T0176843

SOCIAL SYSTEMS ENGINEERING

Wiley Series in Computational and Quantitative Social Science

Computational Social Science is an interdisciplinary field undergoing rapid growth due to the availability of ever increasing computational power leading to new areas of research.

Embracing a spectrum from theoretical foundations to real world applications, the Wiley Series in Computational and Quantitative Social Science is a series of titles ranging from high level student texts, explanation and dissemination of technology and good practice, through to interesting and important research that is immediately relevant to social / scientific development or practice. Books within the series will be of interest to senior undergraduate and graduate students, researchers and practitioners within statistics and social science.

Behavioral Computational Social Science
Riccardo Boero

Tipping Points: Modelling Social Problems and Health
John Bissell (Editor), Camila Caiado (Editor), Sarah Curtis (Editor), Michael Goldstein (Editor), Brian Straughan (Editor)

Understanding Large Temporal Networks and Spatial Networks: Exploration, Pattern Searching, Visualization and Network Evolution
Vladimir Batagelj, Patrick Doreian, Anuska Ferligoj, Natasa Kejzar

Analytical Sociology: Actions and Networks
Gianluca Manzo (Editor)

Computational Approaches to Studying the Co-evolution of Networks and Behavior in Social Dilemmas
Rense Corten

The Visualisation of Spatial Social Structure
Danny Dorling

SOCIAL SYSTEMS ENGINEERING
THE DESIGN OF COMPLEXITY

Edited by

César García-Díaz

Department of Industrial Engineering
Universidad de los Andes
Bogotá, Colombia

Camilo Olaya

Department of Industrial Engineering
Universidad de los Andes
Bogotá, Colombia

Registered Offices
John Wiley & Sons, Inc., 111 River Street, Hoboken, NJ 07030, USA
John Wiley & Sons Ltd, The Atrium, Southern Gate, Chichester, West Sussex, PO19 8SQ, UK

Editorial Office
9600 Garsington Road, Oxford, OX4 2DQ, UK

For details of our global editorial offices, customer services, and more information about Wiley products visit us at www.wiley.com.

Wiley also publishes its books in a variety of electronic formats and by print-on-demand. Some content that appears in standard print versions of this book may not be available in other formats.

Library of Congress Cataloging-in-Publication data applied for

Hardback ISBN: 9781118974452

Cover Image: © Henrik Sorensen/Gettyimages

Set in 10/12pt Times by SPi Global, Pondicherry, India

Printed and bound in Malaysia by Vivar Printing Sdn Bhd

10 9 8 7 6 5 4 3 2 1

Contents

List of Contributors

Ricardo A. Barros-Castro
(Faculty of Engineering, Pontificia
Universidad Javeriana). Email: ricardo-
barros@javeriana.edu.co

Heike I. Brugger
(Department of Politics and Public
Administration, University of Konstanz).
Email: heike.brugger@uni-konstanz.de

William M. Bulleit
(Department of Civil and Environmental
Engineering, Michigan Technological
University). Email: wmbullei@mtu.edu

Jai K. Clifford-Holmes
(Institute for Water Research, Rhodes
University). Email: jai.clifford.holmes
@gmail.com

Bruce Edmonds
(Centre for Policy Modelling, Manchester
Metropolitan University). Email: bruce
@edmonds.name

Grazziela P. Figueredo
(School of Computer Science, University of
Nottingham). Email: Grazziela.Figueredo
@nottingham.ac.uk

César García-Díaz
(Department of Industrial Engineering,
Universidad de los Andes). Email:
ce.garcia392@uniandes.edu.co

John S. Gero
(Krasnow Institute of Advanced Study,
George Mason University and University of
North Carolina, Charlotte). Email: john
@johngero.com

Steven L. Goldman
(Departments of Philosophy and
History, Lehigh University). Email: slg2
@Lehigh.edu

Rafael A. Gonzalez
(Faculty of Engineering, Pontificia
Universidad Javeriana). Email: ragonzalez
@javeriana.edu.co

Zeynep Gurguc
(Digital City Exchange, Imperial College
London). Email: z.gurguc@imperial.ac.uk

Adam Douglas Henry
(School of Government and Public Policy,
University of Arizona). Email: adhenry
@email.arizona.edu

Miwa Hirono
(Department of International Relations,
Ritsumeikan University). Email: hirono-1
@fc.ritsumei.ac.jp

Johann Klocker
(Landeskrankenhaus Klagenfurt). Email:
j.klocker@ipso.at

Chen Liu
(Center for Complex Systems and
Enterprises, Stevens Institute of
Technology). Email: cliu16@stevens.edu

Sandra Méndez-Fajardo
(Faculty of Engineering, Pontificia
Universidad Javeriana). Email: sandra.
mendez@javeriana.edu.co

Jenny O'Connor
(Digital City Exchange, Imperial College
London). Email: j.oconnor@imperial.ac.uk

Mehrnoosh Oghbaie
(Center for Complex Systems and
Enterprises, Stevens Institute of
Technology). Email: moghbaie
@stevens.edu

Peer-Olaf Siebers
(School of Computer Science, University
of Nottingham). Email: Peer-Olaf.Siebers
@nottingham.ac.uk

Camilo Olaya
(Department of Industrial Engineering,
Universidad de los Andes). Email: colaya
@uniandes.edu.co

Carolyn G. Palmer
(Institute for Water Research, Rhodes
University). Email: tally.palmer@ru.ac.za

Michael J. Pennock
(Center for Complex Systems and
Enterprises, Stevens Institute of Technology).
Email: mpennock@stevens.edu

Joseph C. Pitt
(Department of Philosophy, Virginia
Polytechnic Institute and State University).
Email: jcpitt@vt.edu

William B. Rouse
(Center for Complex Systems and
Enterprises, Stevens Institute of
Technology). Email: wrouse@stevens.edu

Martin F.G. Schaffernicht
(Faculty of Economics and Business,
Universidad de Talca). Email: martin@utalca.cl

Markus Schwaninger
(Institute of Management, University of
St. Gallen). Email: markus.schwaninger
@unisg.ch

Anya Skatova
(Warwick Business School, University of
Warwick). Email: anya.skatova@wbs.ac.uk

Jill H. Slinger
(Faculty of Technology, Policy and
Management and the Faculty of Civil
Engineering and Technical Geosciences,
Delft University of Technology and the
Institute for Water Research, Rhodes
University). Email: j.h.slinger@tudelft.nl

Russell C. Thomas
(Department of Computational and
Data Sciences, George Mason
University and University of North
Carolina, Charlotte). Email: russell.
thomas@meritology.com

Koen H. van Dam
(Digital City Exchange, Imperial College
London). Email: k.van-dam@imperial.ac.uk

Chris de Wet
(Institute for Water Research and the
Department of Anthropology, Rhodes
University). Email: c.dewet@ru.ac.za

Zhongyuan Yu
(Center for Complex Systems and
Enterprises, Stevens Institute of
Technology). Email: zyu7@stevens.edu

Preface

We, the editors of this volume, are trained as both engineers and social scientists, and have contrasted the two different mindsets during our careers. Since the time when we were PhD students in the social sciences, we have felt that 'praxis' does not have the same status as theorizing in generating knowledge, as opposed to what happens in engineering. Even the recognition of engineering knowledge as a distinctive kind of knowledge, different from scientific knowledge, seems to remain elusive for both academics and practitioners. In fact, 'praxis' embodies a set of differential elements from pure science. This volume is an effort to bring together elements of engineering thinking and social science into the study of social systems, and more importantly, aims to be a vehicle that emphasizes the necessity of developing practical knowledge – through its proper ways and under its own 'validation' criteria – to provide feasible, yet informed, paths for intervening and improving social systems; that is, systems created and driven by human beings.

Also, through this volume we would like to make explicit the inherent link between systemic thinking and engineering knowledge through the consideration of multiple perspectives and methods. We believe that the merger of both the qualitative and quantitative worlds is essential in order to cope with the complexity of contemporary social systems. Moreover, we believe that engineering thinking, along with its tools and methods, is one of the best chances that we have for designing, redesigning and transforming the complex world of social systems.

This project would not have been possible without the initial motivation provided by Debbie Jupe (commissioning editor at John Wiley & Sons), who was the first editor we met at the Social Simulation Conference in Warsaw (Poland), back in 2013. Debbie invited us to put our ideas in a written proposal, and encouraged us to come up with a book on engineering perspectives to social systems. Thank you Debbie for your encouragement!

We are grateful to all the authors who contributed to this volume for their patience, commitment and understanding over the course of the many months it took to make this book a reality. We are also indebted to the Department of Industrial Engineering of Universidad de los Andes (Colombia) for their continued support and to Claudia Estévez-Mujica, who helped us immensely in putting all the chapters together, checking inconsistencies and assembling the whole book.

Introduction: The Why, What and How of Social Systems Engineering

César García-Díaz and Camilo Olaya

The Very Idea

The expression 'social systems engineering' is not new. As far as we know, its first appearance in the literature dates from the mid-1970s. In 1975, the Proceedings of the IEEE published a special issue on social systems engineering (Chen *et al.*, 1975). Here, Chen and colleagues referred to social systems engineering as the application of systems engineering concepts to social problems. Likewise, the special issue seemed to emphasize that the potential contribution of engineering to social issues was predominantly based on the consideration of quantitative modelling as the workhorse for intervention. Although we concur with some of these points, for us the expression 'social systems engineering' has a broader connotation, not meaning that we advocate exclusively for the application of engineering methods to social issues, but rather that we stand up for the consideration of *design* perspectives as a pivotal way to generate knowledge and transform systems. The intrinsic engineering orientation to *action* and *transformation* as its ultimate goals for improving a system, for meeting needs, for addressing successfully a specific problematic situation that someone wants to improve, etc. are emphases that this book highlights. Such goals demand the recognition of specific engineering considerations and their implications for addressing *social systems*. We want to emphasize the complexity of engineering 'social' (human) systems (as opposed to engineering mechanical systems, electrical systems, etc.), since such systems are then in fact 'social' (formed by purposeful actors that display agency, with diverse, clashing interests and goals) and therefore their design, redesign and transformation, unlike in other engineering domains, cannot be completely determined or planned beforehand. These designs are formal and informal, emergent, always 'in progress', adapting and evolving out of diverse dynamics.

Social Systems Engineering: The Design of Complexity, First Edition. Edited by César García-Díaz and Camilo Olaya.
© 2018 John Wiley & Sons Ltd. Published 2018 by John Wiley & Sons Ltd.

Social systems engineering has a paradoxical status. On the one hand, it is an under-researched topic whose *theoria* has rarely been explored. On the other hand, it is perhaps one of the most common endeavours in society since it concerns the *praxis* that seeks to design, create and transform human organizations. Consequently, we need to understand what engineering thinking means, and how it relates to social systems. Steven Goldman, one of the contributors to this book, stated more than 20 years ago regarding the autonomy of engineering (as distinct from other activities such as science or arts) that 'while engineering has a *theoria*, analogous to, but different from, that of the physical sciences, unlike science, engineering is quintessentially a *praxis*, a knowing inseparable from moral action' (Goldman, 1991, p. 139). The recognition of engineering as an autonomous activity, independent from science (though related in many ways), seems just a recent explicit realization that can be identified with what can be called a 'philosophy of engineering' (Bucciarelli, 2003; Goldman, 2004; Miller, 2009; Sinclair, 1977; Van de Poel and Goldberg, 2010). Perhaps the key word to understand the autonomy of engineering is *design* (Goldman, 1990; Layton, 1984, 1991; Pitt, 2011b; Schmidt, 2012; Van de Poel, 2010). Engineering, being driven by design, shows a distinct *rationality*, as Goldman shows in Chapter 1 of this book. He characterizes engineering design as 'compromised exactness', since its formal apparatus delivers approximate 'solutions' that are subject to their context of application, which means that they are always subjective, wilful and contextual. Social systems, as belonging to the realm of artificial systems, exhibit *design*, which means that they are, and can be, *engineered*, but not in the traditional sense (Remington *et al.*, 2012; Simon, 1996). Traditional engineering, design-based methods, which essentially aim at control and prediction, *cannot* be applied to social systems due to the very nature of these systems – unlike mechanical systems, social systems do not 'obey laws', as Galileo imagined (Galileo Galilei, 1623), but are driven by the agency of human beings. Yet, engineering thinking can be used in several other ways, for instance for *steering* social systems towards a given direction, for influencing action (Pennock and Rouse, 2016), for *opening new possibilities*, for *driving conversations* among its members, for *imagining* different futures, for *learning* about the complexity that social systems entail, etc.

Whenever engineering concerns social systems (i.e., firms, public and private organizations, urban systems, etc.) it implies the design of social artefacts and social constructions such as management structures, incentive schemes, routines, procedures, ways of working (formal and informal, planned and spontaneous), agreements, contracts, policies, roles and discourses, among others (Jelinek *et al.*, 2008; March and Vogus, 2010). Therefore, such types of engineering face a special type of complexity, since these artefacts depend on and are constructed through human action, meaning that not only individuals but also their emotions, language and meanings are involved.

This book seeks to offer an overview of what social systems engineering entails. The reader might hasten to think that this is a mechanistic approach to social systems. However, there is no such thing as optimal design in social systems (Devins *et al.*, 2015). In contrast, the very idea of social systems engineering, although it emphasizes *action*, does not necessarily rely on prediction; it is context-dependent, iterative, builds upon different modelling perspectives and decisively aims at influencing the path of, rather than deliberatively designing, the evolving character of self-organization of human societies. This is a starkly different approach from a purely scientific viewpoint. The book encompasses three sections that follow an intuitive inquiry in this matter. The first section deals with the very idea of what social systems engineering might be and the need for addressing the topic in its own terms. The second section samples illustrative methodologies and methods. The final section illustrates examples of the challenge of designing the complexity that results from systems created through human action.

Epistemic Notions on the Engineering of Social Systems

There are diverse beliefs regarding what engineering is about. Perhaps the most popular is to believe that engineering *is* 'applied science'. However, this would mean assuming that 'scientists generate new knowledge which technologists then apply' (Layton, 1974, p. 31) and therefore would suggest that what makes an engineer an engineer, and what an engineer delivers, is (applied) scientific knowledge, instead of a different type of knowledge (Davis, 2010), which is, at best, misleading (Goldman, 2004; Hansson, 2007; Layton, 1974; McCarthy, 2010; Pitt, 2010; Van de Poel, 2010). The recognition that science and engineering stand on different epistemic grounds (Goldman, 1990; Koen, 2003; Krige, 2006; Layton, 1984, 1987, 1991; Petroski, 2010; Pitt, 2011b; Vincenti, 1990; Wise, 1985) is perhaps the first step in thinking of social systems *engineering* and requires a brief overview.

If it is not 'applied science', what are the defining characteristics of engineering? We can start by realizing that engineering and science usually pursue different goals: scientists, first and foremost, look for systematic *explanations* of phenomena; engineers, on the other hand, pursue the *transformation* of a situation through the design of artefacts that serve as vehicles to solve problems. In short, as Petroski (2010) puts it, scientists seek to explain the world while engineers try to change it. The scientist deals primarily with the question 'what *is* it?' The engineer deals with '*how must this* situation be changed?' and 'what is the *right* action to *do*?' Engineering is concerned 'not with the necessary but with the contingent, not with how things are but with how they might be' (Simon, 1996, p. xii). Such different missions lead to different values, norms, rules, apparatus for reasoning, considerations, type of knowledge, methods, success criteria, standards for evaluating results; in short, different epistemologies.

Engineering knowledge is intrinsic to engineering and different from scientific knowledge. Engineering *know-how* is a distinctive type of knowledge, different from the scientific *know-that* (Ryle, 1945). For example, 'engineering knowledge is practice-generated… it is in the form of "knowledge-how" to accomplish something, rather than "knowledge-that" the universe operates in a particular way' (Schmidt, 2012, p. 1162). Knowledge-how is not concerned with the truth or falsehood of statements, 'you cannot affirm or deny Mrs. Beeton's recipes' (Ryle, 1945, p. 12). Engineers *know how to do* things. It is a type of practical knowledge. Therefore, the resources and information to get the job done can be varied and diverse, in principle they are not rejected under any a-priori principle, 'resolving engineering problems regularly requires the use of less than scientifically acceptable information' (Mitcham, 1994). The scientific 'empirical evidence' might be useful, but it is not a necessary requirement. Such a practical approach requires also that designs must *work* in real life; the effects of friction or air resistance cannot be ignored (Hansson, 2007). Since the task of the engineer is to be effective, to accomplish, then mathematical precision and analytical solutions are not required. Unlike the scientist, the engineer does not assume ideal conditions, s/he knows what to do in imperfect situations.

Engineers address practical problems: their know-how is constructed contingently and for very specific contexts (McCarthy, 2010). Engineering deals with particulars in its particularity, they are not taken as instantiations of a universal (Goldman, 1990). This implies that engineering design faces a variety of constraints related to idiosyncratic values and factors (economic, cultural, political, reliability, viability, ethical) that co-define and specify the design problem, unlike scientific research in which such constraints are absent in the definition of a scientific question (Kroes, 2012). This singularity of each design problem explains why there is no unique solution for an engineering problem: 'an engineer who understands engineering will never claim to have found *the* solution... This is why there are so many different-looking airplanes and automobiles and why they operate differently... they are simply one engineer's solution to a problem that has no unique solution' (Petroski, 2010, p. 54). Moreover, there is usually more than one way to solve an engineering problem. Such diversity of possibilities, methods and solutions contrasts with the goal of scientific communities that typically pursue the one best theory, at any given time, for explaining a phenomenon; when a theory is shown to be erroneous, it can be replaced with a better one.

The activity of engineering does not need epistemic justifications. The intentional creation of artefacts is done by experimental methods that are more fundamental than (and not derived from) any type of *theory* (Doridot, 2008). The origin of design is irrelevant, it does not necessarily have to be *a priori* supported by anything, including theories or data. Design can be freely generated with the help of any procedure, sourced from reason, or guided by previous expectations – 'theoretic' or not (Stein and Lipton, 1989), guided with the help of a model, or just based on imagination, or instincts. 'Empirical evidence', or any other indirect mechanism of representing the world, is just another option, but it is not a requisite. For instance, 'the inventor or engineer... can proceed to design machines in ignorance of the laws of motion... These machines will either be successful or not' (Petroski, 2010, p. 54). Engineering handles a *pragmatic* concept of 'truth' (Doridot, 2008). An artefact or an engineering solution is not false or true (or closer to), simply it works or it doesn't. If it works, engineers succeed. The popular notion of knowledge as '*justified* true belief' means nothing in a pragmatic approach in which knowledge is *unjustified*. In the words of Pitt: 'If it solves our problem, then does it matter if we fail to have a philosophical justification for using it? To adopt this attitude is to reject the primary approach to philosophical analysis of science of the major part of the twentieth century, logical positivism, and to embrace pragmatism' (2011a, p. 173).

We are interested in particular in the engineering of *social systems*. What are the implications of the recognition of such philosophy of engineering for the domain of social systems? Let us consider, for instance, that the predictive logic of scientific causal models relates to the idea that prediction is a requirement of control (Sarasvathy, 2003). A fundamental question is how much prediction, derived from causal explanations, is needed to transform a social system. Before the apparent unpredictability of the behaviour of social systems, one idea is to operate under a different logic and to drop the very idea of prediction in design, as Sarasvathy (2003) puts it. Sarasvathy (2003) claims that, in relation to endeavours of enterprise creation, a design logic highlights the fact that '*to the extent we can control the future, we do not need to predict it*' (Sarasvathy, 2003, p. 208), implying that '*a large part of the future actually is a product of human decision-making*' (Sarasvathy, 2003, p. 209). And yet, the future remains uncertain. How to deal with such uncertainty of social systems? William Bulleit offers a possible answer in this book. The unpredictable and complex nature of human action means to face a special type of uncertainty that is, as Bulleit develops in Chapter 2, much larger than that found in other engineered systems. The uncertainty that engineers usually confront resembles an explorer in a jungle with unknown dangers; this explains why engineers consider *as part of* their design considerations, elements such as 'safety factors', 'safety barriers', 'unforeseen factors', etc. (Doorn and Hansson, 2011; Hansson, 2009a,b). However, unlike probabilistic risk analysis, the design of social systems deals with true uncertainty under *unknown* probabilities. As Hansson (2009a) pictures it, such uncertainty is unlike that which a gambler faces at the roulette wheel. Social systems represent perhaps the extreme case, whose design and maintenance requires a distinct mind-set that brings together bottom-up and top-down solutions, along with the recognition of the adaptive nature of social systems, as Bulleit suggests.

How to engineer problem-solving designs in such unpredictable social systems? The recognition of adaptive and evolutionary dynamics leads us to think of the possibility of producing designs without 'knowing' beforehand the way in which the system to be designed or transformed 'works'. Perhaps the main contribution of Charles Darwin is in the realm of philosophy, indicating a way to produce a design without a 'designer' (Ayala, 2007; Dennett, 1995; Mayr, 1995, 2001). Evolution already shows how and why the selection of blind variations explains the success of any system that adapts to changing and unknown environments (Campbell, 1987; Harford, 2011; Popper, 1972). Perhaps we must resist the apparent requisite of having knowledge beforehand for doing something. Bruce Edmonds makes an analogy in Chapter 3 that compares social systems engineering with farming. Since there is no such thing as 'designing' a farm, farmers instead know that they must continuously act on their farms to achieve acceptable results. Edmonds underlines that, since we are far from even having a minimal and reliable understanding of social systems, then engineers of social systems must recur to system farming. Edmonds emphasizes that traditional design-based engineering approaches are simply not possible to be applied to social systems; a systems farming lens should rely more on experience rather than on system control, should operate iteratively rather than as a one-time effort, and should make use of partial rather than full understanding, among other considerations.

Yet, the notion of evolution challenges the very idea of whether humans can deliberately improve social systems. Is it possible to control, manage or at least direct an evolutionary process? Martin Schaffernicht deals with this question in Chapter 4. Like Edmonds, Schaffernicht questions whether deliberate social system designs can actually be made and if they can really be translated into improvement. Schaffernicht rather suggests that engineering can contribute to influence the pace of the evolutionary nature of social systems through *policy* engineering. He underlines that *collective policies* are evolving artefacts that drive behaviours – they are never definitive but in constant revision and adaptation – and become the central elements for developing an interplay between evolution and engineering that ends up shaping open-ended social systems.

These brief ideas indicate the immense challenge in 'engineering' (designing and redesigning, that is) social systems, or as put by Vincent Ostrom, it means a problem of 'substantial proportions... In Hobbes's words, human beings are both the "matter" and the "artificers" of organizations. Human beings both design and create organizations as artifacts and themselves form the primary ingredient of organizations. Organizations are, thus, artifacts that contain their own artisans' (Ostrom, 1980, p. 310). Human beings co-design the social systems that they form, this is why those designs might be intentional up to some point but they are also emergent, dynamic, incomplete, unpredictable, self-organizing, evolutionary and always 'in the making' (Bauer and Herder, 2009; Garud *et al.*, 2006, 2008; Kroes, 2012; Krohs, 2008; Ostrom, 1980). The ultimate challenge is to address the complexity posed by the relations between human beings. Joseph Pitt illustrates this concern with a concrete example: what does it mean to be a friend of someone? This question will lead us to challenge the very possibility of designing a social system. In Chapter 5, Pitt suggests that we can only design an environment in which a social system emerges and evolves, a suggestion that is in line with the first part of this book that calls for the need to recognize the experimental, evolving and open-ended nature of social systems. This is the first requisite for anyone aspiring to transform a social system.

Using Engineering Methods

How to engineer social systems? The second part of this book introduces different methods for engineering social systems. Engineers proceed in a distinctive way. Billy Vaughn Koen in his book *The Discussion of the Method* (2003) defines engineering by its method. For him, the engineering method is any 'strategy for causing the best change in a poorly understood situation with the available resources' (p. 7). Engineers call such strategies 'heuristics'. 'A heuristic is anything that provides a plausible aid or direction in the solution of a problem but is in the final analysis unjustified, incapable of justification, and potentially fallible' (Koen, 2010, p. 314). Koen highlights the distinctive nature of heuristics as opposed to other ways of facing the world; in particular, he considers the differences from scientific theories. A heuristic does not guarantee a solution, it may contradict other heuristics (Koen 2009); it does not need justification, its relevance depends on the particular situation that the heuristic deals with and its outcome is a matter of neither 'truth' nor generalizability. The engineering method – as opposed to the scientific method – is a heuristic; that is, unjustified, fallible, uncertain, context-defined and problem-oriented. Hence, the second part of this book can be seen as a small sample of heuristics that in particular share a common preferred strategy of engineers: modelling.

Engineering design requires the capacity to 'see' and imagine possible (both successful and unsuccessful) futures. Zhongyuan Yu and her colleagues show in Chapter 6 how policy flight simulators may help to address 'what if...' questions through model-based interactive visualizations that enable policy-makers to make decisions and anticipate their consequences. Policy flight simulators drive the exploration of management policies according to possible factors that contribute to an existing or potential state of a system. Through two detailed cases, the chapter shows how such simulators can be developed and how groups of people (rather than individuals) interact with them. These interactions are the central piece of the method, since the involved stakeholders and policy-makers bring conflicting priorities and diverse preferences for courses of action. The chapter illustrates with practical cases the mentioned idea of Schaffernicht: the centrality of the evolution of 'collective policies' for transforming social systems and the way in which such evolution can be enhanced through learning. Yu and her colleagues underline that the key value of their models and visualizations lies in the insights that they provide to those intending to engineer their own social systems.

Models are powerful tools for supporting design activities (Dillon, 2012; Dodgson *et al.*, 2007; Elms and Brown, 2012; Will, 1991). Unlike scientific models that are usually built for analysis of observations and generating 'true' explanations (Norström, 2013), engineering models are judged against their usefulness for specific, diverse (Epstein, 2008) purposes. For engineers, they serve as focal points 'for a story or conversation about how a system behaves and how that behaviour can be changed. It is by mediating in this process – acting to focus language by stressing some features of the real system while ignoring others – that models contribute to new shared understandings in a community of engineering practice' (Bissell and Dillon, 2012, p. vi). Chapter 7 by Peer-Olaf Siebers and colleagues introduces a structured framework for guiding such conversation processes through model development, from conceptual design to implementation. In particular, this framework organizes both the process of building and using agent-based models and the way in which the resulting simulation models can be used as decision-support tools for exploring the application of policies. Being a heuristic, they adapt what they consider appropriate for developing their framework; in particular, they borrow ideas from software engineering for tackling problem analysis and model design. The chapter uses international peacebuilding activities in South Sudan as an example to illustrate the practical possibilities of their proposal.

There are diverse ways of building models. Sandra Méndez-Fajardo and colleagues show, in Chapter 8, how social systems engineering can employ (social) science through a methodological framework that uses actor-network theory as a heuristic for designing and building agent-based models. They use an applied case in waste of electrical and electronic equipment management as an illustrative example. Their proposal presents a way to overcome the distinction between human and non-human actors, and underlines the centrality of 'actor-networks' (rather than just actors) in social systems. Although these theoretic contributions stand on their own as valuable results, they unmistakably underline the engineering character of their proposal, which concerns the pragmatic usefulness of modelling rather than its theoretical validity. They frame the application of actor-network theory as a heuristic for intervening social systems through the use of simulation models to enact policy changes.

Engineering may use scientific theories but may also contribute to science. Computational modelling can complement diverse theoretic approaches, for instance it is useful for supporting theory building in social science (e.g., Schwaninger and Grösser, 2008). To complete the second part of the book, in Chapter 9 Russell Thomas and John Gero use social theory to explore the process of institutional innovation and how to influence innovation trajectories in pre-paradigmatic settings (which the authors call 'contested territories'), where there are rival worldviews regarding the nature of problems and innovations. The authors illustrate their methodological approach with the case of cyber security and the problem of quantifying security and risk under two rival worldviews: the 'quants' (for whom cyber security and risk can and should be quantified) and the 'non-quants' (who believe that cyber security and risk either cannot be quantified or its quantification does not bring enough benefits). The chapter frames the process of institutional innovation in Boisot's theory of the social learning cycle and the role of knowledge artefacts during the cycle. A computational model helps to explore how knowledge artefacts of different characteristics affect innovation rate and learning. The chapter makes provocative suggestions regarding not only how social science can contribute to social systems engineering but also the other way around: how this latter approach can contribute to deal with scientific questions, such as the assessment of the scientific merit of each school of thought (in terms of explanatory coherence) and the possibility of addressing further theoretic issues of social dynamics such as legitimization, power struggles and structuration, among others.

Into Real-World Applications

Since social systems engineering is praxis, then real-world applications become perhaps the true way to depict this type of engineering. The last part of the book places the emphasis on practical applications that illustrate the richness and possibilities that the first two parts suggest.

Chapter 10 by Adam Douglas Henry and Heike Brugger deals with developing strategic scenarios for the adoption of environmentally friendly technologies. Through agent-based computational modelling, they inspect non-trivial policy answers to two simultaneously desirable outcomes regarding sustainable technologies: the speed of their adoption and the guarantee of equal access to them. Chapter 11 by Clifford-Holmes and colleagues combines ethnographic data collection with participatory system dynamics modelling in the design of potential strategies in water resource management in South Africa. Clifford-Holmes and colleagues emphasize the 'muddled middle' between policy and implementation, and propose new directions in participatory modelling. In Chapter 12 Markus Schwaninger and Johann Klocker provide an account of the 30-year evolution of the oncological care system in Klagenfurt, Austria, exposing the threat of organizational over-specialization in patient treatment and highlighting the importance of holistic approaches to healthcare system design by using causal loop diagrams and organizational cybernetic concepts. Last but not least, in Chapter 13 Jenny O'Connor and colleagues explore four case studies of smart city projects in the United Kingdom and highlight the importance of understanding the unpredictability of individual and societal behaviour when confronted with new sustainable-related policies derived from technical aspects only. O'Connor and colleagues explicitly call for the inclusion of the social dimension in the engineering of social systems.

In summary, social systems engineering goes beyond the application of engineering methods to social problems. In different instances there has been a tendency to equate engineering a social system with a traditional, mechanistic, one-shot undertaking that attempts to reach optimality according to some well-pre-established objective (Devins *et al.*, 2015). That is not what social systems engineering is about. In contrast, we aim to highlight the importance of trial and error, failure, iteration, adaptability and evolution as salient features of any design-oriented process. Stimulating self-organization (as opposed to direct intervention) as a way to foster growth of desirable properties (e.g., adaptability and resilience) is also intrinsic to any design-oriented endeavour. Engineering a social system implies 'steering' a system towards a desirable state (Penn *et al.*, 2013), even if such a state is not completely understood and is subject to different interpretations (e.g., a sustainable community), and even if the journey towards it is filled with unexpected occurrences. We hope that this book will provide a broader, multidisciplinary, conceptual approach to social systems design, and stimulate the growth of ideas towards solution-oriented perspectives (Watts, 2017) in dealing with social systems issues.

References

Ayala, F.J. (2007) Darwin's greatest discovery: Design without designer. *Proceedings of the National Academy of Sciences*, **104**(1), 8567–8573.

Bauer, J.M. and Herder, P.M. (2009) Designing socio-technical systems, in A. Meijers (ed.), *Philosophy of Technology and Engineering Sciences*, North Holland, Amsterdam, pp. 602–630.

Bissell, C. and Dillon, C. (2012) Preface, in C. Bissell and C. Dillon (eds), *Ways of Thinking, Ways of Seeing. Mathematical and other modelling in engineering and technology* (Vol. **1**), Springer-Verlag, Berlin, pp. v–vii.

Bucciarelli, L. (2003) *Engineering Philosophy*, Delft University Press, Delft.

Campbell, D.T. (1987) Evolutionary epistemology, in G. Radnitzky and W.W. Bartley III (eds), *Evolutionary Epistemology, Rationality, and the Sociology of Knowledge*, Open Court, La Salle, IL, pp. 47–73.

Chen, K., Ghaussi, M. and Sage, A.P. (1975) Social systems engineering: An introduction. *Proceedings of the IEEE*, **63**(3), 340–344.

Davis, M. (2010) Distinguishing architects from engineers: A pilot study in differences between engineers and other technologists, in I. Van de Poel and D.E. Goldberg (eds), *Philosophy and Engineering. An emerging agenda*, Springer-Verlag, Dordrecht, pp. 15–30.

Dennett, D.C. (1995) *Darwin's Dangerous Idea*, Penguin Books, London.

Devins, C., Koppl, R., Kauffman, S. and Felin, T. (2015) Against design. *Arizona State Law Journal*, **47**, 609.

Dillon, C. (2012) Models: What do engineers see in them?, in C. Bissell and C. Dillon (eds), *Ways of Thinking, Ways of Seeing. Mathematical and other modelling in engineering and technology* (Vol. **1**), Springer-Verlag, Berlin, pp. 47–69.

Dodgson, M., Gann, D.M. and Salter, A. (2007) The impact of modelling and simulation technology on engineering problem solving. *Technology Analysis & Strategic Management*, **19**(4), 471–489.

Doorn, N. and Hansson, S.O. (2011) Should probabilistic design replace safety factors? *Philosophy & Technology*, **24**, 151–168.

Doridot, F. (2008) Towards an 'engineered epistemology'? *Interdisciplinary Science Reviews*, **33**(3), 254–262.

Elms, D.G. and Brown, C.B. (2012) Professional decisions: The central role of models. *Civil Engineering and Environmental Systems*, **29**(3), 165–175.

Epstein, J.M. (2008) Why model? *Journal of Artificial Societies and Social Simulation*, **11**(4), 12.

Galileo Galilei (1623) The assayer, in S. Drake (ed.), *Discoveries and Opinions of Galileo* (trans. S. Drake 1957), Anchor Books, New York, NY.

Garud, R., Kumaraswamy, A. and Sambamurthy, V. (2006) Emergent by design: Performance and transformation at Infosys Technologies. *Organization Science*, **17**(2), 277–286.

Garud, R., Jain, S. and Tuertscher, P. (2008) Incomplete by design and designing for incompleteness. *Organization Studies*, **29**(3), 351–371.

Goldman, S.L. (1990) Philosophy, engineering, and western culture, in P.T. Durbin (ed.), *Broad and Narrow Interpretations of Philosophy of Technology*, Kluwer, Amsterdam, pp. 125–152.

Goldman, S.L. (1991) The social captivity of engineering, in P.T. Durbin (ed.), *Critical Perspectives on Nonacademic Science and Engineering*, Lehigh University Press, Bethlehem, PA.

Goldman, S.L. (2004) Why we need a philosophy of engineering: A work in progress. *Interdisciplinary Science Reviews*, **29**(2), 163–176.

Hansson, S.O. (2007) What is technological science? *Studies in History and Philosophy of Science*, **38**, 523–527.

Hansson, S.O. (2009a) From the casino to the jungle. Dealing with uncertainty in technological risk management. *Synthese*, **168**, 423–432.

Hansson, S.O. (2009b) Risk and safety in technology, in A. Meijers (ed.), *Philosophy of Technology and Engineering Sciences*, North Holland, Amsterdam, pp. 1069–1102.

Harford, T. (2011) *Adapt. Why success always starts with failure*, Picador, New York, NY.

Jelinek, M., Romme, A.G.L. and Boland, R.J. (2008) Introduction to the special issue. Organization studies as a science for design: Creating collaborative artifacts and research. *Organization Studies*, **29**(3), 317–329.

Koen, B.V. (2003) *Discussion of the Method*, Oxford University Press, Oxford.

Koen, B.V. (2009) The engineering method and its implications for scientific, philosophical, and universal methods. *The Monist*, **92**(3), 357–386.

Koen, B.V. (2010) Quo vadis, humans? engineering the survival of the human species, in I. Van de Poel and D.E. Goldberg (eds), *Philosophy and Engineering. An emerging agenda*, Springer-Verlag, Dordrecht, pp. 313–341.

Krige, J. (2006) Critical reflections on the science–technology relationship. *Transactions of the Newcomen Society*, **76**(2), 259–269.

Kroes, P. (2012) Engineering design, in P. Kroes (ed.), *Technical Artefacts: Creations of mind and matter*, Springer-Verlag, Dordrecht, pp. 127–161.

Krohs, U. (2008) Co-designing social systems by designing technical artifacts, in P.E. Vermaas, P. Kroes, A. Light and S.A. Moore (eds), *Philosophy and Design*, Springer-Verlag, Dordrecht, pp. 233–245.

Layton, E.T. Jr. (1974) Technology as knowledge. *Technology and Culture*, **15**(1), 31–41.

Layton, E.T. Jr. (1984) Science and engineering design. *Annals of the New York Academy of Sciences*, **424**(1), 173–181.

Layton, E.T. Jr. (1987) Through the looking glass, or news from lake mirror image. *Technology and Culture*, **28**(3), 594–607.

Layton, E.T. Jr. (1991) A historical definition of engineering, in P.T. Durbin (ed.), *Critical Perspectives on Nonacademic Science and Engineering*, Lehigh University Press, Bethlehem, PA, pp. 60–79.

March, S.T. and Vogus, T.J. (2010) Design science in the management disciplines, in A. Hevner and S. Chatterjee (eds), *Design Research in Information Systems*, Springer-Verlag, New York, NY, pp. 195–208.

Mayr, E. (1995) Darwin's impact on modern thought. *Proceedings of the American Philosophical Society*, **139**(4), 317–325.

Mayr, E. (2001) The philosophical foundations of Darwinism. *Proceedings of the American Philosophical Society*, **145**(4), 488–495.

McCarthy, N. (2010) A world of things not facts, in I. Van de Poel and D.E. Goldberg (eds), *Philosophy and Engineering. An emerging agenda*. Springer-Verlag, Dordrecht, pp. 265–273.

Miller, G. (2009) London calling philosophy and engineering: WPE 2008. *Science and Engineering Ethics*, **15**, 443–446.

Mitcham, C. (1994) *Thinking Through Technology. The path between engineering and philosophy*, University of Chicago Press, Chicago, IL.

Norström, P. (2013) Engineers' non-scientific models in technology education. *International Journal of Technology and Design Education*, **23**(2), 377–390.

Ostrom, V. (1980) Artishanship and artifact. *Public Administration Review*, **40**(4), 309–317.

Penn, A.S., Knight, C.J., Lloyd, D.J., Avitabile, D., Kok, K., Schiller, F. *et al.* (2013) Participatory development and analysis of a fuzzy cognitive map of the establishment of a bio-based economy in the Humber region. *PLoS ONE*, **8**(11), e78319.

Pennock, M.J. and Rouse, W.B. (2016) The epistemology of enterprises. *Systems Engineering*, **19**(1), 24–43.

Petroski, H. (2010) *The Essential Engineer. Why science alone will not solve our global problems*, Vintage Books, New York, NY.

Pitt, J.C. (2010) Philosophy, engineering, and the sciences, in I. Van de Poel and D.E. Goldberg (eds), *Philosophy and Engineering. An emerging agenda*, Springer-Verlag, Dordrecht, pp. 75–82.

Pitt, J.C. (2011a) *Doing Philosophy of Technology*, Springer-Verlag, Dordrecht.

Pitt, J.C. (2011b) What engineers know, in J.C. Pitt (ed.), *Doing Philosophy of Technology*, Springer-Verlag, Dordrecht, pp. 165–174.

Popper, K. (1972) *Objective Knowledge. An evolutionary approach*, Oxford University Press, Oxford.

Remington, R., Boehm-Davis, D.A. and Folk, C.L. (2012) Natural and engineered systems, in *Introduction to Humans in Engineered Systems*, John Wiley & Sons, Hoboken, NJ, pp. 7–13.

Ryle, G. (1945) Knowing how and knowing that. *Proceedings of the Aristotelian Society, New Series*, **46**, 1–16.

Sarasvathy, S.D. (2003) Entrepreneurship as a science of the artificial. *Journal of Economic Psychology*, **24**(2), 203–220.

Schmidt, J.A. (2012) What makes engineering, engineering?, in J. Carrato and J. Burns (eds), *Structures Congress Proceedings*, American Society of Civil Engineers, Reston, VA, pp. 1160–1168.

Schwaninger, M. and Grösser, S. (2008) System dynamics as model-based theory building, *Systems Research and Behavioral Science*, **25**, 447–465.

Simon, H.A. (1996) *The Sciences of the Artificial* (3rd edn), MIT Press, Cambridge, MA.

Sinclair, G. (1977) A call for a philosophy of engineering. *Technology and Culture*, **18**(4), 685–689.

Stein, E. and Lipton, P. (1989) Where guesses come from: Evolutionary epistemology and the anomaly of guided variation. *Biology and Philosophy*, **4**, 33–56.

Van de Poel, I. (2010) Philosophy and engineering: Setting the stage, in I. Van de Poel and D.E. Goldberg (eds), *Philosophy and Engineering. An emerging agenda*, Springer-Verlag, Dordrecht, pp. 1–11.

Van de Poel, I. and Goldberg, D.E. (2010) *Philosophy and Engineering. An emerging agenda*, Springer-Verlag, Dordrecht.

Vincenti, W.G. (1990) *What Engineers Know and How They Know It*, The Johns Hopkins University Press, Baltimore, MD.

Watts, D.J. (2017) Should social science be more solution-oriented? *Nature Human Behaviour*, **1**, 0015.

Will, P. (1991) Simulation and modeling in early concept design: An industrial perspective. *Research in Engineering Design*, **3**(1), 1–13.

Wise, G. (1985) Science and technology. *Osiris*, **1**, 229–246.

Part I

Social Systems Engineering: The Very Idea

1

Compromised Exactness and the Rationality of Engineering

Steven L. Goldman

1.1 Introduction

In the spring of 1929, on the occasion of the Gifford Lectures at Edinburgh University, John Dewey asked: 'Are there in existence the ideas and the knowledge that permit experimental method to be effectively used in social interests and affairs?' (Dewey, 1988, p. 218). By 'experimental method', Dewey meant systematic reasoning about effective means for achieving a specified end. This was problem-solving reasoning *par excellence* for Dewey, because it was reasoning that was reflexively shaped by its consequences in a cognitive positive feedback loop characteristic of applied science and engineering. It was just this 'experimental method', Dewey argued, that by uniting the results of experiment-validated scientific knowledge with the objectives of engineering practice had enabled the society- and culture-transforming accomplishments of nineteenth-century technological innovations. What Dewey was asking in the Gifford Lectures, then, was: Do we know enough, not *in* science and engineering, but *about* the methodologies employed in applied science and engineering, to apply those methodologies to 'social interests and affairs'?

Here we are, eighty-six years later, asking the same question: Is there, in the kind of reasoning routinely employed so successfully by engineers to solve technical problems, a model for the design of more effective social systems? Do we, today, know enough about engineering practice – specifically engineering practice rather than the practice of science – to help us formulate more effective public policies, create more effective organizational structures and develop better social systems: educational systems, health-care systems, judicial systems, financial systems, even political systems?

A first step towards answering these questions would be clarifying the distinctiveness of engineering reasoning *vis-à-vis* scientific reasoning. This would help us to understand why it

Social Systems Engineering: The Design of Complexity, First Edition. Edited by César García-Díaz and Camilo Olaya.
© 2018 John Wiley & Sons Ltd. Published 2018 by John Wiley & Sons Ltd.

is that we ask about a model for developing better public policies and better social systems in engineering and not in science.

A second step would be to describe the centrality to engineering reasoning of the design process, and within that process, of trade-off decision-making, in order to assess its transposability to the design of public policies and social systems. Even if it seems transposable in principle, however, the roles of two fundamental features of engineering practice in the reasoning underlying the design process must be taken into account, namely, its experimental and its evolutionary character.

Of course, people have been designing and redesigning social systems, and implementing public policies, for all of recorded history, and for a long time before that. As nearly as we can tell, they did this without asking for help from the 'engineers' of the day, that is, from contemporary possessors of craft know-how. What seems new in assessing social systems today is a perception on the one hand of manifest expertise in applied science and engineering and on the other hand of a problematic situation confronting humanity for which science and engineering clearly bear some responsibility: a technology-enabled, globalized social, political and economic life causing a threatening reaction of the physical world to our science- and technology-based action upon it. The distinctive re-engineering of human being in the world that has taken place over the past 200 years has clearly contributed to this threatening situation, so, we ask, can engineering show us how to ameliorate, if not resolve, its most threatening features?

That question will be explored here, beginning with putting the perception that science and engineering could play a role in designing new social systems into a historical context. As a matter of fact, this perception *has* a history, one extending back some 400 years. That it does, implies that the turn to science and/or engineering today for guidance in formulating public policies and designing social systems is not uniquely a response to today's technology-caused problematic world situation. Long before that situation arose, people had proposed basing social systems on science or engineering. Exposing the history of such proposals may thus shed light on the motives, and prospects, for turning for help now to engineering practice.

1.2 The Historical Context

Claude-Henri de Saint-Simon and Auguste Comte were perhaps the first people to propose a wholesale reorganization of society – in truth a re-engineering of society – in order to put scientists and engineers in leadership roles, alongside industrialists and financiers. New forms of science-informed, technology-driven industrialization were then just beginning to be recognized as constituting an 'industrial revolution' that was creating a new basis for prosperity for society. This was consistent with Adam Smith's vision in *The Wealth of Nations* (1776), and with Alexander Hamilton's *Report on Manufactures* (1791) to the US Congress, but it was in sharp contrast to the views of the Physiocrats, who in mid-eighteenth-century France had formulated the first holistic economic theory. This theory was propounded by Francois Quesnay in his *Tableaux Economique* (1759) and developed further by Baron Turgot, Minister of Finance to Louis XVI from 1774 to 1776. The Physiocrats argued that national wealth derived solely from agriculture, or more generally from extractive activities, including mining, fishing and arboriculture, so that the only productive class in society was made up of those citizens working in the extractive sector. Merchants, including factory owners and industrialists, artisans and even wealthy landowners, were 'sterile' in that they generated no net wealth themselves, but only repackaged and redistributed the wealth created by the extractors.

Adam Smith's economic vision, by contrast, was one in which trade and industry did indeed create wealth. Within twenty-five years of the publication of *The Wealth of Nations*, wealth created by technologically transformed industries in England and Western Europe was beginning to transform society. In his *Letters to an Inhabitant of Geneva* (1802), Saint-Simon argued that France's future prosperity depended on optimizing industrial production. To do that, French society needed to be reorganized so as to reap the benefits from an industry-driven, rather than an agriculture-driven, economy. Saint-Simon proposed the creation of a European committee of twelve scientists and nine artists to guard 'civilization' as this transition was made. He also called for a Council of Newton, composed of twenty-one scientists, the 'Elect of Humanity'. These men and women were to be nominated by, and supported by contributions from, the public to ensure that they were apolitical. Their task was to do research and to oversee a new scientific religion that Saint-Simon saw as central to a 'modern' society.

Seventeen years later, in his book *Social Organization*, Saint-Simon called for the creation of a new parliament comprising three chambers: a chamber of invention, composed of scientists, artists and engineers; a chamber of examination, composed of scientists only; and a chamber of execution, composed of leaders of industry. In a society whose well-being depended on optimizing industrial production, scientists and industrialists would be, for Saint-Simon, the 'natural leaders' of the working class, and a new political system was needed that reflected that reality. It followed that industrialists should replace the feudal and military classes in government, because the business of government was ensuring a social order that allowed for an optimal industrial–entrepreneurial environment.

Like Francis Bacon, who had argued early in the seventeenth century for mass education in the technical arts informed by his version of natural philosophy, Saint-Simon had no significant knowledge of science and even less of engineering. Both men were social reformers who had visions that science and technology could drive a nation's wealth and security. Unlike Bacon, however, Saint-Simon's vision was informed by his witnessing the beginnings in his lifetime of what he correctly foresaw would become a flood-tide of industrial development, driven by successive technological innovations that drew on scientific knowledge. If France were to manage the shift to such an industrial economy successfully, traditional politics, social organization, social systems and personal as well as social values needed to be replaced by new ones that reinforced science serving the needs of industry.

Saint-Simon could do no more than sketch what those new systems and values would be like, but his ideas reached a wide audience, especially among students at the École Polytechnique. Some of these students did subsequently influence French industrial policies, playing roles in the creation of the Credit Mobilier bank to finance industrialization and in the promotion of grand engineering projects like the Suez Canal, but the direct impact of Saint-Simon's ideas was modest. Their indirect impact, however, through the writings of Auguste Comte, was considerable.

Comte was thoroughly literate in the science and engineering of his day. Before serving as Saint-Simon's private secretary and assimilating his ideas, Comte had been a student at the École Polytechnique. Although he was ultimately expelled for his political activities, Comte received an excellent education in science-based engineering at a time when the Polytechnique was the leading engineering educational institution in the world, pioneering engineering education based on science, mathematics and the laboratory instead of the machine shop and field experience. Especially in his *Course of Positive Politics* (1851–1854), Comte presented by far the most detailed plan yet for creating a new society whose well-being would be driven by science applied to industry, with the parallel goals of order and progress.

Progress, Comte claimed, would come from the continuing growth of 'positive', that is, empirical, fact-based, knowledge via Bacon's experimental method, and its application by engineers to newer and better products and production technologies. Order would be preserved by a political system anchored in a science-reinforcing 'rational' religion. The values of this religion, inculcated into citizens from birth, would constrain not only the behaviour of the worker masses, but also the behaviour of the governors of society, ensuring that their decision-making would always be in the best interests of society rather than of the industrial elite alone. The key was for the governors to use the methods of 'positive' science – which for Comte always meant applied science, science stripped of metaphysical, merely theoretical, pretensions – to make political decisions.

Comte's new political system was openly hostile to parliamentary democracy, as was Saint-Simon's. For both, transposing the 'infallibility' of science to society entailed creating an elite ruling class. This was justified for them by the promise of continually improving the well-being of all citizens by optimizing the production of wealth by industry. This, in turn, required optimizing the ability of engineers to apply the growing body of scientific knowledge, and employing the methods of science and engineering in politics and in ethics. Scientists and engineers were to be core members of the ruling class, whose task was to maintain social order while enabling technology-driven progress given the continual social changes that progress entailed.

These plans of Saint-Simon and Comte did not erupt out of a vacuum. They were an extension of a broad eighteenth-century perception that the 'new philosophie' of nature created in the seventeenth century, what we call early modern science, could be, and needed to be, an agent of social reform. Already in the seventeenth century Hobbes, in *Leviathan* (Chapter 30), had called for using Galileo's method to make political philosophy scientific, and John Graunt, William Petty and Edmond Halley had begun the collection of social statistics for the express purpose of developing more effective public policies in the areas of trade and of government-issued annuities. In Part IV of *The Art of Conjecture* (1713), Jakob Bernoulli introduced the application of probability theory to 'Civil, Moral, and Economic Matters', including judicial reasoning and public policies. Both Turgot and his protégé the Marquis de Condorcet promoted probability theory as a rational basis for the design of social institutions. Turgot prepared for Louis XVI a 'rational' taxation policy (which led to his dismissal as Finance Minister!) and in 1785 Condorcet published *Essai sur l'application de l'analyse a la probabilite des decisions rendue a la pluralite des voix*, containing important theorems on jury and political voting procedures. Condorcet's 'voting paradox' is an example of his use of mathematics to determine public policy, namely, the voting process he was to recommend for the Revolutionary constitution. He was able to show that under quite plausible circumstances, voters ranking as few as three candidates would generate equally objective but non-unique outcomes, that a two-stage voting process could easily lead to an outcome the majority of voters did not prefer, and that voting was sensitive to the order in which choices were posed. (The so-called Impossibility Theorem of the mid-twentieth-century economist Kenneth Arrow is a more sophisticated version of Condorcet's voting paradox, one that lends itself to showing the inevitably non-unique character of ostensibly objective engineering design trade-off decisions.)

These are particular instances of a more general intellectual development in the eighteenth century: the claim that the new science of Newton *inter alia* was comprehensively applicable to human affairs. There is something paradoxical about this claim in that the new science, whether in its Galilean, Cartesian, Newtonian or Leibnizian form, was rigorously deterministic. But if reality, personal and social no less than physical, is deterministic, how can it be deliberately, wilfully, *reformed*? Nevertheless, the perception that the new science was successful because it epitomized reason, and that it was applicable to human affairs, was the foundation on which claims of Enlightenment and the Age of Reason rested.

For example, although Comte introduced the term 'sociology', Montesquieu had already effectively founded social 'science' in *The Persian Letters* (1721) and *The Spirit of the Laws* (1748). In these books, Montesquieu naturalized social institutions and values, bringing them within the scope of scientific inquiry and explanation. Between 1772 and 1791, Gottfried Herder developed a naturalistic, comparative cultural anthropology, reintroduced to a more receptive audience Giambattista Vico's earlier *New Science* (of cultural evolution) and published a *Dissertation on the Reciprocal Influence of Government and the Sciences* (1780). Naturalizing the economy as a reality amenable to scientific and eventually mathematical analysis was another eighteenth-century innovation. It is reflected in the essays on economic issues of David Hume, in the programme of the Physiocrats, in Adam Smith's *Theory of Moral Sentiments* (1759) as well as in his *Wealth of Nations*, and early in the nineteenth century in the very influential writings of David Ricardo, contemporary with Saint-Simon.

Hume followed Hobbes and Locke in formulating a naturalistic theory of knowledge, based on a scientific understanding of how the mind works. For Hume, this meant seeking a 'Newtonian physics' of the mind. The Abbe Condillac's writings triggered a movement whose members pursued a scientific psychology, after millennia of speculative philosophical psychology. Politics and ethics, it was argued, needed to reflect the reality of human beings in the physical and social world, as revealed by science. This was a founding principle of the American and French Revolutions, and of the English Reform movement. Denis Diderot's *Encyclopedie* project was, as the authorities well understood, subversive, and intentionally so. Diderot wanted to provoke social change by making technological no less than scientific knowledge available to all, hence the inclusion in the *Encyclopedie* of eleven volumes of illustrations of technical processes. Diderot was motivated to emphasize technology by his concern that the new science and the new mathematics were, by the late eighteenth century, becoming too abstract and too esoteric, at the expense of being useful to society.

This was the context out of which the social re-engineering programmes of Saint-Simon and Comte sprang. In fact, Western European, American and, from 1868, Japanese societies *were* transformed in the course of the nineteenth century, but the transformation was not along Saint-Simonian or Comtean lines, and not at all the product of thoughtful re-engineering. Changes in policies, institutions, organizations and values resulted from piecewise accommodations to the needs of new vested interests created by new forms of production and commerce. The new form of greatest impact, perhaps, was the centrally administered, hierarchically organized, vertically integrated industrial corporation. The explosive growth of these enterprises was wholly dependent on new production, communication, transportation and information technologies. These technologies, in turn, created the need for unprecedented numbers of science- and mathematics-trained engineers as employees, and for scientists and mathematicians to train them. It was in the context of optimizing the operation of these corporations that efficiency became a core value for business, and through business for society.

Frederick Winslow Taylor's 'scientific' analysis of work was the approach of a modern engineer to optimizing operations and workflow in a factory treated as a system (Kanigel, 1997). The obvious conclusion to be drawn from Taylor's studies was that engineers should be in charge of industrial corporations, not financiers and lawyers. Taylor's rationale for industrial leadership by engineers was preceded by Thorstein Veblen. In his *Theory of the Business Enterprise* (1904), Veblen had argued that capitalist-driven business was dominated by the pursuit of profit, with no concern for the well-being of society. New technologies, however, had the potential to improve all of society, if only they were developed and their benefits distributed to that end. If engineers were in leadership positions, Veblen wrote, they would use their commitment to specifically engineering values – accuracy, precision, efficiency and successfully meeting goals – to benefit society rather than to make profit the primary goal of industrial operations.

Veblen repeated these ideas in *The Engineers and The Price System* (1921) and in his last work, *Absentee Ownership and Business Enterprises in Recent Times* (1929). The economy, he argued, needed to be managed as one would manage an efficient machine. This entailed putting engineers in charge of the economy, not profit-driven businessmen. On a more abstract plane, Max Weber's *The Protestant Ethic and the Spirit of Capitalism* (1905), but even more so his *The Theory of Social and Economic Organization* (English translation 1947, but only part of a larger work that evolved between 1915 and 1925) independently reinforced Taylor's and Veblen's focus on the overriding value of efficiency in modern societies. Weber argued that rationality based on efficiency, defined techno-scientifically, was *the* core value of modern, industrial capitalist society. There was, therefore, no practical possibility that capitalists would cede control of wealth-creating businesses to anyone other than managers made in their own image. Jacques Ellul extended Weber's point in his highly influential *La Technique ou l'Enjeu du Siecle* (1954; in English, *The Technological Society*, 1964). Ellul argued the fundamentally anti-human character of technical rationality and his book stimulated a broad critique in the United States and in Europe of the roles of scientists and engineers as enabling contemporary social policies by providing technical expertise to political and economic elites.

In parallel with these intellectual/theoretical analyses, the publics in societies that were being revolutionized by technology-driven industry were acutely aware of how their personal and social lives were being transformed, and by whom. From the second half of the nineteenth century right through the Great Depression of the 1930s and into the post-war period, hundreds of popular novels, plays and films explored the roles of engineers and scientists in driving social change, typically by serving the interests of entrepreneurs and financiers (Goldman, 1989). Some of these works treated engineers and scientists heroically, some as pawns manipulated by capitalists. Some works were utopian, some were dystopian. Some, like the film version of H.G. Wells' *Things to Come* (1936), began as engineer-led utopias but ended as dystopias. Earlier, Fritz Lang's film *Metropolis* (1925) depicted an attempt by capitalists to annihilate the human working class in favour of robots developed for the capitalists by a 'mad' scientist. Lang's film echoed Karl Capek's play *R.U.R.* (1920), which introduced the term 'robot'. In *R.U.R.*, and in his novel *War with the Newts* (1936), humans are conquered by sentient servants created for humanity by scientists and engineers. Rene Claire's film *A Nous la Liberte* (1931) had a screen-play that might have been written by Veblen, and Charlie Chaplin's *Modern Times* (1936), which borrowed from Claire's film, mocked the inhumanity of technologically defined modernity.

With the onset of the Great Depression, Veblen's ideas about industrial leadership by engineers were taken up by Howard Scott and M. King Hubbert. In 1931, they created

Technocracy Incorporated, an organization whose goal was to put the economy under the control of scientists and engineers who would implement a thermodynamics-based econometric model keyed to energy efficiency, rather than to money (Meynaud, 1968). Although popular for a while, the movement that Scott and Hubbert hoped to precipitate never formed. William Akin, in *Technocracy and the American Dream* (1977), attributed this to the failure of Scott and Hubbert to articulate a believable political plan for accomplishing their goal, but it is also the case that every attempt to organize American engineers politically has failed.

What can we learn from this very brief survey that is relevant to our question of whether engineering reasoning can help us to design more effective social systems and public policies?

First of all, that the idea of applying to what Dewey called 'social interests and affairs' the scientific method and knowledge generated by that method is as old as early modern science itself, long preceding the experience of industry-caused social change. It follows that looking to science and/or engineering for more effective, because more rational, social systems and policies is not uniquely a reaction to the world situation today. Rather, the situation today has resurrected a recurring perception that the successes of science and of engineering are the result of subjecting experience to systematic, logical reasoning. The argument is as follows: science and engineering are undeniably extraordinarily successful in solving the problems they choose to address; this success is the result of following a well-defined, logical reasoning process; so, we as a society should be able to solve social problems by adopting that same process.

It is not at all clear, however, that there *is* a single method in accordance with which scientists reason. On the contrary, a great deal of evidence has accumulated that there is no single form of reasoning, no one method used by all scientists even within a single discipline, let alone across all disciplines. Furthermore, it has become increasingly clear that engineering reasoning is different from scientific reasoning, though engineering practice has been subject to much less study by historians, philosophers and sociologists than the practice of science.

A second lesson is that it was only in the nineteenth century that engineering was recognized as necessary, *along with science*, in order to realize the idea of making social systems 'rational'. By the early twentieth century, however, engineering increasingly seemed the *primary* model for rational management of social institutions, even as engineering increasingly incorporated knowledge generated by scientists. It was through engineering that inventions were turned into innovative technologies, as engineers selectively exploited scientific knowledge and stimulated the production of new engineering know-how, in order to solve the problems posed by bringing innovations into the marketplace. The result is that it is reasonable today to conclude that it is the way that engineers solve *problems* that society should adopt in order to develop more rational social systems, and not the way that scientists generate *knowledge*. The world we live in is directly the result of the form in which technological innovations were introduced into society and disseminated, a process in which engineering problem-solving played a pivotal, enabling role. Understanding how engineers play this role seems much more likely to provide a model for solving social problems than the process by which scientists produce idealized, universal theories explicitly removed from their social context.

Finally, we can learn from history the fruitlessness of any attempt at shifting political or managerial power to engineers or scientists and away from entrepreneurs (who may themselves be engineers or scientists), financiers and non-technically trained politicians, managers and bureaucrats. History and common sense strongly suggest that is not going to happen: power has more inertia than matter!

Our question thus becomes: Is there a method of reasoning employed in engineering practice that can be abstracted and applied by non-engineers to develop more effective social systems and public policies? Engineers would surely play a consultative role in such efforts, but as methodology mentors only.

1.3   Science and Engineering: Distinctive Rationalities

Scientists pursue understanding of natural and social phenomena. In the process, they generate what they call 'knowledge'. For the overwhelming majority of scientists, the object of this knowledge, what it is about, exists independently of the cognitive process. What scientists claim to know is the way that things are, independently of human experience and, in principle, independently of the human mind, individual or collective. This is the essence of scientific realism: a scientific knowledge claim is true, is validated as knowledge, by its correspondence with the way things 'really are, out there'. Even the minority of scientists who claim not to be realists claim that what they know transcends experience: there may not, for them, *be* black holes 'really', but there really *is* a stable pattern 'out there' that is revealed in experience and that corresponds to what we call a 'black hole'. These patterns are facts that are *revealed* in experience, but are not *produced* by experience.

From its seventeenth-century beginnings to the present day, science has been quite explicit that it was simultaneously empirical and, so to speak, trans-empirical; that it was rooted in experience but sought to disclose an unexperienced, indeed an unexperienceable, reality underlying experience and causing it. What its founders claimed distinguished modern science from Renaissance magical nature philosophy, which made the very same claim, was its use of a subject-neutral, objective method that reliably – fallibly and corrigibly, but reliably nevertheless – linked experience to its unexperienced causes.

Francis Bacon formulated an inductive, experimental, non-mathematical version of this method. Descartes formulated a deductive, mathematical version in which experiment played a very limited role. Galileo formulated a third version, modelled after his reading of Archimedes, that was intensively experimental but also intensively mathematical and deductive. Newton called himself a Baconian but was closer to Galileo and even to Descartes, protestations to the contrary notwithstanding. Newton described his method as analysis/ decomposition followed by synthesis/recomposition, success in the latter alone qualifying as understanding. But Newton also understood that the method a scientist used – the term 'scientist' is anachronistic until the 1830s, but easier to use than 'natural philosopher' – resulted in idealizations that did not correspond to the way things 'really' were, which is, of course, not revealed in experience but only in the mind, through calculation, which brings empiricism perilously close to rationalism.

What Newton and all his fellow scientists *did* share, methodologically, was a deep commitment to the non-subjective and value-free character of scientific reasoning. The logic of this reasoning was, in practice, fuzzy: a mix of induction, deduction, intuition, guesswork, serendipity and even some wishful thinking. Nevertheless, the self-professed goal of modern science has been and remains to describe nature as it is and to understand natural phenomena by providing an explanation of how nature works as a consequence of what its fixed properties are.

At the same time, and from its inception, modern science has been deeply conflicted about the status of its knowledge claims, and this conflict affects the usefulness of applying scientific reasoning to human affairs. As revealed by the rhetoric of science, from Bacon to the present day, scientific theories are said to describe a universal reality that transcends human experience, but which somehow is accessible to reason, even though reason seems limited to experience, which is particular. Scientific knowledge is said to be universal, necessary (because constrained by an independently existing reality) and capable of certainty (if it gets reality right), but it is dependent on experience, which is contextual, contingent and never certain.

The conflict lies in scientists recognizing that logic precludes claims that scientific knowledge is universal, necessary and certain, even in principle, and precludes having as its object an experience-transcending reality, while at the same time wanting precisely that to be the case. The only route to certainty is by reasoning deductively, but reasoning that begins with experience is ineluctably inductive, and it is no secret that there is no way of logically bridging the gulf between induction and deduction. That the experimental method cannot bridge this gulf was well known in the seventeenth century, as committing what some logicians call the 'Fallacy of Affirming the Consequent' and others the 'Fallacy of Affirming the Antecedent'. Christian Huyghens, for example, in the preface to his *Traite de la Lumiere* (1690), wrote that logically, experiments can never establish a necessary connection to an underlying cause, but when an experiment confirms a theory's predictions, especially in surprising ways, how can we not conclude that truth has been revealed to us! In this regard, Bacon was just as insistent as Descartes that following his method would lead to knowledge of reality. Galileo claimed that what we knew about nature by the deductive Archimedean method we knew in the same way that God knew, though of course God knew infinitely more. For Newton, the goal of his physics was to discover the 'true causes' of natural phenomena, not merely hypothetical, logically possible causes.

This continuing conflict within modern science over the nature of scientific knowledge, implicating the status of its knowledge claims and its trans-experiential object, echoes an ancient philosophical conflict. In his dialogue *The Sophist* (246a *et seq.*), Plato wrote of a perennial battle between the Sky Gods and the Earth Giants over the nature of knowledge. For the Sky Gods, the object of knowledge was an ideal, universal, timeless reality and knowledge was universal, necessary and certain, fundamentally different from belief and opinion. For the Earth Giants, the object of knowledge was the material world encountered in concrete, mutable experience and knowledge was a species of belief, thus particular, contingent and probable only. The (Platonic) philosopher was an ally of the Gods and pursued their version of knowledge. The allies of the Giants and their version of knowledge called themselves philosophers, but for Plato they were merely sophists, *knowing* nothing. Modern scientists clearly allied themselves with the Gods and Plato's philosophers in pursuing idealized knowledge for its own sake, even as they fraternized with Giants to collect data and perform experiments. Engineers, however, are wholly on the side of the Giants and the sophists. They are not interested in knowledge for its own sake, but in knowing for the sake of doing (Goldman, 1990).

Engineering and science are both intentional activities, but the intentionalities could not be more different. The intended object for science is understanding; the intended object for engineering is action. The rationality of science is the reasoning that results in claims of understanding and providing explanations. This reasoning is explicitly value-free, though implicitly it incorporates what philosophers of science call epistemic values ranging from honesty to employing assumptions that are not logical consequences of data. The object of scientific reasoning is assumed to be independent of the reasoning process, antecedent to the reasoning process and unaffected by that process.

The rationality of engineering, in contrast, is the reasoning that leads not to abstract knowledge, but to a kind of knowledge-enhanced know-how, the reasoning employed to solve a particular problem. As such, engineering reasoning is explicitly and inescapably valuational. Values are intrinsic both to the definition of an engineering problem and to what will constitute an acceptable solution to that problem. The rationality of scientific reasoning is measured by the conformity of the results of that reasoning to a supposed antecedent and independent state of affairs, called nature or reality. The rationality of engineering reasoning is measured by whether the results of that reasoning 'work', as judged relative to a set of highly contingent and very subjective value judgements. Furthermore, the object of engineering reasoning does not exist independently of that reasoning process, and it is very directly affected by that process because it is created by it!

Paraphrasing Theodor von Karmann, if science aims at revealing what is, engineering aims at introducing into the world what never was, for the sake of acting on the world in new ways. As Ortega Y Gassett argued in *Towards a Philosophy of History* (1941), this is not reducible to responding to physical necessities only. Today, as in remote antiquity, engineering the world is overwhelmingly a matter of projecting imagined possibilities onto a world that lacks them, and then making those possibilities actual. Engineering the world begins with simple tools and weapons, but advances to controlling fire and creating shelters and clothing, metals and ceramics, rafts and boats, new, unnatural kinds of animals and plants, fortified cities, monumental buildings, canals, irrigation systems. By the time writing was invented in the fourth millennium BCE, thousands of years of complex yet reliable know-how had accumulated and afterwards continued to accumulate, growing in complexity and scope, largely indifferent to writing.

Where is the dividing line between craft know-how and the kind of know-how that we recognize, after the fact, as representative of engineering? How did people acquire know-how? How did know-how cumulate and evolve over centuries and millennia without writing? We have a common-sense idea of how know-how is taught, by apprenticeship, but this does not explain such accomplishments as breeding dogs from wolves, wheat from wild grasses, or cultivable rice from wild rice, projects that took centuries of sustained, intentional effort. Animals have know-how, but human know-how is unique for being cumulative, adaptive and imaginative. Finding answers to questions about know-how, and thereby illuminating the engineering dimension of human beings in the world, has been ignored by Western philosophers, including natural philosophers, in favour of an obsession with abstract knowledge (Goldman, 2004). (What I mean by know-how here is very different from the philosopher Gilbert Ryle's term 'knowing-how'. Ryle's 'knowing-how' serves an epistemological function, fitting into his theory of knowledge. Know-how for me is ontological. It is a highly specific, and typically complex, form of doing or making, for example, making glass or breeding animals and plants or building a sailing ship. Historically, engineering as we use that term emerged when knowledge was applied to know-how. Perhaps the earliest explicit reference to doing this is in the opening chapter of Vitruvius' *On Architecture* (approximately 15 BC). The idea was revived in the Renaissance and was consciously pursued by self-styled engineers through the seventeenth and eighteenth centuries to the momentous convergence of engineering and the newly powerful scientific theories in the nineteenth century that generated the world we live in today.)

Elias Canetti put the subordination of engineering to science in Western culture this way: 'Among the most sinister phenomena in [Western] intellectual history is the avoidance of the concrete. [Thinkers] have had a conspicuous tendency to go first after the most remote things, ignoring everything that they stumble over close by' (Canetti, 1984, p. 14). But 'remote', abstract, knowledge, as Aristotle who embraced it acknowledged, is incapable of determining action, because action is always concrete, contextual and temporal. Aristotle concluded that action is determined by desire, desire shaped by experience-based beliefs and opinions, and at best informed by knowledge, but always and only initiated by desire. Engineering, which overlaps what Aristotle called 'praxis' and what he called 'art' (action consciously informed by established know-how), is decisively inferior to science/knowledge on this view, precisely because it is action-focused and lacks the universality, necessity and certainty of knowledge.

1.4   'Compromised Exactness': Design in Engineering

It seems clear enough, then, why the reasoning scientists employ in producing scientific knowledge is not going to be a model for formulating action plans. The knowledge itself may be useful, but not the reasoning process by which it was produced, whatever that process is, precisely. Scientific reasoning explicitly excludes the kinds of value judgements that are integral to action. And acting on the basis of scientific knowledge must be qualified by recognizing that it is corrigible and fallible. Scientific theories are continually evolving, continually being modified as new data are acquired, as new experiments are performed, as new instruments are invented, leading to new explanations of the same phenomena and to new theories based on new assumptions. But what about the reasoning process that engineers employ to produce solutions to engineering problems?

Engineering reasoning *always* subserves action and thus is intrinsically value-laden. On the face of it, then, engineering reasoning may well be a model for designing social systems and formulating public policies, both of which aim at action. Note, however, that the distinctiveness of scientific and engineering reasoning is ideal. As practices, science and engineering overlap in employing similar problem-solving modes of reasoning. Scientists, especially experimental scientists, look very much like engineers as they design, adapt and integrate equipment to create projected experimental situations that will generate anticipated data. Inevitably, individual pieces of equipment and the experimental setup as a whole require 'tinkering' that is pulled by the intended object of the experiment, its intended outcome, which thus acts like an Aristotelean final cause: the experimental setup exists in order to produce a specific outcome. That is, the anticipated outcome 'pulls' the conception of the experiment, its design and its execution, especially the judgement that it 'worked'.

The recent announcement that the Large Hadron Collider had revealed the existence of the Higgs boson, a logical consequence of the prevailing Standard Model of quantum physics, is an example of this. The machine was designed and built as dictated by the prevailing theory in order to produce the Higgs boson, and it did. At least, its operation was interpreted via specialized computer-based analysis as having produced Higgs bosons, albeit very few of them. Had they not been produced, scientists would either have had to modify their theory, search for a new theory, or decide that the Large Hadron Collider was not working properly or not up to the job. The latter conclusion was reached in the case of the failure of the $400 million Laser Interferometer Gravity Observatory to detect the gravity waves predicted by the General

Theory of Relativity and so was modified, at considerable additional expense, and then it did do the job!

In its experimental mode, science as a practice implicitly *seems* very much like engineering, but without the explicit wilfulness and extra-technical value judgements that are central to engineering. For their part, engineers are often said to employ scientific reasoning when they engage in experimental testing of designs, acquire and keep systematic records of data, and modify designs in light of that data. This is, however, a popular and misleading use of the term 'science', as technologists even in remote antiquity did the very same thing as they created bodies of sophisticated know-how. There is thus a form of problem-solving reasoning on which both science and engineering build, but which precedes both, logically and historically.

The distinctiveness of engineering *vis-à-vis* science is most clearly revealed in the design process, understanding 'design' as that facet of engineering practice that produces a specification of the terms of a problem together with criteria of *acceptable* solutions to that problem. Both of these require multiple complex value judgements centred on what someone wants the outcome of the design process to *do*. The outcome of the design process is not the revelation of a pre-existing state of affairs as in science, but an act of creation. What engineers enable is an outcome determined by the wilfulness motivating a desire for what some outcome must do and how it needs to do it. The problems that engage scientists are 'there', waiting to be recognized. Engineering problems, by contrast, are created by people who *want* to do something specific and are constrained in various ways, to a degree by what nature will allow, but primarily by highly contingent factors that, from a logical as well as a natural perspective, are arbitrary: time, money, markets, vested interests and social, political and personal values.

It follows that engineering reasoning in the design process is in a sense 'captive' to the wilfulness underlying the specification of engineering problems and their solutions (Goldman, 1991). This makes the reasoning employed in engineering problem-solving profoundly different from that in scientific problem-solving. Since the early nineteenth century, however, engineering has become increasingly dependent on scientific knowledge, lending support to the claim that engineering is 'merely' applied science. But engineers deploy scientific knowledge selectively, on terms dictated by the criteria of the engineering problem and the solution specifications. These criteria reflect complex, contingent and competing value judgements that have nothing to do with technical knowledge and are beyond the control of the engineers engaged in the design process. There is, therefore, a profound gulf between engineering and science, as profound a gulf as the one in logic between inductive and deductive inference. Engineers selectively employ scientific knowledge on terms dictated by the requirements of solving engineering problems as scientists selectively deploy mathematical knowledge on terms dictated by the requirements of solving scientific problems. Engineering is no more *just* applied science than science is *just* applied mathematics.

In the course of creating a new science-based engineering curriculum for the University of Glasgow in 1855, one that justified distinguishing engineering education from the existing science education while still linking engineering to science, William Rankine, eminent in science and in engineering, characterized the reasoning underlying engineering practice as 'compromised exactness' (Channell, 1982). The exactness in engineering came from its utilization of experimentally validated scientific knowledge. Such knowledge was exact because it was universal, abstract and deductive, as in the then prominent Lagrangian version of Newtonian mechanics, Fourier's mathematical theory of heat and the new science of thermodynamics.

But what makes something engineering, and not merely applied science, is that the *application* of exact technical knowledge was 'compromised' by commercial considerations of economy and efficiency. If scientific knowledge was, as it claimed to be, objective, then engineering, in spite of utilizing scientific knowledge, was not objective in the same sense, and could even be said to be subjective, because it incorporated person- and situation-dependent value judgements. What Rankine meant by 'compromised exactness', then, goes to the heart of the distinctiveness of engineering *vis-à-vis* science.

Engineering cannot be derived from science because application, by its nature, is constrained by explicitly contingent value judgements. That's what Rankine meant by characterizing engineering as 'compromised' exactness; not that engineering is simply approximate, while science is exact, but that engineering is inescapably wilful, valuational and contextual. Engineers are challenged to build *this* bridge, for *this* purpose, in *this* place, within *this* budget and in *this* time frame. Engineers need to design an airplane engine, an automobile transmission, a computer that satisfies a specification that is virtually never driven by what the engineers consider the best possible design given their technical knowledge, but by one among many possible interpretations of what is judged best in a particular situation by individuals committed to highly contingent commercial or political agendas.

The IBM PC, for example, back in 1981 was in no way the best 'personal' computer design possible then, from an engineering perspective, but its commercial success justified its 'compromised' character: selectively applying available technical knowledge on behalf of IBM management's interpretation of projected market opportunities. Tracy Kidder (1981) exposed the very same 'compromised' character of engineering in the design of a highly successful mini-computer that saved the Data General Corporation (for a while). The same kind of 'compromising' is revealed in the designs of the US Space Shuttle (Logsdon, 2015), the International Space Station (Logsdon, 1998), and indeed of everything from weapons systems to consumer products that engineers enable bringing into existence. Sometimes the 'compromised' design is successful and sometimes not, but success or failure has less to do with the technical knowledge employed than with the value judgements constraining the application of that knowledge. The Segway is a case in point: technically brilliant, but a marketplace failure, relative to its inventor's, and its investors', expectations (Kemper, 2003). In the case of Apple's iPod, iPad, iPhone and iWatch, marketplace success followed not from technical excellence, but from managerial judgements of what the finished product needed to look like, what it would do and most importantly, *how* it would do what it would do compared with what competing products were doing.

Rankine's characterization of engineering as distinctive because it is 'compromised exactness' thus finds its clearest expression in the engineering design process, because that is where problems and acceptable solutions are defined by the introduction of contingent value judgements. But if the design process is the essence of engineering, then the essence of the design process is trade-off choices among the many natural and contingent parameters that comprise the technical specification of the design problem and its acceptable solution, with an emphasis on 'acceptable'. The term 'technical' here disguises a host of very non-technical and subjective value judgements. In the case of physical products, for example, these would include trade-offs among manufacturing cost (which implicates the materials to be used and the suppliers), size, reliability, safety, longevity, serviceability, performance, appearance, ease of use, compatibility with related products and services, compliance with relevant laws and regulations, scale of production, scalability of product and its modifiability, time to market, and market niche and marketability (given existing products and services). Services and software designs would generate a similar list of trade-off choices.

It is in trade-off choices that the wilfulness underlying engineering problem-solving emerges. What engineers know, and what they can adopt and adapt from relevant pools of engineering know-how and scientific knowledge, certainly influences what they can do as engineers to solve technology-related problems for people (Vincenti, 1990). This knowledge influences the negotiations eventuating in a well-defined problem and its acceptable solution, but it does not determine that outcome. From the perspective of engineering, problems and possible solutions are wildly over-determined: there are too many possibilities, and the ones that engineers would prefer on engineering grounds typically are not the ones that management prefer!

Attempts at reducing the subjectivity of trade-off decisions by means of an objective, and thus 'scientific', formal methodology have been numerous. They include: Analogy-Based Design; Analytic Hierarchy Process; Suh's Axiomatic Design; Decision Matrix Techniques; General Design Theory; Quality Function Deployment and the House of Quality; IDEF; PDES-STEP; PERT charts; Pugh Concept Selection; Taguchi's Theory of Robust Design; DYM; PYM; TRIZ; and many more. The sheer number of theories and commercial products that claim to make trade-off choices objective alone suggests that none of them do the job. For any given design team, however, one or more of these will be helpful, and some tool for managing a large number of mutually implicating decisions is close to a necessity. In the end, however, wilfulness and subjectivity are ineluctable features of trade-off decision-making, and therefore of the design process, and therefore of engineering reasoning and practice. This conclusion is supported by arguments that it is not possible to generate a unique ranking of three or more parameters, each of which has three or more possible states or values. Among these arguments are Condorcet's voting paradox mentioned earlier and Kenneth Arrow's 'Impossibility Theorem' (Arrow, 1951). The ranking process may be quantitative and thus apparently objective, but there are no unique outcomes, and at least some of the ranking weights must be subjectively assigned. How much weight should be assigned to size, weight, reliability, specific materials, certain performance features compared to others, cost of production, ... [An analogous problem exists with assigning prior probabilities in Bayesian probability theory: once assigned, the probabilities generated are strictly objective, but no one has yet come up with a process for assigning prior probabilities objectively, though many have tried.]

1.5   Engineering Social Systems?

The upshot of all this is that there is a strong similarity between the engineering design process with its trade-off choices and the process by which social institutions are created and public policies are formulated. Both are political processes in the sense of being framed and determined by irreducibly subjective, highly contingent and competing value judgements. If this is indeed the case, then how could engineering reasoning contribute to designing better social systems and formulating better public policies? Not, obviously, by transforming the latter into an objective process. But the 'logic' of the engineering design process shares with the social system design process the goal of a functional accommodation among multiple competing and value judgements. In the case of engineering, however, all of the value judgements that determine the specification of problems and solutions are explicit and subordinated to the functionality of the projected outcome. As a result, and in spite of the irreducible subjectivity of the process, the process generates objective metrics for assessing whether the outcome works as anticipated.

This clearly has not been the case historically with social systems and public policies. These, too, are the outcomes of negotiated design processes, but the outcomes of those processes typically are keyed to satisfying holders of the contending value judgements that their interests have been served. And the contending value judgements are often disguised or misrepresented as facts. In the end, the historical process results in an outcome that satisfies the interests driving the trade-off process and is only loosely coupled to the functionality of that outcome for the public as a whole.

This is the reverse of Dewey's 'experimental method', the methodology of engineering, in which means have no value in themselves but only insofar as they subserve explicit ends and are measured by their effectiveness in doing so. In social system and public policy design, the value judgements underlying the selection of means do have value in and of themselves. They have value to the special interests shaping the design process, and they even shape the end so that it reflects those interests. If the design of social systems and public policies required making explicit the criteria for assessing the outcome as successful, along with the trade-off judgements made to achieve that outcome, then there is every reason to believe that systems and policies would be improved.

But adopting the form of its design and trade-off processes is not all that engineering can contribute to creating improved social systems and public policies. The design process as described here captures the essence of what makes engineering *reasoning* distinctive *vis-à-vis* scientific reasoning. But reasoning is only one dimension of engineering *practice*, which is also essentially experimental and evolutionary. The outcome of the design process needs to be implemented, it needs to be produced and it needs to be used. Implementation reveals that the engineering design process is open-ended. Unlike the deductive inferences drawn from a scientific theory, design solutions can never be guaranteed to work successfully, as judged either by the engineers themselves or by their employers. In this sense, engineering practice is experimental (Petroski, 1992, 2012).

Implementation of design problem solutions generates ongoing feedback into both the solution and the problem. The feedback may reveal weaknesses in the solution that require reassessment of the trade-off choices made and a redesign. But feedback from practice may require reassessment of the problem itself, along with reconsideration of what *now* constitutes an acceptable solution to it. Especially when the design problem involves a new technology, use typically stimulates unanticipated applications that may require modification of the original solution, but always generates new design problems in order to exploit new action opportunities enabled by these applications, or to prevent or manage them. In this sense, engineering practice is evolutionary in the biological sense of having an unpredictable dimension to it. We are witnessing this phenomenon today in the continuing evolution of semiconductor-based technologies and their coupling to the evolution of the Internet, from its initial implementation as an esoteric DARPA-funded project to its extraordinary global penetration of personal, social, economic, political, cultural and professional life. But the Internet is, in this evolutionary respect, typical of engineering design projects. Consider the reciprocally influencing technical and social impacts of the implementations of the steam engine, the gaslight industry, the telegraph and telephone, the automobile and electric power (Hughes, 1983; Hunter, 1985; Starr, 2005; Tomory, 2012; Volti, 2006).

The processes by which social systems and public policies have been designed in the past lack the open-ended, experimental and evolutionary character of engineering design driven by feedback from their implementation. For reasons that are external to engineering *per se*, that is, external to engineers simply as possessors of technical knowledge, engineering designs succeed or fail and are forced to evolve. Often this evolution is driven by factors that to engineers are irrational, for example, marketplace factors that value form over function, mere fashionability, advertising-manipulated style for its own sake, a company's need to distinguish its products from competitors, or a strategy of driving sales by generating change for its own sake, claiming to be 'new and improved'. Often, however, the evolution of engineering designs is driven by the 'space' a new design opens for various forms of improvement and by the innate unpredictability of technological innovations.

Once adopted, innovations typically inspire applications that had not been anticipated, analogous to mutations in biological evolution, and these drive the need for newer and new kinds of designs. Again, the Internet is a perfect instance of this evolutionary character of engineering, not at all 'blind variation' but unpredictable nevertheless. [And thus different from at least Donald Campbell's conception of evolutionary epistemology (Campbell, 1960, but see Polanyi, 1962 and Popper, 1972).] The ubiquity today of music streaming, video streaming, social media, photo and video file sharing, and their implications for hardware and software design are obvious. But this was just as true of Edison's phonograph and moving-film projector. The implementation of innovations also can, and typically does, have consequences that are undesirable as well as unpredictable. Death and injury caused by the automobile, its impact on mass transit and the design of cities, and its socio-political and environmental impacts illustrate this. So does the increasingly threatening 'dark side' of the Internet, including pornography, gambling, identity theft, hacking, cybercrime, cyberwarfare and terrorist networks.

By contrast with engineering's explicitly experimental and evolutionary character, social systems and public policies are promulgated *as if* they were definitive solutions to social problems to which personal and social action must conform, rather than tentative, pragmatic solutions that are expected to be modified by the consequences of their implementation. They do not openly incorporate feedback mechanisms that make learning from the implementation process a normal feature of the social systems and public policy design processes. Instead, the politics of these processes virtually precludes incorporating into policy implementation legislation clear metrics for assessing when a policy is meeting its proclaimed goals together with a process for modifying the policy in light of its actual outcomes. In the United States, instances of this would include the Prohibition amendment to the Constitution, the 'War on Drugs', mandatory federal prison sentences, the original Medicare act and its prohibition of negotiating the cost of prescription medications with pharmaceutical companies, the federal student loan programme with no provision for monitoring the consequences of open-ended lending, the Cuba embargo with no provision for assessing its impact on the Cuban people rather than on the Cuban government, the exemption of the fracking industry from the Clean Drinking Water Act, the commitment to NATO with no post-Cold War exit strategy, and many others.

Implicitly, social policies, like engineering designs, *are* open-ended; they are also implicitly experimental and in principle capable of modification in response to feedback from outcomes. The Prohibition amendment, for example, was repealed, but the 'war' on drugs drags on. As the other policy instances listed above illustrate, the political dimension of social policy-setting precludes keeping the policy-setting process open to modification by feedback from accumulating experience with the policy's implementation. In the world of engineering, it is expected that designs will evolve in light of experience, while in the world of social policy, actors behave *as if* policies were definitive. To be sure, there is a difference between the timelines of engineering design evolution and social policy evolution. It would be counterproductive for policies to be modified too quickly in response to outcomes. Social policies need to promote a sense of stability, a sense that they are not going away, if they are to force, and enforce, changes in behaviour, laws and investment strategies. And there is often a moral dimension to social policies, for example in the case of civil rights legislation, that complicates the process of revising a policy based on its short-term consequences, which included numerous and sustained acts of violence, for example, over school desegregation and busing. It is nevertheless suggestive that modelling social system and public policy design after a *comprehensive* understanding of engineering methodology, one that includes the distinctiveness of engineering reasoning together with the experimental and evolutionary character of engineering practice, has the potential to significantly improve the performance of social systems and the effectiveness of public policies.

References

Arrow, K. (1951) *Social Choice and Individual Values*, Yale University Press, New Haven, CT.

Campbell, D.T. (1960) Blind variation and selective retention in creative thought as in other knowledge processes. *Psychological Review*, **67**(6), 380–400.

Canetti, E. (1984) *The Conscience of Words* (trans. J. Neugroschel), Farrar, Straus and Giroux, New York, NY.

Channell, D. (1982) The harmony of theory and practice: The engineering science of W.J.M. Rankine. *Technology and Culture*, **23**(1), 39–52.

Dewey, J. (1988) *The Quest for Certainty*, Southern Illinois University Press, Carbondale, IL.

Goldman, S.L. (1989) Images of science and technology in popular films. *Science, Technology and Human Values*, **14**(3), 275–301.

Goldman, S.L. (1990) Philosophy, engineering and western culture, in P. Durbin (ed.), *Broad and Narrow Interpretations of Philosophy of Technology*, Kluwer, Amsterdam.

Goldman, S.L. (1991) The social captivity of engineering, in P. Durbin (ed.), *Critical Perspectives on Nonacademic Science and Engineering*, Lehigh University Press, Bethlehem, PA.

Goldman, S.L. (2004) Why we need a philosophy of engineering: A work in progress. *Interdisciplinary Science Reviews*, **29**(2), 163–176.

Hughes, T. (1983) *Networks of Power: Electrification in western society 1880–1930*, Johns Hopkins University Press, Baltimore, MD.

Hunter, L.C. (1985) *Steam Power: A history of industrial power in the United States, 1780–1930*, University of Virginia Press, Charlottesville, VA.

Kanigel, R. (1997) *The One Best Way: Frederick Winslow Taylor and the enigma of efficiency*, Viking Press, New York, NY.

Kemper, S. (2003) *Code Name Ginger: The story behind the segway and Dean Kamen's quest to change the world*, Harvard Business Review Press, Cambridge, MA.

Kidder, T. (1981) *The Soul of a New Machine*, Little Brown, Boston, MA.

Logsdon, J.M. (1998) *Together in Orbit: The origins of international participation in the space station*, NASA, Washington, D.C.

Logsdon, J.M. (2015) *After Apollo? Richard Nixon and the American Space Program*, Palgrave Macmillan, New York, NY.

Meynaud, J. (1968) *Technocracy* (trans. P. Barnes), The Free Press, New York, NY.

Petroski, H. (1992) *To Engineer is Human: The role of failure in successful design*, Vintage Press, New York, NY.

Petroski, H. (2012) *To Forgive Design: Understanding failure*, Belknap Press, New York, NY.

Polanyi, M. (1962) *Personal Knowledge. Towards a Post-Critical Philosophy*, Routledge & Kegan Paul, London.

Popper, K. (1972) *Objective Knowledge. An Evolutionary Approach*, Oxford University Press, Oxford.

Starr, P. (2005) *The Creation of the Media: Political origins of modern communication*, Basic Books, New York, NY.

Tomory, L. (2012) *Progressive Enlightenment: The origins of the gaslight industry 1780–1820*, MIT Press, Cambridge, MA.

Vincenti, W. (1990) *What Engineers Know and How They Know It*, Johns Hopkins University Press, Baltimore, MD.

Volti, R. (2006) *Cars and Culture: The life story of a technology*, Johns Hopkins University Press, Baltimore, MD.

2

Uncertainty in the Design and Maintenance of Social Systems

William M. Bulleit

2.1 Introduction

Engineering design must be performed under a wide range of uncertainties. The uncertainties arise from a number of sources, leading to aleatory uncertainties (those related to inherent randomness) and epistemic uncertainties (those arising from limited knowledge). Systems where prototypes can be built and tested during the design process (e.g., automobile engines) generally exhibit a lower level of uncertainty than systems where prototypes cannot be used (e.g., bridges) (Bulleit, 2013). Social systems are non-prototypical, since we cannot build a prototype of the system and test it during the design process. Most engineers today typically deal with systems that are *simple* or *complicated*. A simple system is generally a single component in a larger system. A complicated system is a combination of simple systems that exhibit component interactions, but generally the interactions do not lead to non-intuitive system behaviour. Simple and complicated systems can be either prototypical or non-prototypical.

Social systems are *complex*, and more likely a subset of complex referred to as *complex adaptive*. Complex and complex adaptive systems have components that are highly interconnected, with strong interactions. In the case of complex adaptive systems, they also include agents, such as humans, who can adapt their behaviour to each other and to the environment in which they live. These types of complex systems present special problems to engineers, because they can exhibit non-intuitive behaviours through self-organized criticality and emergence (e.g., Bak, 1996; Holland, 1998; Miller and Page, 2007). Furthermore, the level of uncertainty that must be dealt with in the design of these systems is greater, potentially far greater, than for simple or complicated systems, whether they are prototypical or not.

Since complex systems can exhibit chaotic behaviour, self-organized criticality and emergence of unusual events, even system collapse (Crutchfield, 2009), the techniques used for *design prediction* of simple and complicated systems will generally not be adequate for

Social Systems Engineering: The Design of Complexity, First Edition. Edited by César García-Díaz and Camilo Olaya.
© 2018 John Wiley & Sons Ltd. Published 2018 by John Wiley & Sons Ltd.

full-scale social systems. Complex systems produce behaviour that is highly variable, sometimes even over short time frames. Thus, one issue that must be addressed for design is the *prediction horizon* (i.e., at what point in the future does the probability of a useful prediction become so small that it is no longer valid). For simple/complicated systems, the prediction horizon is often measured in years for mechanical systems and decades for civil structures. For instance, design life in the United States is 50 years for most civil structures. The prediction horizon is likely to be quite a bit smaller than that for complex systems, particularly social systems. Estimating the prediction horizon will be a necessary part of any design of social systems. Techniques to model complex and complex adaptive systems to estimate the prediction horizon might include information-theoretic approaches (Bais and Farmer, 2007), stochastic process approaches (Bhat, 1972; Williams, 2003), simulation (Bulleit and Drewek, 2011; Epstein and Axtell, 1996; Gilbert and Troitzsch, 2005) and system dynamics (Forrester, 1995). These techniques might prove useful in the future, but it is not clear at the present time whether any of them will eventually prove useful in social systems engineering.

Events that are far outside the realm of experience are sometimes referred to as *black swan* events (Taleb, 2007). Since, by definition, it is not possible to design for black swan events, social systems will need to be designed such that their details increase the chances that the system will respond to unexpected events in a *resilient* manner. This requirement means that when social systems are designed, engineers will need to consider *robustness* (the ability to withstand an event) and *rapidity* (the ability to recover from an event). The combination of robustness and rapidity is often referred to as *resilience*.

The type of system that is chosen in a social system design will affect its behaviour. Distributed systems generally are more robust to external stressors than are systems where the component behaviours are tightly coupled or highly correlated. In those cases, an event that affects only one component can lead to system-wide damage or even collapse. Furthermore, top-down control of a distributed system will tend to increase the correlation between components, thus likely reducing robustness. Design of social systems requires an appropriate level of top-down control (usually fairly small), balanced by bottom-up design and maintenance. This balance is difficult to find and difficult to maintain once it is found, and the balance will likely change as the system evolves. In the case of social systems, some of the uncertainty may even be beneficial to the evolution of the system.

Design of an effective social system requires accounting for human and other living agents, a potentially complicated environment, and strong agent-to-agent and agent-to-environment interactions, which cause the prediction horizon to be potentially very short. Any system with these limitations requires that feedback become an important aspect of the design and maintenance. In many ways the design becomes the maintenance. Thus, the term 'maintenance' becomes somewhat misleading, because it refers to maintaining the behaviour of the entire system, even though the system is evolving and changes are being made to local regions of the system, both of which will alter the system.

In the remainder of this chapter, we will delve more deeply into uncertainty and its effects on the engineering of systems. The uncertainties inherent in a given system will affect the types of heuristics, colloquially 'rules of thumb', that can be used in the design. Thus, design of simple systems with low uncertainty may be vastly different from design and maintenance of complex adaptive systems such as social systems.

2.2 Uncertainties in Simple and Complicated Engineered Systems

Uncertainty increases as we move from simple, prototypical systems (e.g., an automobile frame) to complicated, non-prototypical systems (e.g., long-span bridges), to complex adaptive systems (e.g., the U.S. economy). The uncertainties can generally be separated into aleatory and epistemic uncertainties (Der Kiureghian and Ditlevsen, 2009). Consider the flipping of a fair coin. Flipping a fair coin is usually thought of as aleatory uncertainty, because the uncertainty appears to be related to chance. But it might be possible to model the coin-flipping process so well – including coin imbalance, air resistance, hand behaviour, etc. – that the results of the flip would be nearly predictable. If a prediction such as that were possible, then the uncertainty in coin flipping would be primarily epistemic rather than aleatory, because an increase in knowledge about coin flipping, better modelling of the process, reduced the uncertainty. In this section we will examine some sources of uncertainty in terms of the contribution to each from aleatory and epistemic uncertainties. The manner in which each of these sources should be dealt with in the design of engineered systems is often affected by whether the contributing uncertainties are primarily aleatory or epistemic (e.g., Bulleit, 2008). There are five major sources of uncertainty in these systems: temporal uncertainties, randomness, statistical uncertainties, model uncertainties and human influences.

Temporal uncertainties arise from actions such as using past data to predict future occurrences (e.g., using snow load data over the past 50 years or so to predict the snow load over the next 50 years), changes of loadings due to societal change (e.g., bigger and heavier trucks being allowed on the road), changes in material properties (e.g., soil properties can change over time through physical and chemical processes in the ground) and modifications to design standards due to evolving engineering knowledge (e.g., it is possible that a system being designed today is being under-designed because present design knowledge is inadequate). As will be discussed later, complex and complex adaptive systems exhibit even more temporal uncertainties. Since we can never know enough about the future to remove all temporal uncertainties, most of them are primarily aleatory in nature.

Uncertainties due to *randomness* are ubiquitous in engineering. All material properties are random variables, varying about a central value, such as the mean value. All loads (e.g., wind and earthquake) are highly variable. These uncertainties are generally viewed as aleatory, but some portion of them is likely to be epistemic. For example, wind interacts with a structure to produce the loads. Such effects as flutter and vortex shedding increase the forces that the structure experiences. Since engineers have been able to learn more about these effects through theory and testing, such as wind tunnel tests, we have reduced the uncertainty of the wind loads on the structure. In the past, much of that uncertainty would have been attributed to randomness, since the wind is a stochastic process. Thus, although uncertainty from randomness is aleatory, what appears today to be randomness may be limited knowledge; and if that is the case, then that portion of the uncertainty is epistemic. Complex and complex adaptive systems will exhibit uncertainties that may appear to be due to randomness but may prove to be able to be reduced by modelling in the future. Examples are chaotic behaviour, self-organized criticality and emergence.

Statistical uncertainties arise when we try to reduce the effects of randomness by gathering appropriate data. One way to help reduce the uncertainty due to material property randomness is to obtain material data, such as taking soil borings before designing a foundation, or taking concrete samples to obtain concrete strength properties for the concrete going into a structure. In either case, only a small sample of specimens can be tested or a small number of soil borings obtained. So, although the data give some information about the future system, there is uncertainty about whether the small-sample data set is representative of the soil under the foundation or the concrete in the structure. This uncertainty is statistical in nature. It is primarily epistemic since we can, in theory, reduce the uncertainty by using a larger number of samples. From a practical standpoint though, we will be limited by the cost of obtaining and testing the samples necessary to significantly reduce the uncertainty. Thus, although much of the statistical uncertainty is epistemic, we can never obtain enough data to reduce that variability significantly. As a result, the uncertainties here include a combination of aleatory and epistemic contributions. In social systems, data will need to be collected to examine whether changes that have been made to the system are having the desired effect. Statistical uncertainty may prove to be significant when examining social system data.

The next source of uncertainty is associated with the models used in the analysis of the systems being designed. In the design of non-prototypical engineered systems, models are the primary way to examine the behaviour of the entire system prior to its construction. Thus, *model uncertainty* will be a larger contributor to the uncertainty in the system analysis of non-prototypical systems, including social systems, than it is for engineered systems where prototypes can be used. There are two basic types of model uncertainty. The first is related to how well a prediction equation or algorithm models test data and the second is related to how well a system model (e.g., a finite element model or agent model) predicts the actual behaviour of the system. The first type can be dealt with in design by developing a model bias factor that accounts for the difference between the predicted behaviour and the test data. Because the bias factor is itself a random variable, the uncertainty in this context seems to be aleatory; but, as in the flipping of the coin, if we can develop better prediction models then we will reduce the bias between the prediction and the tests. Another way to reduce the uncertainty in the bias factor is to perform more tests for comparison to the prediction model. It is possible, though, that the additional tests may show that the prediction model is worse than believed, which would either falsify the model or increase the uncertainty in the model. Thus, this first type of model uncertainty is a combination of aleatory and epistemic uncertainties. The second type of model uncertainty is also a combination of epistemic and aleatory. This uncertainty is affected by the engineer's conceptual understanding of the modelling technique being used, how much effort the engineer can afford to refine the model and the accuracy of the modelling technique when it is used to its fullest. Given enough time and money, a model could be developed that would closely represent the built system, but the amount of time and money necessary would be so great that we could not afford to build the system. Thus, model uncertainty is a combination of both aleatory and epistemic uncertainties, because no matter how carefully we model our system we can never account for all behaviours in the final version, except of course the 'model' we create when we construct the system. Modelling will be at least as important for design and maintenance of social systems, but due to the present limited knowledge about appropriate modelling techniques, large-scale interactions of agents, system evolution and the possibility of emergence (in the system and the model), the model uncertainty may be a significant contributor to the uncertainties that must be dealt with in social systems.

The last source of uncertainty is *human influences* on the system. For simple and complicated engineered systems, human influence, in the form of human error, is a major contributor to the uncertainty in the design and behaviour, and many failures are primarily due to it

(Petroski, 1992, 1994). The uncertainties produced by human errors are primarily epistemic, since increases in modelling skills will help reduce conceptual errors. For social systems we will have human error in modelling and design, but we will also have human actions that will influence the system behaviour in possibly significant ways (e.g., terrorist attacks) and may also have other agent influences, such as the effects of living beings on an ecological system. Human and other agent influences will likely become the primary source of uncertainty in social systems.

2.3 Control Volume and Uncertainty

Engineers use a broad range of *heuristics* in design. According to Koen (2003), 'A *heuristic* is anything that provides a plausible aid or direction in the solution of a problem but is in the final analysis unjustified, incapable of justification, and potentially fallible' (p. 28, italics in original). Heuristics allow engineers to solve problems that would otherwise be intractable from a purely mathematical and scientific standpoint. Koen (2003) goes as far as to say that the engineering method is to use heuristics. The problem, from the standpoint of this chapter, is that these heuristics have been developed for systems that are simple, complicated or, at most, complex. Consider a simple system, a single component in a larger system, such as a beam in a building, a crankshaft in an engine, a weir in a water system or a pipe in a sewage system.

A complicated system is a combination of simple systems that exhibits component interactions, but the interactions do not lead to chaotic or emergent behaviour. Certain complicated systems can exhibit 'domino behaviour' (Harford, 2011), in which one part of the system fails and leads to a collapse of the entire system or a significant portion of it. One example is progressive collapse in structures, such as occurred at the Ronan Point apartment buildings in London, when a gas explosion in an apartment near the top of the building caused all the apartments above and below it to collapse (Delatte, 2009). Examples of complicated systems are automobiles, aircraft, buildings and large bridges.

In all these examples, both simple and complicated, the system, or a portion of it, is considered for design to be inside a *control volume* (Vincenti, 1990). For instance, considering a weir in a water system, when the engineer designs the weir, only the weir itself is considered directly. The water flowing in is an input and the water flowing out is an output. The weir is isolated from the entire water system. The assumption is that the weir will not have an effect on the entire system, except in as much as it changes the input flow to the output flow. A tall building, an example of a complicated system, is also considered inside a control volume during design. The environment applies loads to the building and once we know those, we no longer need to worry about how the building affects anything else. Once we have isolated the force diagram 'we can forget about the rest of the world' (Bucciarelli, 2003, p. 50). This approach to design has worked well for engineers in the past, and will continue to work well in the future, *except* when the simple or complicated system being designed is part of a *complex* or *complex adaptive system* (Crutchfield, 2009; Holland, 1995, 1998; Miller and Page, 2007). Pool (1997) and Perrow (1999) discuss this dichotomy in the context of the nuclear power industry. Designing social systems will require that a wide range of control volumes be used, always keeping in mind that the use of control volumes that do not comprise the entire system may lead to unintended changes to the overall system.

What if the buildings, the control volume, under consideration were the World Trade Center towers on September 11, 2001? In that case, the towers withstood the plane crashes, didn't topple over and stayed up long enough to allow most of the occupants to escape. The original design of the towers included the possibility of a slow-moving Boeing 707 crashing into a tower during landing. This design criterion was suggested by the crash of a B-25 into the Empire State Building in 1945 while it was flying in fog (Delatte, 2009). The high-speed crash into the towers of Boeing 767s being flown by terrorists far exceeded the design forces from a slow-moving 707, but each building withstood the crash. Even though the towers eventually collapsed, they exhibited robust behaviour with respect to the initial crash. The attack would have been much worse if the towers had toppled over under the impact of the aircraft.

The crashing of the planes into the towers was, in many ways, a so-called *black swan* event (Taleb, 2007) – that is, the consequences were far outside the realm of experience. ('Black swan' refers to the long-held belief that all swans were white until black swans were discovered in Australia.) Karl Popper (2002) used the black swan example when discussing the problem of induction. Taleb (2007) discussed black swan events from the perspective of his experience in the investment community. As far back as 1921, Knight (1948) considered various types of uncertainty in the business environment. He referred to them as *measurable uncertainties* and *unmeasurable uncertainties*. In a business environment, if you can determine the probability of an event, it is a measurable uncertainty that can be managed with insurance. An unmeasurable uncertainty cannot be managed using insurance, because its probability cannot be determined due to its unpredictability. Knight's unmeasurable uncertainties are black swan events. In simple and complicated systems, if we have measurable uncertainties we can deal with them using safety factors and characteristic values. A characteristic value is a value selected for design from the statistics of a load or material property. For example, a 100-year flood is a characteristic flood level that has a probability of 1/100 of occurring in any given year. Then, for design, the engineer might decide to use 2.0 times the 100-year flood level as the design flood. That would mean a safety factor of 2.0 on the 100-year flood. There are other definitions of safety factor, but this one is a reasonable example for our purposes. In complex and complex adaptive systems, social systems, we will have both measureable and unmeasurable uncertainties and the methods to deal with each will be different due to the nature of the uncertainties. At the present time, we have few heuristics to help with unmeasurable uncertainties in social system design and maintenance. Vitek and Jackson (2008) have suggested that we should deal with system uncertainties such as this by being humble and accepting our ignorance. From a design standpoint, it appears that ignorance is simply another way of describing the uncertainties exhibited by a complex adaptive system.

Black swan events emerge out of complex and complex adaptive systems. The complex adaptive system from which the Twin Towers disaster emerged is the world society. The Twin Towers interacted with certain factions of society to drive two terrorist attacks on the structures. The towers survived the first one, but not the second. This type of interaction is only thought of in passing for the design of many engineered structures. As society becomes more complex, a wide range of interactions will become more important in design. Engineers have few heuristics to deal with these interactions, and the science of complex and complex adaptive systems is only beginning to reach a point where these interactions can be addressed (Bulleit and Drewek, 2011; Crutchfield, 2009). Furthermore, as Crutchfield (2009) discusses, the complex systems we are developing in modern society will exhibit fragilities that presently cannot be predicted. Thus, black swan events may become more common in the future, and engineers will need to develop ways to think about, design for and mitigate them. Doing this will require using control volumes that are much larger than typically used for the design of simple and complicated systems, and will introduce uncertainties and ignorance much greater than engineers are presently used to handling. Design and maintenance of social systems will require adaptation, both by the social system engineer and by the system itself.

The methods and heuristics developed by engineers for the design of simple and complicated systems, prototypical and non-prototypical, occurred in a reductionist mindset where the system was isolated from the broader environment in a fairly small control volume. Social system engineering will require enhanced methods and heuristics to design and maintain complex and complex adaptive systems under significant amounts of uncertainty. The choice of control volume will affect the types of analysis models that should be used, and will affect the uncertainty with respect to how changes based on those analyses will alter the behaviour of the entire system. The design methodologies and mental models of engineers will need to be enhanced to include the ability to consider black swans, complexity and evolution/adaptation. Although any changes to the design/maintenance process for engineered systems will still need to be mathematical and predictive, the consideration of possibilities that come from outside the chosen control volume or emerge from the system dynamics in a non-intuitive manner (e.g., black swan events) will provide the impetus to include concepts from disciplines such as cognitive science and psychology and likely will increase the uncertainty for decision-makers. A system design paradigm that works across disciplines to develop design methodologies will necessarily need to include a range of engineering disciplines, cognitive sciences, psychology, complexity science, information theory and likely some aspects of philosophy. Also, complex and complex adaptive systems, because they are far from equilibrium systems, require continual maintenance of the system and continual learning about the system (i.e., adaptation). The engineers and others who design and maintain such systems need a mindset that allows them to be continually learning about the system with which they are working.

2.4 Engineering Analysis and Uncertainty in Complex Systems

Engineers have a range of methods that are presently used for the design of simple and complicated systems. The design methods use various techniques that allow the engineer to predict or anticipate the behaviour of the future system. These methods include calculations that determine a characteristic value of the capacity of the member or system, often encoded in a structural building code (Bulleit, 2008). The methods also include characteristic values of the loads on the structure. For environmental loads, such as snow and wind, the in-time aspect of the variability in the loads is accounted for using extreme value theory, which makes the in-time behaviour implicit in the design (Bulleit, 2008). Similar approaches are used for many water resource designs and geotechnical engineering problems. In some complicated engineering systems, actual in-time behaviour is treated in a more rigorous fashion (e.g., fatigue design in aircraft and modern response history analysis in seismic design). These types of design predictions need to be made for control volumes in complex and complex adaptive systems, but the techniques are generally not adequate for full-system analysis due to the in-time evolutionary behaviour of complex and complex adaptive systems.

Since complex and complex adaptive systems can exhibit chaotic behaviour, or self-organized criticality, and emergence of black swan events, the techniques used for design prediction of simple and complicated systems will generally not be adequate for complex and complex adaptive systems unless a small subsystem, control volume, of the entire system is being designed. Some of the methods used in the design of simple and complicated systems may work as reasonable approximations in subsystems of complex and complex adaptive systems, but the criteria necessary to decide which ones will work are not presently available. The development of engineering design prediction methods suitable for complex and complex adaptive systems, including which existing techniques are adequate, will be an important part of the future of social systems engineering. Thus, at this time, the model uncertainty in engineering of social systems will be a significant portion of the overall uncertainty.

Complex and complex adaptive systems produce behaviour that is clearly stochastic in nature and may also be non-intuitive. Thus, one issue that must be addressed for design and maintenance is the prediction horizon: the point in the future where the probability of a useful prediction becomes so small that it is no longer useful. A prediction horizon describes a control volume in time rather than in space; we choose to examine the system over a limited time. We may also examine a small portion of the system, a subsystem, over a relatively short period of time (i.e., a control volume in space and time). For simple and complicated systems, the prediction horizon is often measured in years and decades. For complex and complex adaptive systems, the prediction horizon may be quite a bit smaller than that. Estimating the prediction horizon will be a necessary part of the design and maintenance of social systems. Techniques that may prove helpful for estimating the prediction horizon include information-theoretic approaches (Bais and Farmer, 2007), stochastic process approaches (Bhat, 1972; Williams, 2003), simulation (de Neufville and Scholtes, 2011; Epstein and Axtell, 1996; Gilbert and Troitzsch, 2005) and system dynamics (Forrester, 1995). Pool (1997, p. 271), referring to the 1979 near meltdown of a nuclear reactor at Three Mile Island near Harrisburg, PA, stated:

> Most of the causes of the accident could be traced back, directly or indirectly, to a mindset and an organizational structure inherited from a time when the utility operated only fossil-fuel plants.

It seems that today's engineers are in an analogous position with respect to simple and complicated systems versus complex and complex adaptive systems. We have a mindset based on simple and complicated systems that we are trying to apply to the design and maintenance of complex and complex adaptive systems. We have the wrong mindset for social systems engineering, and that mindset will for some time increase the uncertainty in the results of our decisions made under that mindset.

Social systems engineers will need to develop mental models, analytical techniques and heuristics that will allow engineers and others to design and manage complex and complex adaptive systems in a way that reduces the dangers from external events, emergent events and overconfidence in the ability to predict or anticipate the future of such systems. Furthermore, techniques to enhance resilience (robustness and rapidity) of systems will be of key importance.

It should be becoming evident that the design and maintenance of complex systems will require that feedback from the behaviour of the system will need to be used to decide how next to alter the system to maintain it. Traits of the system will need to be considered. One possible trait is the probability distribution of outputs from the system. For instance, if the outputs of the system can be modelled using a normal distribution, then the system is likely simple and possibly complicated. Output data with a skewed distribution (e.g., lognormal) will indicate that the system is likely to be complicated and possibly complex. As we move into the realm of complex and complex adaptive systems, the distribution of outputs becomes highly skewed, often power law distributed (Mitzenmacher, 2004; Newman, 2005). Furthermore, lognormal distributions are connected to power-law distributions in a natural way (Mitzenmacher, 2004). The appearance of highly skewed lognormal distributions is an indicator of a shift to more complexity, and an indication that fat-tail distributions are becoming important to the prediction of system behaviour. Power-law distributions indicate a high level of complexity, including complex adaptive (Miller and Page, 2007, p. 166). Correlation between elements of the system may indicate that the system is about to change in a significant way. For instance, correlation lengths tend to increase before a phase transition (Barabasi, 2002, p. 74). Other possible indicators might come from system identification analysis of the system output, unusual outliers in the data that cannot be written off as errors, and outputs from the system that seem to show fragilities in the system behaviour.

2.5   Uncertainty in Social Systems Engineering

It should be apparent from the above discussion that there is a wide range of possible uncertainties that will affect the decisions that need to be made in the design and maintenance of social systems. The uncertainties when engineering social systems will be produced by all the sources of uncertainty for typical engineered systems, particularly non-prototypical systems, and will include other significant sources. Since social systems will be complex adaptive, they will evolve even without any intentional perturbations (i.e., maintenance). Engineering and maintenance alterations to the system will change the evolutionary path in some way, and the results of that perturbation will need to be evaluated in order to make further design alterations (Simon, 1996). Even without intentional perturbations, the system will adapt to stressors from inside and outside the system. Social systems are adaptive, so variation and selection are occurring all the time (Harford, 2011). The concern with the engineering of social systems is that the engineering and maintenance perturbations must be made so that the system does not move in a potentially dangerous direction, or even collapse, due to the consequences of the changes. The engineering decisions that must be made will be made under levels of uncertainty that far exceed even that for large-scale engineered systems. Engineers will need to use large control volumes, develop new heuristics and deal with the variation and selection that occurs, whether the system is engineered or not.

The uncertainties in engineering social systems, ignorance according to Vitek and Jackson (2008), will require that the complex nature of the system be included in deciding how to engineer the system. For instance, it might prove useful to model the entire system, but engineering decisions will likely need to be made working with a smaller control volume. The size of the control volume will be directly related to the prediction horizon. An engineering model of the system will only be able to make useful predictions between the present time and the prediction horizon. Then, once the engineering decision has been made, feedback from the system can be used to examine how well the model predicted the behaviour of the system. This information can then be used to enhance the model in order to make the next perturbation to the system. Olsen (2015) discusses the use of this approach, referred to by him as the

observational method, as a possible way for civil engineers to deal with climate change in the
design of large projects.

> When it is not possible to fully define and estimate the risks and potential costs of a project and
> reduce the uncertainty in the time frame in which action should be taken, engineers should use
> low-regret, adaptive strategies, such as the observational method to make a project more resilient
> to future climate and weather extremes.

Social systems, being complex adaptive systems, exhibit large uncertainties due to a range
of agent types (e.g., humans, animals, corporations and government agencies) interacting
together in an environment where the agents and the environment are strongly connected. This
type of system exhibits emergence of unforeseen events, some of them leading to catastrophic
consequences (i.e., black swan events). It is the nature of social systems that causes the prediction horizon to be highly uncertain and potentially very close to the present. Furthermore,
the strongly interconnected nature of the system makes it possible for local changes to lead to
dangerous system-wide situations. It seems, under our present ability to model complex
adaptive systems, that any engineering of social systems must be done on a local level over a
fairly short time frame. Assuming that this is true, engineering of social systems will need to
be mostly a bottom-up process. Yes, efforts to model the long-term behaviour of the overall
system will be useful, but only as an aid in developing models to increase the time to the prediction horizon and enhance the base system model. The enormous levels of uncertainty
inherent in the system should preclude the use of long-term, system-wide models to make
top-down changes to the system. Clearly, these ideas imply that many of our present top-down
approaches to social system engineering may lead to potentially serious consequences for
the system.

The last aspect of system behaviour that is important to social systems engineering is
resilience. The portion of resilience that will be discussed here is robustness, the ability of the
system to withstand shocks to the system, whether the shocks come from outside the system
or emerge from inside the system. The primary trait of a social system that makes it robust is
that the system is broadly distributed and has relatively weak correlations between elements
of the system. A distributed system is able to withstand local events without the consequences
of the event spreading to other elements, both because of the distributed nature of the system
and because of a lack of correlation between elements. Here, variability between elements
allows a form of redundancy, so that some elements of the system behave differently enough
that events that damage one element do not necessarily damage another. Again, from the
standpoint of engineering, bottom-up and local modifications help minimize correlations that
can make even a well-distributed system sensitive to relatively small perturbations. So, once
again, top-down control of a system will reduce the distributed nature of the system and
increase the correlation between elements. The system then becomes more sensitive to system
collapse from relatively local events. Consider two examples of the effects of correlation. The
first is the near nuclear reactor meltdown at Three Mile Island, referred to previously (Perrow,
1999; Pool, 1997). The failure started with a moisture leak into an instrumentation line, which
unnecessarily shut down the turbines. From there a series of events led to nearly exposing the
reactor core, which would likely have caused a reactor meltdown if it had occurred. In this
case the correlation of components, including safety devices, was caused by what Perrow
(1999) refers to as 'tight coupling'. This simply means that a local failure cascades throughout

the system, possibly causing a system failure. The cascade at Three Mile Island occurred in about 13 seconds, yet took the operators over 2 hours to remedy. The second is correlation produced by top-down control. A tightly controlled franchise system forces each of the franchisees to follow in lock step the directions from above. This tight control means that the behaviour of each of the franchise owners is very similar, and thus correlated. If the top-down decision is inappropriate, then the resulting financial problems will be felt by all or most of the franchisees, and likely the entire franchise system. More loose control, while possibly increasing the risk of problems for some local franchise owners, makes the system more distributed, reducing the possibility of system-wide problems. If the franchise owners are allowed to make some changes to each of their individual franchises, then that produces some variation in the system, which might allow adaptation of the system. The possible benefits of this type of variation will be discussed subsequently.

Thus, the mindset for good social systems engineering and maintenance will be one that recognizes that the system should be perturbed only in local areas over time spans that allow feedback from the system to drive the next engineered perturbation. The mindset will mean that for large-scale social systems, like the types of human–natural systems discussed in Vitek and Jackson (2008), it may be necessary to allow the system to go its own way for a while before an engineering change is attempted. This approach allows variation and selection that is inherent to the system to occur. It will also mean that large-scale perturbations, such as full-earth geo-engineering, should not be attempted. Large-scale, top-down perturbations produce changes that are potentially dangerous to the system, and the system uncertainty is too great to make reasonable predictions about what the consequences of the top-down perturbation will be. As has been pointed out by Petroski (2006, 2010), engineering advances by failures. If we are going to push the envelope in engineering design, we are going to have failures. For social system engineering, we will also get failures, and the only safe way to advance is to engineer only small portions of the system over short time spans, because that limits, as much as possible, the consequences of a failure. Large-scale, top-down perturbations will eventually lead to a system collapse. The question is not if there will be a collapse, but when.

Harford (2011, p. 243) has suggested three principles (heuristics) for adapting, which seem appropriate for social system engineering: first, 'try new things, expecting that some will fail'; second, 'make failures survivable: create safe places for failure or move forward in small steps'; and third, 'make sure you know when you've failed, or you will never learn'. Engineers working on simple and complicated systems generally have little difficulty knowing when failures have occurred. This is not necessarily true for social systems, such as government programmes. Consider the Affordable Care Act in the United States. As a large federal government healthcare programme it certainly tries something new, but it is not a small step and has no safe place for failure. Furthermore, those who developed it are highly unlikely to admit failure. More consistent with the three principles of adaptation described above would have been to allow the states to develop healthcare programmes. Then there would have been 50 experiments, leading to a range of variation. The steps to change would have been smaller, and failure would have been more apparent, since comparisons could have been made between states. These comparisons would then have allowed selection. Certainly this is not the only possible approach, but it would arguably have been a better engineering approach.

A somewhat counter-intuitive aspect of social systems engineering is that a system that allows high variability in behaviour among the agents of the system will change in ways that will permit the system to evolve better. The various behaviours will increase the variability in the system, but the various behaviours will act as experiments, much like crossover and mutation produce experiments in natural evolution. Thus, increased uncertainty in the system is likely good for the system in the long run, because it increases variation. In many ways the system becomes a learning system. Social systems engineers will need to be tolerant of a system that is able in some cases to select its own path from inherent variation.

2.6   Conclusions

Social systems engineering will involve dealing with a wide range of uncertainties, much larger than even those in large-scale non-prototypical engineered systems. The level of uncertainty will be large enough that engineers who design and maintain social systems will require a mindset that is different from the mindset of the vast majority of today's engineers. To quote Dennett (2003, p. 280):

> Engineers, like politicians, are concerned with the art of the possible, and this requires us, above all, to think realistically about what people actually are, and how they got that way.

Social systems engineers will find that the variability, and thus uncertainty, in the systems that they design and maintain will be both a hindrance and, if allowed to be, a help. A mindset that leads an engineer to make changes to the system mostly in a bottom-up, local manner, and at times allows the system to choose its own path, will be required for an effective social system engineer. Social systems are adaptive systems, and thus variation and selection are vital to their health, including engineered variation and selection as well as variation and selection inherent in the system. Engineering of modern engineered systems often entails reducing uncertainty to its lowest level and then dealing with what remains using time-tested heuristics. To a great extent, this approach will not work for social systems, except insofar as control volumes can be used to analyse and design small portions of the overall system over a short time frame. Heuristics to deal with a wide range of measurable and unmeasurable uncertainties will be required for social systems engineering. Engineering of social systems will need to be a form of adaptation, occurring within an environment of large uncertainty, done in such a way as to prevent large-scale failures.

References

Bais, F.A. and Farmer, J.D. (2007) The physics of information. Working paper, Santa Fe Institute, Santa Fe, NM. Available at: www.santafe.edu/media/workingpapers/07-08-029.pdf.

Bak, P. (1996) *How Nature Works: The science of self-organized criticality*, Copernicus, New York, NY.

Barabasi, A. (2002) *Linked: How everything is connected to everything else and what it means for business, science, and daily life*, Perseus Publishing, Cambridge, MA.

Bhat, U.N. (1972) *Elements of Applied Stochastic Processes*, John Wiley & Sons, New York, NY.

Bucciarelli, L.L. (2003) *Engineering Philosophy*, Delft University Press, Delft.

Bulleit, W.M. (2008) Uncertainty in structural engineering. *Practice Periodical on Structural Design and Construction*, **13**(1), 24–30.

Bulleit, W.M. (2013) Uncertainty in the design of non-prototypical engineered systems, in D. Michelfelder, N. McCarthy and D. Goldberg (eds), *Philosophy and Engineering: Reflections on practice, principles, and process*, Springer-Verlag, Dordrecht, pp. 317–327.

Bulleit, W.M. and Drewek, M.W. (2011) Agent-based simulation for human-induced hazard analysis. *Risk Analysis*, **31**(2), 205–217.

Crutchfield, J.P. (2009) The hidden fragility of complex systems – consequences of change, changing consequences. *Essays for cultures of change/changing cultures*, Barcelona, Spain. Available at: www.santafe.edu/media/workingpapers/09-12-045.pdf.

Delatte Jr, N.J. (2009) *Beyond Failure: Forensic case studies for civil engineers*, American Society of Civil Engineers, Reston, VA.

de Neufville, R. and Scholtes, S. (2011) *Flexibility in Engineering Design*, MIT Press, Cambridge, MA.

Dennett, D.C. (2003) *Freedom Evolves*, Penguin Books, New York, NY.

Der Kiureghian, A. and Ditlevsen, O. (2009) Aleatory or epistemic? Does it matter? *Structural Safety*, **31**, 105–112.

Epstein, J.M. and Axtell, R. (1996) *Growing Artificial Societies: Social science from the bottom up*, Brookings Institution Press, Washington, D.C.

Forrester, J.W. (1995) Counterintuitive behaviour of social systems. Available at: web.mit.edu/sysdyn/road-maps/D-4468-1.pdf.

Gilbert, N. and Troitzsch, K.G. (2005) *Simulation for the Social Scientist*, Open University Press, Milton Keynes.

Harford, T. (2011) *Adapt: Why success always starts with failure*, Picador, New York, NY.

Holland, J. (1995) *Hidden Order: How adaptation builds complexity*, Basic Books, New York, NY.

Holland, J. (1998) *Emergence: From chaos to order*, Perseus Books, Cambridge, MA.

Knight, F.H. (1948) *Risk, Uncertainty, and Profit*, Houghton Mifflin, Boston, MA (originally published in 1921).

Koen, B.V. (2003) *Discussion of the Method: Conducting the engineer's approach to problem solving*, Oxford University Press, Oxford.

Miller, J.H. and Page, S.E. (2007) *Complex Adaptive Systems: An introduction to computational models of social life*, Princeton University Press, Princeton, NJ.

Mitzenmacher, M. (2004) A brief history of generative models for power laws and lognormal distributions. *Internet Mathematics*, **1**(2), 226–251.

Newman, M.E.J. (2005) Power laws, Pareto distributions, and Zipf's law. *Contemporary Physics*, **46**(5), 323–351.

Olsen, J.R. (2015) *Adapting Infrastructure and Civil Engineering Practice of a Changing Climate*, ASCE, Reston, VA. Available at: ascelibrary.org/doi/pdfplus/10.1061/9780784479193.

Perrow, C. (1999) *Normal Accidents: Living with high risk technologies* (2nd edn), Princeton University Press, Princeton, NJ.

Petroski, H. (1992) *To Engineer is Human. The Role of Failure in Successful Design*, Vintage Books, New York, NY.

Petroski, H. (1994) *Design Paradigms: Case Histories of Error and Judgment in Engineering*, Cambridge University Press, Cambridge.

Petroski, H. (2006) *Success Through Failure: The paradox of design*, Princeton University Press, Princeton, NJ.

Petroski, H. (2010) *The Essential Engineer: Why science alone will not solve our global problems*, Vintage Books, New York, NY.

Pool, R. (1997) *Beyond Engineering: How society shapes technology*, Oxford University Press, Oxford.

Popper, K.R. (2002) *The Logic of Scientific Discovery* (15th edn), Routledge, London.

Simon, H.A. (1996) *The Sciences of the Artificial* (3rd edn), MIT Press, Cambridge, MA.

Taleb, N.N. (2007) *The Black Swan: The impact of the highly improbable*, Random House, New York, NY.

Vincenti, W.G. (1990) *What Engineers Know and How They Know It*, Johns Hopkins University Press, Baltimore, MD.

Vitek, B. and Jackson, W. (2008) *The Virtues of Ignorance: Complexity, sustainability, and the limits of knowledge*, University Press of Kentucky, Lexington, KY.

Williams, R.H. (2003) *Probability, Statistics, and Random Processes for Engineers*, Brooks/Cole, Pacific Grove, CA.

3

System Farming

Bruce Edmonds

3.1 Introduction

Consider farmers. They may know their animals, crops and land in some detail, but are under no illusion that *designing* a farm in advance would contribute more than a small part to its success. Rather, they understand that they have to be acting upon the elements of their farm constantly to try to get acceptable results – less an exercise in careful planning than disaster avoidance using constant monitoring and continual maintenance. In such a situation, new ideas cannot be assessed on the grounds of reason and plausibility alone (even those suggested by scientific theory), but have to be tried out *in situ*. Solutions are almost never permanent and universal, but rather a series of tactics that work for different periods of time in particular circumstances. Techniques that are seen to provide real benefit (even if that benefit is marginal) are adopted by others nearby so that, in the longer run, a community of specific and local practices evolves. This chapter suggests that in order to effectively produce complex sociotechnical systems (CSS) that work, we have to learn to be more like farmers and less like mathematicians or traditional engineers.[1]

This is a difficult lesson to adsorb. As students, we are taught the value of good design – constructing the system sufficiently carefully so that we know *in advance* that it will work well once built. We know that bugs are bad and we can limit them by careful specification and implementation. However, CSS may not be very amenable to such an approach, requiring considerable and continual *post-construction* adaption, where the system is only ever partially understood. Here bugs (in the sense of unintended effects) may not only be inevitable, but also the very things that make the CSS 'work'. In other words, learn to 'farm' the systems that we need – *system farming*.

---

[1] Cook (2008) compared engineering to a garden, noting that a garden is not an entirely natural construct, but is also a human artefact. Here we use a similar analogy, but in the opposite direction, that engineering of CSS needs to be more like gardening.

Social Systems Engineering: The Design of Complexity, First Edition. Edited by César García-Díaz and Camilo Olaya.
© 2018 John Wiley & Sons Ltd. Published 2018 by John Wiley & Sons Ltd.

This chapter argues that a *design-centric* engineering approach to producing CSS will not work. Such an approach requires either: (1) a robust and *well-validated* understanding of social systems upon which a design-based engineering approach may be built *or* (2) an approach which focuses on the post-implementation stages of system development. Since we are far from anything *even approaching* a reliable understanding of social systems, approach (1) is infeasible. We conclude that the only approach that is currently feasible is the second approach, system farming. In particular, we argue that an approach based on abstract simulations that are *not* well validated, but merely plausible, has no hope of obtaining its goal, however suggestive the results. In other words, a 'foundationalist' approach, which tries to base the design of such systems upon some abstract principles, will fail – especially if humans interact with it.

3.2 Uncertainty, Complexity and Emergence

In this section we discuss and define unpredictability, complexity and emergence in complex systems (including CSS). In particular, we want to show that just because a particular emergent feature is *caused* by the mechanisms and setup of a complex system and each micro-step is completely understandable in a deterministic way, this does not mean that the emergent feature is *reducible* to the mechanisms and its setup. It may be that the assumption that 'predictability' scales up from the micro to the macro is what lies behind much confusion with respect to CSS. Rather, it seems that, as Philip Anderson put it, 'More is Different' (Anderson, 1972).

Complexity has many different definitions and meanings. This is because the complexity of a system is relative to the type of difficulty that concerns it, as well as the frame/descriptive language in which that system is represented (Edmonds, 1999). In this case, we could characterize the *syntactic complexity* as the 'computational distance' from the setup to the resultant behaviour at a later point in time. That is, the minimum amount of computation necessary to determine a certain aspect of the outcomes given the initial conditions, setup, plans, programs, etc.[2] If an easy-to-calculate shortcut to do this exists, then we say that this aspect of the system's behaviour is simple. In contrast, if the shortest way to determine this is by running the system up to that point then it is (syntactically) complex (this is equivalent to Bedau's definition of 'weak emergence'; Bedau, 1997).

Clearly, syntactic complexity can make it infeasible for an actor/agent with computational limitations to predict future behaviour, even if they have the full details of the initial setup and any subsequent environmental inputs. In particular, if we wish to be able to predict the behaviour of a class of such setups without simulating each one, then the presence of such complexity makes this infeasible to do directly. Thus, syntactic complexity can be a cause of effective unpredictability. Pseudo-random number generators are an example of this in practice – their syntactic complexity makes their output unpredictable and arbitrary in practice.

Emergence occurs when some significant behaviour occurs that (a) is not reducible to the details of the system setup (otherwise it would not be new) but yet (b) is totally consistent with those details (otherwise it would not be from the system). Clearly, both (a) and (b) being the case is impossible within a simple formal system. Rather, what tends to happen within a

---

[2] This is similar to 'logical depth' (Bennett, 1988) but without the condition that the program needs to be the shortest possible.

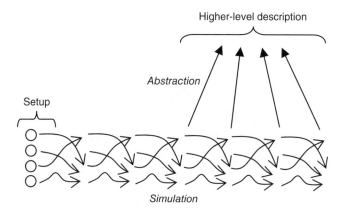

Figure 3.1   Emergence resulting from syntactic complexity plus abstraction.

system of high syntactic complexity is as follows: since the behaviour is not predictable from the setup, the observed behaviour (that appears significant to an observer) is described in a different type of language to that of the detailed interactions in the system. The fact that this description represents the observed behaviour of the system ensures that it will be consistent with its implementation, but because it is in a different language, it will not be reducible to descriptions of the implementation. For example, in Schelling's model of racial segregation (Schelling, 1978), the implementation is in terms of counters on a checkerboard and when they move, but the emergent behaviour that results is an observed global clustering of like colours, even at high levels of tolerance (Edmonds and Hales, 2005). This is illustrated in Figure 3.1.

Emergence and unpredictability are inevitable features of complex systems, even if they are deterministic at the micro-level and are perfect implementations of the programmer's intentions. This can be seen by looking at some formal systems which, although simpler than most CSS, can easily be mapped into them (i.e., they can be seen as a simplification of some CSS).

Wolfram (1986) exhibits a cellular automaton (CA), which produces a seemingly random binary sequence in its central column, in a deterministic manner from a given initial state (Figure 3.2). Of course, since this is a deterministic system and one has 'run' it before with a particular initial state, one knows (and hence in a sense can 'predict') what sequence will result, but if one only knows what resulted from similar (but not identical) initial states then there seems to be no way of knowing what will result beforehand. Of course, this is true of any pseudo-random-number-generating program; the point is that, in this case, it would be easy to design a CSS with the same properties using essentially the same mechanism, and in that CSS what had happened in the past would be no guide as to how it behaved in the future.

Of course, it is very difficult to prove a negative, namely that there is no 'shortcut' to determining such a system's behaviour without doing the whole simulation. Therefore, despite the available evidence, it is always possible for people to simply assert that there must be some way of doing this. However, there is formal evidence against such a hope in the form of Gregory Chaitin's proof that there are mathematical truths of the following kind: those whose simplest proof is as complicated as the truth itself (Chaitin, 1994). For these truths there is no shortcut – no underlying, simpler reason why they are true. In fact, Chaitin's construction

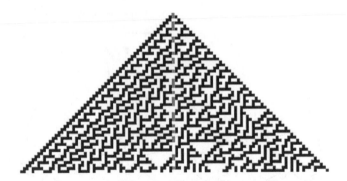

Figure 3.2 CA rule 30 (Wolfram, 1986). This is iteratively applied to an all-dark line with a single white cell (time is downwards, so the development is clear). What state the original cell (shown with a blue line) is in after a number of iterations seems to be only retrievable by calculating the whole.

shows that all but a vanishing number of such truths are like this, so that those which are explainable by a simpler construction are the tiny exception. In other words, for most formal constructions, there is no simpler explanation of them – they are not amenable to any simplification.

The consequences of these kinds of complexity within CSS means that approaches that abstract away from the detail of the systems are unlikely to be very useful. In particular, abstract mathematical models (such as global-level system dynamics models) will miss key emergent features of CSS and their consequences. In these cases, the only kind of model that might help would be individual-based simulation models (i.e., ones where the entities are represented separately in a heterogeneous fashion).

3.2.1 The Double Complexity of CSS

CSS are often complex in the ways just described, but they also have additional sources of complexity, making them 'doubly complex'. Here we discuss only three of these additional sources: downwards causation, intention recognition and context-dependency.

Whilst many distributed physical systems have been shown to display emergence, this is always an 'upwards' process whereby the interaction of the parts determines their aggregate behaviour. In social systems, in which humans play a part, there is also the action of 'downward causation' where the properties of the *whole* effect and/or constrain the behaviour at the micro-level. A classic example of this is the action of social norms (Xenitidou and Edmonds, 2014); when a standard of behaviour becomes common within a society and is perceived as useful, it may become a social norm and be recognized throughout. Once this happens, the social norm will be enforced across that society, even by those who would otherwise behave differently. Thus, a social conformity results that would not have arisen as a result of only 'upwards convergent' processes because a perception of what is generally pertinent across a society is co-developed by its members. This feature is key to how socio-technical systems evolve – as soon as a new system (such as Facebook, Wikipedia, twitter) appears, and people start using it, they develop norms as to what one then does. These norms are often essential to the success and working of the system as a whole – the systems would not function without them, and would function very differently if different norms had evolved.

Another key feature of human interaction is that its members (successfully) understand each other's intentions and goals. That is, they apply the 'Intentional Stance' (Dennett, 1987) as a strategy. This is essential to the way they interact – people do not just react to each other's observable behaviour but to their knowledge of what the goals of the others are. For example, if an individual is perceived to have entirely selfish goals, then they will be afforded less assistance and tolerance than one who is perceived to recognize some social obligation, *even if their actions are essentially the same*. This stance relies upon a suitably rich social interaction between its members to function well. Systems that are essentially anonymous (agents do not recognize other individuals' identity), such as early P2P file-sharing systems, were not successful due to a tragedy of the commons process, whereby selfish individuals could get away with not contributing to the common good and simply exploit the system. Systems where individuals are known and where there is a richer interaction (such as Facebook) can be self-organizing and foster cooperative enterprises (charities, campaigns, etc.).

The third added complexity is context-dependency. Whilst many models of human action presume that there are some identifiable general patterns (e.g., following economic incentives) combined with a lot of unexplained variation from these, this variation is usually dealt with as 'noise' and represented as an essentially random process (Edmonds, 2009). However, human behaviour is rarely random, but rather is highly context-dependent. That is, humans recognize different kinds of situation (a lecture, a party, commuting, etc.) and apply *different* rules, knowledge, norms, etc. to these. In other words, they decide how to behave in completely different ways in different kinds of situation – they do not have many uniform behavioural rules or patterns. This supports a powerful mechanism for social coordination, because different coordination mechanisms can then evolve for different kinds of situation. For example, there may be one set of norms and expectations for a set of fishers in their normal business, but this might change completely if a storm appears. This is important for how socio-technical systems are used, since recognizable 'contexts' are created within these systems which can then allow particular kinds of cooperation and norms to develop, such as a memorial page on Facebook.

A simulation model of a distributed social system that includes humans as key actors and does not take into account pertinent features of human interaction, such as social norms, intention recognition and context-dependency, will not predict how such systems will work in practice. An engineered CSS where people are involved is likely not to work as predicted by its design. Humans adapt to the tools they have, they don't just use them as a designer intended but invent new ways of using the systems they inhabit and some of the ways they do this are complex and depend upon rich social interaction. Many simulations of CSS do not adequately capture the kinds of responses that people will display, and hence diverge critically from any implemented system.

3.3 Science and Engineering Approaches

'Science' and 'engineering' are different; however, they are often confused or conflated. The difference between science and engineering is in terms of their goals: science aims to understand phenomena, whilst engineering aims to produce systems that perform well when deployed. Engineering uses what reliable science is available. A lot of the production of goods and structures that we rely on today relies on the understanding of science in their design. Indeed, we are so used to this that we often forget the large number of scientific models and laws that go into their construction, but that is because these models and laws have been tested extensively and shown to be reliable. Historically, it was often the case that engineering did not have much reliable science to go on, and proceeded instead via a 'trial and error' approach. For example, many mediaeval cathedrals collapsed until the invention of 'flying buttresses'. In the opposite direction, the development of science often requires a more advanced engineering (in terms of instrumentation for observation and experiments).

To summarize, there are only two broad approaches to reliable engineering:

1. Using models based on science that have been validated extensively against real-world data.
2. Using an extensive period of trial and error to achieve systems that work within a particular context.

The first of these would permit a kind of design-based approach, starting with well-known scientific principles and constructing designs and models based on these. However, in the case of CSS these are missing, at least in any system in which there is distributed human agency. Thus, at the moment, it is not a case of building first approximations that can be refined later via iterative testing and validation; we do not have the foundations to start building them at all. The second approach can, over time, result in the development of a number of heuristics and 'rules of thumb' that can form the basis for an engineering approach, in the widest sense (Simon, 1996; Koen, 2003).[3] This could be considered as a kind of design approach, but only in the very limited sense that the range of what one might try is limited to the menu of heuristics to hand. This approach tends to be very specific to the kinds of situation it has been developed within and is not systematic. With such heuristics, one has no idea of their applicability within a different kind of situation – in other words, the second approach does *not* support generalization or general principles.

Historically, the second approach has tended to come first, with the first approach only emerging once there has been a body of trial and error experience to build upon. The transition tends to be marked by a period of systematic experimentation – the testing of ideas and theories against observations of the world. Before this happens, theory can be worse than useless in the sense that it can be misleading and hamper the emergence of a science. Before Vesalius started dissecting human bodies, the (wrong) theories of Aristotle concerning human anatomy were believed. Those working on CSS face a similar danger, that of following a plausible theory that lacks an extensive empirical basis.

Of course, a lot of engineering uses a mixture of the above two methods – the design of a large bridge is never scientifically 'proved', but rather the components have known properties, the general principles are known and many calculations are made to check the adequacy of key

---

[3] The purest, most exploratory form of this is when the choice for what is tried is not 'pre-filtered' by what is expected to work (based on current knowledge) – the 'blind variation and selective retention' of Campbell (1960).

parts, but the ultimate test is when it is built and tested *in situ*.[4] This chapter argues that with CSS we will need more of the 'trial and error' approach because there is no reliable science of CSS. In particular, basing the design of CSS on simulations that are built on plausible rules but are not extensively validated against real-world data is not only foolish but also dangerous. We would never authorize the use of medicines simply on the basis that it is plausible that they *might* be helpful!

In particular, there is one engineering strategy that is emphasized within a lot of the academic literature – what might be called the 'specify and design' strategy. Roughly, this goes as follows: (1) decide goals for the system; (2) write down a formal specification for a system that would meet those goals; (3) design a system that meets that specification; and (4) implement a system that corresponds to that design. In other words, this seeks to formalize the stages of system construction. This does have many advantages; principally, one can check each stage and eliminate some of the errors that might have crept in. However, the 'double complexity' of CSS blows big holes in this strategy and limits its usefulness.

3.3.1 The Impossibility of a Purely Design-Based Engineering Approach to CSS

Even apparently simple distributed systems can display complex behaviour. To illustrate this, Edmonds (2005) describes an apparently simple multi-agent social system. In this, agents have a fixed library of very simple fixed plans that determine how they pass unitary tokens between themselves. However, this turns out to be equivalent to a Turing machine, and hence is unpredictable in general (i.e., its outcomes cannot be completely determined in advance of their appearance). Despite it being *far* simpler in nature (at the micro-level) than any existing CSS and most abstract social simulations of socio-technical systems, one can prove that it is impossible (in general) to either predict the outcomes from the setup or design a setup to achieve any specification *even when one knows that such a setup exists* (Edmonds and Bryson, 2004)! That is, there will be no general effective method for finding a program that meets a given formal specification (even if we know one exists), or a general effective method of checking whether a given program meets a given formal specification. In effect, this system implements integer arithmetic that puts it beyond the scope of formal methods. Since most CSS involve processes such as arithmetic, there will be no general and effective methods for predicting many of the properties of any of these CSS.

These sorts of results are versions of Gödel's results (Gödel, 1931). In a sense, Gödel's results went further, they showed that (for most varieties of CSS) there will be true properties of such systems that are not provable at all! That is, one might (correctly) observe properties of such a system that are not obtainable in terms of a formal approach. In CSS terms that means there may well be some properties of deterministic systems that emerge as they run that cannot be proved within *any* particular logic or formal system. Similarly, Wooldridge (2000) shows that, even for finite MAS, the design problem is intractable (PSPACE complete).

---

[4] More recently, the bridge would be simulated to explore how it might behave under various circumstances, but this does not completely prevent undesirable behaviour, as the case of the Millennium Bridge in London showed (bbc.co.uk/education/clips/zvqb4wx).

Of course, the situation is even worse in the real world, where there is a lot of 'noise' – essentially, non-deterministic factors from a variety of sources, including actions from actors/agents of unknown composition and goals, random inputs, chunks of legacy code which have insufficient documentation but are still used, bugs and machine limitations. This noise is unlikely to be random, and hence will not cancel out or be very amenable to statistical approximations (as we will see in a following section). That computer systems of any complexity have unpredictable and emergent features, even isolated and carefully designed systems, is part of our everyday experience. That it is even more difficult to get complicated MAS systems to behave in a desirable way than traditional isolated and deterministic code is also part of our everyday experience.

In support of the engineering picture of programming reliable multi-agent systems, Nick Jennings explicitly suggested that we stop these systems becoming too complicated in order to try and maintain the effectiveness of the design tactic. For example, in Jennings (2000) his advice includes (among others):

- do not have too many agents (i.e., more than 10);
- do not make the agents too complex;
- do not allow too much communication between your agents.

These criteria explicitly rule out the kind of CSS that are described in this book, as well as all those in any of the messy environments characteristic of the real world where they may be used. These rules hark back to the closed systems of unitary design that the present era has left behind. What is surprising is not that such systems are unpredictable and, at best, only partially amenable to design-based methods, but that we should ever have thought they were.

3.3.2 Design vs. Adaptation

As we have discussed, there are two main ways that engineering achieves its goal: by using reliable science or via an extended trial and error process. As illustrated in Figure 3.3, for any system there are processes that seek to ensure its reliability and suitability before and after its initial construction. *In general*, the stage before construction is associated with *engineering* and that after with *maintenance*. The former is associated with design, the latter with adaption and monitoring. Both stages involve iterations of improvement/development and comparison against some standard of desirable behaviour. Both kinds of process are similar in many ways, involving considerable experimentation and observation in each.

The former is the 'specify and design' strategy mentioned above, which has gained status in today's world due to the good scientific foundations of how to make physical objects. The person who designs a machine has far greater pay and status than the person who maintains it.[5] However, this focus needs to change for CSS. Facebook was not particularly distinguished from similar systems when it was launched, but its social context[6] and the way it adapted to

---

[5] The wish to maintain the status of system 'authorship' is one of the factors that impede the emergence of new ways of working that are more suitable to CSS.

[6] A 'facebook' was a Harvard University mechanism that the electronic Facebook mimicked. The service was first launched to the U.S. young upper classes there and then at similar universities.

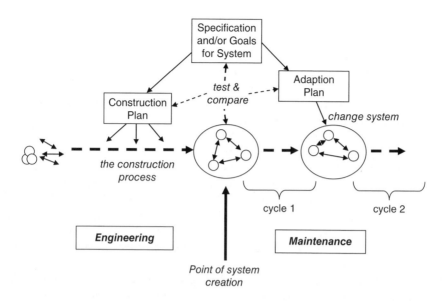

Figure 3.3 The engineering and maintenance phases before and after system creation.

the needs of its users, and the ways they were using the system, were very different. These days, Facebook has a department that researches how Facebook is used and its properties – doing experiments upon its users to *discover* the properties of their own system (e.g., Kramer *et al.*, 2014). Here the maintainers, constantly trying to understand and adapt the system, are more important than the original designer. There have been several occasions when adaptive decisions by social networking systems have been far more important than the original designs: friendster.com was the first social networking system as we understand it, but it tried to prevent musicians using it to promote their music, which myspace.com encouraged.

Engineering is, of course, not precisely defined and encompasses a variety of approaches. These approaches differ in many ways, including the formality of their stages, the extent to which looping or iteration is used to embed the results of experience into the systems being produced, the extent to which adaptive intelligence is incorporated into the system and the extent to which users are involved in the design and adaption of the system. These will be discussed below. However, none of these change the need for a fundamental shift from design-like activities to maintenance-like activities (and many of them are part of this shift). This does not mean that there will be no role for design activities, but that they need to be more empirically based and play a lesser role.

3.3.3   *The Necessity of Strongly Validated Foundations for Design-Based Approaches*

For a design-based engineering approach to work (i.e., where the focus of the effort is put into the pre-construction phases of system development), there has to be a reliable source of knowledge concerning how such systems work. The above reasons show why a general science is not currently available, and hence there are no fundamental micro-foundations for constructing designs. This accords with simple experience with trying to design and predict complex social systems. This has profound consequences for how we might seek to produce complex social systems.

Here we need to distinguish what might make for a *reliable* science of social systems – something a traditional engineering approach could be based upon. Key to such reliability is the replicable validation of models – that is, the models encapsulating theory compare well and repeatedly against observational data. Vague reproduction of patterns is not enough to establish reliable foundations, because these are so open to interpretation.[7]

In particular, plausibility of models is insufficient for such foundations. In these, the model is used more as an analogy than a reliable model (Edmonds, 2001). An analogy seems to fit reality, but because we have interpreted it to do so – in other words, the relationship between an analogy and what it models is not precise. With an analogy, we are free (indeed we have) to reinvent its relationship to what is observed each time. Analogies are useful as a way of thinking, but they do not provide reliable understanding – they appear hopeful but do not give scientific knowledge that is only established by rigorous and *systematic* comparison with data over many different cases.

People often assume that a simple model will give *some* guide to a more complicated version – that a more abstract model will somehow be more general, but this is not the case. To see this, consider taking a key variable out of an equation; the result is simpler, but then it fails to produce the correct outputs. In a CSS, and in the absence of a well-validated general science of CSS, one does not know which aspects one can safely leave out. The result might be a model of a social system based merely on plausible principles, not validated against how humans behave at the micro-level (see Section 3.2.1 above) and not systematically validated against real-world data.

The alternative to a fundamentalist modelling approach, building upon abstract and general principles, is a bottom-up, empirically driven approach. This approach seeks to represent and understand a specific observable system. Modelling an existing system has lots of advantages: one can observe how the humans and technology in the system are currently behaving (rather than relying upon abstract models such as utility comparison) so as to build an appropriate model, and it is possible to extract good data from the system at all levels to enable detailed validation of outcomes. This kind of modelling results in complicated and specific models – more like a computational description than anything like a general theory. To summarize, a lot of evidence and/or knowledge, either from observed systems or constructed systems, will need to be applied for the project of engineering social systems to be effective.

3.4 Responses to CSS Complexity

3.4.1 Formal Methods

So, given this situation, what can we do to try to get CSS systems to behave within desirable constraints? We consider some of the possibilities below.

The formalist answer is to attempt to use formal methods (i.e., proof) to make the engineering of CSS 'scientific'. Edmonds and Bryson (2004) proved that formal methods are

---

[7] Roughly, the more complex the model the more validation is needed, e.g. multiple aspects at many levels.

insufficient for the design of any but the simplest of CSS (e.g., those without arithmetic). Hence, complete formal verification is only possible for the very simplest CSS components and almost no CSS that would help solve real-world problems. However, formality can help in some more mundane ways, namely:

1. Providing a precise *lingua franca* for engineers, for specifications and programs (allowing almost error-free communication).
2. Allowing for specifications and programming to be manipulated in well-defined and automatic ways.
3. Facilitating the inclusion of consistency checks within code to prevent or warn of some undesirable outcomes.
4. Providing a stable and expressive framework/language for developers (or communities of developers) to gain experience and expertise.

What is important is to abandon the delusion that formal proof will ever be a major component in generating or controlling the complex system-level behaviour that we see in real-world problems.

3.4.2   Statistical Approaches

The statistical approach is another way of getting at apparently disordered systems. The first step is to assume that the system can be considered as a set of central tendencies plus essentially arbitrary deviations from these. The idea is that although one might not be able to predict or understand all the detail that emerges from such a system, this does not matter if there are some broad identifiable trends that can be separated from the 'noise'. Thus far is fairly uncontroversial, but more problematic is the next step typically taken in the statistical approach – that of making assumptions about the nature of the noise, usually such as its independence, randomness or normality. That these are suspect for CSS is indicated by systems which exhibit self-organized criticality (SOC) (Bak, 1996). Jenson (1998) lists some criteria which indicate when SOC might occur. These are:

- Agents are metastable – i.e., they do not change their behaviour until some critical level of stimulus has been reached.
- Interaction among agents is a dominant feature of the model dynamics.
- Agents influence but do not slavishly imitate each other.
- The system is driven slowly, so that most agents are below their critical states a lot of the time.

Clearly, this includes many CSS. In such systems, one cannot make the usual assumptions about the nature of any residual 'noise'. For example, when one scales up Brian Arthur's 'El Farol Bar' model (Arthur, 1994) to different sizes and plots the variation of the residuals, it does not obey the 'law of large numbers' as it would if it were essentially random. That is, the proportion of the variation in system size does not reduce with increasing system size, as would happen if the residuals were random, but a substantial residual variation remains. This is shown by Edmonds (1999), as suggested by the results of Kaneko (1990). In this model a

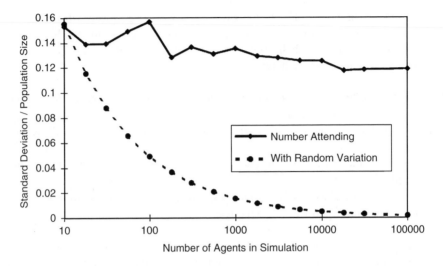

Figure 3.4 A plot of scaled standard deviation against different population sizes averaged over 24 runs over 500 cycles for each point in the El Farol Bar model.

fixed number of individuals have to decide whether or not to go to the El Farol Bar – basically they want to go if others do, but not if many others want to go. They make their decision in a variety of ways based upon the history of attendance numbers. This sort of system results in a sharp SOC attendance pattern around the 'break-even' point. The variance in this attendance is plotted in Figure 3.4 – one can see that this shows no evidence that the variation around the central tendency is dropping as a proportion of system size as the system gets larger. This means that the 'noise' is not random and its distribution may well have undefined moments (which can invalidate many standard statistical techniques such as regression).

The solid line connects the observed values; the dashed line what one would expect were the deviations random. This is exactly the same sort of lack of law of large numbers that was found by Kramer *et al.* (2014), showing that in a globally coupled chaotic system what appeared to be noise did not diminish (as a proportion of the whole signal) with ever larger samples.

Here, care should be taken to distinguish between descriptive and generative statistics. In descriptive statistics, one is simply describing/summarizing a set of known data, whilst in generative statistics, a data-generating process is encapsulated which is supposed to be a model of an observed data stream. Thus, in the latter case, there must be some sense in which the statistics are a model of the source of the observed data. So, for example, if one does have a SOC system which is producing a stream of data with no defined second moment, then positing a distribution with a defined second moment would be a fundamental misrepresentation, whereas any finite set of data obtained from this source will have a defined second moment, which might be a meaningful description of the data for some purposes (e.g., in comparison with a different set of data from the same source and with the same length). However, it would be a mistake to interpret that statistic as representing anything about the underlying system, since it is merely an artefact of the fact that you are dealing with a finite data set.

The fact that one cannot rely on something like the law of large numbers means that one cannot rely on the Monte Carlo method of averaging over multiple randomized runs of the system.

3.4.3   Self-adaptive and Adaptive Systems

Self-adaptive systems (Babaoglu *et al.*, 2005), which include self-designing systems (Garud *et al.*, 2006, 2008) are those that monitor their own performance and then adapt or fix themselves in some way. In other words, a meta-level process is built in that can act upon the functional processes, changing or fixing themselves as needed. This follows a long tradition of exploring the use of meta-level mechanisms in the field of artificial intelligence. Adaptive systems (Beer, 1990; Holland, 1992) are ones which include a learning or evolutionary process in their decision processes so as to change the decisions or even the decision-making process. The difference between self-adaptive and adaptive systems is not entirely well defined. However, self-adaptive systems tend to have a reasoning or learning level that reflects upon a lower, more functional level (and hence is somewhat separate), whilst in adaptive systems, the reflection is more in the mind of the programmer and the adaptive process embedded within the functional decision-making process itself (although there are various mixtures of these approaches).

The temptation is to assume that systems that incorporate either self-adaptive or adaptive components are more general than ones that are more straightforward. However, this is not generally the case for any situations where such algorithms adapt or change correctly to meet environmental change; there will also be cases where they react wrongly, since any signals for change can be misleading. Self-adaptive and adaptive mechanisms typically have a range of circumstances that they can cope with and adapt to successfully, but outside this range they can be more catastrophically wrong than a simpler mechanism. For example, in Edmonds (2002), in which 'clever' agents build models of stockmarket trends and 'dumb' agents just react to the market, neither made greater profits in the long run (although they made and lost money in different patterns). Also, the more complex the mechanism, the greater the resources required in terms of computational time, memory and data.

This lack of a *general* advantage for meta-mechanisms is summed up by the 'No Free Lunch' theorems (Wolpert, 1996; Wolpert and Macready, 1997), which show that over *all* problem spaces no mechanism (adaptive, self-adaptive or otherwise) is better than any other. However, these general results do not hold when the target (environmental) problem spaces are constrained to be those of a particular type. In other words, the mechanism needs to exploit knowledge of its environment in order to gain advantage. If one knows certain facts about an environment, then one can use this to design a more effective algorithm. However, with complex and/or co-evolving systems whose nature may change fundamentally (or even antagonistically), there may be little or no certain knowledge of the environment to exploit. In such systems, there is no *a priori* guarantee that a cleverer mechanism will do any better than a simpler one.

3.4.4   Participatory Approaches and Rapid Prototyping

Travelling in the opposite direction to formalist and design-centric approaches are a set of directions that might be characterized as 'participatory' (Bok and Ruve, 2007) or involving 'rapid prototyping' (Connell and Shafer, 1989). Instead of a relatively linear 'specify and design' process involving a long and careful process of system production by engineers, this employs an open and interactive dialogue with users by producing systems in a relatively 'quick and dirty' manner then getting feedback from users about how this should be adapted or developed. Thus, the engineers get an early steer from users concerning the system, so that quick feedback from what is working in practice can be incorporated. In other words, moving the date of first system creation to as early as possible and shifting effort from the specify and design-style engineering phase to the post-construction maintenance phase (see Figure 3.3). The more that this shift occurs in various approaches, the more it coincides with the suggestions and arguments expressed in this chapter.

3.5   Towards Farming Systems

The infeasibility (and even impossibility) of using formal or statistical techniques to predict what a CSS will do, based on fundamental principles, leaves us with a problem, namely: what *can* we do to understand and manage CSS? As argued above, even a shift to a 'clever' self-adaptive or adaptive system does not get around this problem.

Dealing with CSS rather than simple systems will call for a change in emphasis. Below are some of the 'axes' along which such a change may have to occur. This is a shift of viewpoint and approach – looking at the CSS as a given whole system. These shifts cannot be *proved* in general, more than the indications from abstract cases (such as those above) indicate.

3.5.1   Reliability from Experience Rather Than Control of Construction

At worst, many of the systems we will have to manage will not be those where we have had any control over construction. These systems may be composed of a variety of nodes or components, which participate within the overall system but whose origin is out of our control (as with many P2P systems). Even if all the components are *theoretically* under our control in the construction phases, this does not mean that this can ensure a reliable system in total, if the system is complex and adaptive.

In either case for CSS, trying to achieve system reliability by means of careful construction will have limitations, and some of the desired reliability will have to be gained via management as a result of us watching and learning how to intervene to keep the system going. That is using observation and experience of the system as it happens to be, rather than what it *should* be according to its design or construction. Ferguson (1992) highlighted the importance of feedback from practice to design – with CSS the trend towards incorporating experience needs to have even greater prominence. In a way, this is a shift from a reductionist thinking about computational systems to a more holist one, as a result of a recognition that *in practice* (and often in theory), the detailed construction of a CSS is only a guide to resulting system behaviour. The construction may provide a good *hypothesis* for the system behaviour, but no more than this.

3.5.2   Post-Construction Care Rather Than Prior Effort

As a consequence of the above, a lot of the effort in managing CSS has to shift from before the moment of construction to after. No amount of care before a system is constructed will eliminate the need for substantial post-construction care and management. Instead of viewing our software constructions like a bridge whose parts can be largely relied upon and forgotten once in place, we may have to think of them more like a human organization whose parts need substantial management once it is going.

3.5.3   Continual Tinkering Rather Than One-Off Effort

Since parts of a CSS are constantly changing and adapting, they will need continual adjustment. There may well be no permanent configuration that will be sufficient for the conditions it is to encounter. Instead of a one-off 'heroic' effort of engineering, the more mundane activity of management and adjustment will be needed – re-engineering, adjusting and rebuilding parts as needed.

3.5.4   Multiple Fallible Mechanisms Rather Than One Reliable Mechanism

Since all mechanisms are fallible with CSS, multiple and overlapping systems can increase reliability, such as is found in biological systems. Thus if, for some unforeseen reason, one fails to work, another might. While we cannot foresee that more than one mechanism will definitely be needed, we cannot tell that parallel and overlapping mechanisms will not be helpful. Rather, a variety of mechanisms is more likely to cope with a variety of unforeseen circumstances.

3.5.5   Monitoring Rather Than Prediction

CSS do not lend themselves well to prediction. If we could predict what they would do, we would be able to *tell* if our design was the correct one beforehand. It might well be that the best we can do is to merely catch trends and developments within a system as soon as possible after they start to develop. In this way, ameliorative action can be taken as quickly as possible. The inherent unpredictability of many CSS means that we cannot be sure to detect it before it occurs (i.e., predict it).

3.5.6   Disaster Aversion Rather Than Optimizing Performance

The large uncertainties in CSS mean that one has no hope of optimizing its performance (by any measure). Rather, a more attainable target is simply the aversion of system breakdown. For example, preventing the invasion of a file sharing by uncooperative users, or distrust breaking a market-based system.

3.5.7   Partial Rather Than Full Understanding

CSS mean that we are never going to have a full understanding of what is happening, but will have to be satisfied with partial or incomplete understanding. Such partial understandings may be fine for a certain set of conditions or context, but break down completely in another set.

Gaining a working partial understanding in a particular set of circumstances might be more important than attempting to achieve a full understanding. A continual partial re-understanding of CSS may just be more effective than spending a lot of time attempting a fuller one.

3.5.8   Specific Rather Than Abstract Modelling

The fact that some CSS are susceptible to sharp 'phase changes' with changes of situation means that general and abstract models (or *theories*) of their behaviour may simply not be applicable. Rather, we can get a greater handle on their brittleness and kinds of behavioural traits by modelling them in very specific and detailed ways – eschewing abstraction. This takes a lot more effort in terms of model construction but is more straightforward, since less abstraction is required – with fewer decisions of what to include and what not to. A detailed simulation can be both a prototype of a system (so the resulting global behaviour can be checked as well as its construction) as well as a (fallible) diagnostic tool once it is going. Engineering, at its more practical end, has always been more concrete, singular and context-dependent than science (Bouma *et al.*, 2010; Goldman, 1990, 2004). However, there is a tendency at the academic end of engineering to yearn for some of science's abstraction and generality; this will be unattainable in CSS until the emergence of well-validated models (as discussed above).

3.5.9   Many Models Rather Than One

Knowledge concerning the CSS may be incorporated in a number of simulations at different levels of abstraction, the lowest level being the detailed descriptive simulation described immediately above and higher levels modelling aspects of that simulation. Indeed, there will typically need to be a series of simulations at different levels of abstraction. This multi-layered simulation approach was attempted by Hales and Edmonds (2005), with a sequence of simulation models going from the abstract to real applications. Similarly, due to the changing nature of CSS with circumstance, there will almost inevitably need to be a sequence of such models (or a sequence of model chains) as the system evolves and develops.

3.5.10   A Community Rather Than Individual Effort

Any effective working information about CSS will necessarily be detailed and specific to a particular set of circumstances. Pinelli (2001) pointed out the importance of engineering knowledge communities. This means gathering many more examples, case studies and evidence about the behaviour of CSS than it is feasible for an individual to collect. Thus, those engaged with similar CSS being used in similar situations will need to pool their knowledge, spreading what does and does not work.

3.6   Conclusion

The wish for a 'shortcut' to the production and control of predictable CSS is strong, almost as strong as the wish for a 'proper engineering' of CSS with firm foundations in logical, formal or simulation methods. But wishing does not make things true. We should accept that producing and managing CSS is fundamentally different from simple computational systems. For CSS careful design, whether at the micro- or macro-level, is not enough; adding adaptive or self-adaptive features may just lead to more subtle difficulties or more catastrophic failures.

There is a place for specification and design methods in the production and management of CSS, but it is not a prominent one – rather, the bulk of the progress will rely on trying out techniques and seeing which ones work, but only once a specific system in a particular human and cultural context is deployed. There may be few principles for producing CSS that can be relied upon in general, and these more in the nature of heuristics that evolve over time with experience from practice.

In particular, we call upon those in the distributed systems and simulation community to explicitly and loudly reject those principles and approaches that are not applicable to the systems they are working with (even though they may be applicable for other, simpler systems). Namely to *reject* that:

- abstract 'proof of principle' simulations will contribute significantly to their production or management;
- simple simulations will give general (albeit approximate) guidelines for CSS construction;
- there is likely to be any 'magic bullet' techniques with universal applicability for designing CSS;
- the validation, management and adaptation of CSS are secondary matters that can be significantly ameliorated by good design;
- the 'specify and design' methodology is the only real way to proceed.

We argue that the community will need to take a different path to that pursued in the past, putting it in the vanguard of the 'software revolution' detected by Zambonelli and Parunak (2002).

The nub is that we need to accept a lower-status role, the software equivalent of a farmer. Less prior 'heroic' individual design and more mundane, collective *post-hoc* management. Less abstract and general theory used to predict system properties and more specific and context-dependent modelling used to guide system monitoring and fault diagnosis. Less neat understanding and more of a messy 'community of practice' using rules of thumb and models for particular circumstances and situations. Less assurance from good design and more from a history of having used and worked with such systems in similar situations in practice.

A great deal of work in software engineering has been to discover how such farming might be reduced and/or eliminated as a result of intelligent design. This chapter is a reminder that with CSS, such efforts will be limited in their efficacy and that, if we are to develop *effective* means of managing CSS, we might have to concentrate on the more mundane business of *system farming*. Working CSS are not formally neat systems but messy and complicated evolved systems, like organisms, and have to be treated as such.

References

Anderson, P.W. (1972) More is different. *Science*, **177**(4047), 393–396.

Arthur, W.B. (1994) Inductive reasoning and bounded rationality. *American Economic Association Papers and Proceedings*, **84**(2), 406–411.

Babaoglu, O., Jelasity, M., Montresor, A., Fetzer, C., Leonardi, S., van Moorsel, A. and van Steen, M. (2005) The self-star vision, in O. Babaoglu, M. Jelasity, A. Montresor, C. Fetzer, S. Leonardi, A. van Moorsel and M. van Steen (eds), *Self-star Properties in Complex Information Systems*, Springer-Verlag, Berlin, pp. 1–20.

Bak, P. (1996) *How Nature Works: The science of self-organized criticality*, Copernicus, New York, NY.

Bedau, M.A. (1997) Weak emergence. *Noûs*, **31**(s11), 375–399.

Beer, R. (1990) *Intelligence as Adaptive Behavior*, Academic Press, New York, NY.

Bennett, C.H. (1988) Logical depth and physical complexity, in R. Herken (ed.), *The Universal Turing Machine, A Half-Century Survey*, Oxford University Press, Oxford, pp. 227–257.

Bok, B.M. and Ruve, S. (2007) Experiential foresight: Participative simulation enables social reflexivity in a complex world. *Journal of Future Studies*, **12**(2), 111–120.

Bouma, T.J., de Vries, M.D. and Herman, P.M. (2010) Comparing ecosystem engineering efficiency of two plant species with contrasting growth strategies. *Ecology*, **91**(9), 2696–2704.

Campbell, D.T. (1960) Blind variation and selective retention in creative thought as in other knowledge processes. *Psychological Review*, **67**(6), 380–400.

Chaitin, G.J. (1994) Randomness and complexity in pure mathematics. *International Journal of Bifurcation and Chaos*, **4**(1), 3–15.

Connell, J.L. and Shafer, L. (1989) *Structured Rapid Prototyping: An evolutionary approach to software development*, Yourdon Press, Raleigh, NC.

Cook, S.D.N. (2008) Design and responsibility: The interdependence of natural, artifactual, and human systems, in P. Kroes, P.E. Vermaas, A. Light and S.A. Moore (eds), *Philosophy and Design: From engineering to architecture*, Springer-Verlag, Dordrecht, pp. 259–273.

Dennett, D.C. (1987) *The Intentional Stance*, MIT Press, Cambridge, MA.

Edmonds, B. (1999) Modelling bounded rationality in agent-based simulations using the evolution of mental models, in T. Brenner (ed.), *Computational Techniques for Modelling Learning in Economics*, Kluwer, Dordrecht, pp. 305–332.

Edmonds, B. (2001) The use of models – making MABS actually work, in S. Moss and P. Davidsson (eds), *Multi Agent Based Simulation*. Lecture Notes in Artificial Intelligence, vol. **1979**, Springer-Verlag, Berlin, pp. 15–32.

Edmonds, B. (2002) Exploring the value of prediction in an artificial stock market, in V.M. Butz, O. Sigaud and P. Gérard (eds), *Anticipatory Behavior in Adaptive Learning Systems*. Lecture Notes in Artificial Intelligence, vol. **2684**, Springer-Verlag, Berlin, pp. 262–281.

Edmonds, B. (2005) Using the experimental method to produce reliable self-organised systems, in *Engineering Self-organising Systems: Methodologies and applications* (ESOA 2004). Lecture Notes in Artificial Intelligence, vol. **3464**, Springer-Verlag, Berlin, pp. 84–99.

Edmonds, B. (2009) The nature of noise, in F. Squazzoni (ed.), *Epistemological Aspects of Computer Simulation in the Social Sciences*. Lecture Notes in Artificial Intelligence, vol. **5466**, Springer-Verlag, Berlin, pp. 169–182.

Edmonds, B. and Bryson, J.J. (2004) The insufficiency of formal design methods – the necessity of an experimental approach for the understanding and control of CSS, in *Proceedings of the 3rd International Joint Conference on Autonomous Agents and Multiagent Systems (AAMAS 2004)*, IEEE Computer Society, New York, NY, pp. 938–945.

Edmonds, B. and Hales, D. (2005) Computational simulation as theoretical experiment. *Journal of Mathematical Sociology*, **29**(3), 209–232.

Ferguson, E.S. (1992) Designing the world we live in. *Research in Engineering Design*, **4**(1), 3–11.

Garud, R., Kumaraswamy, A. and Sambamurthy, V. (2006) Emergent by design: Performance and transformation at Infosys Technologies. *Organization Science*, **17**(2), 277–286.

Garud, R., Jain, S. and Tuertscher, P. (2008) Incomplete by design and designing for incompleteness. *Organization Studies*, **29**(3), 351–371.

Gödel, K. (1931) Uber formal unentscheidbare sätze der principia mathematica und verwandter system i. *Monatshefte für Mathematik und Physik*, **38**(1), 173–198.

Goldman, S.L. (1990) Philosophy, engineering, and western culture, in P.T. Durbin (ed.), *Broad and Narrow Interpretations of Philosophy of Technology*, Kluwer, Amsterdam, pp. 125–152.

Goldman, S.L. (2004) Why we need a philosophy of engineering: A work in progress. *Interdisciplinary Science Reviews*, **29**(2), 163–176.

Hales, D. and Edmonds, B. (2005) Applying a socially-inspired technique (tags) to improve cooperation in P2P networks. *IEEE Transactions in Systems, Man and Cybernetics*, **35**(3), 385–395.

Holland, J.H. (1992) Complex adaptive systems. *Daedalus*, **121**(1), 17–30.

Jennings, N.R. (2000) On agent-based software engineering. *Artificial Intelligence*, **117**(2), 277–296.

Jenson, H.J. (1998) *Self-organized Criticality: Emergent complex behavior in physical and biological systems* (vol. **10**), Cambridge University Press, Cambridge.

Kaneko, K. (1990) Globally coupled chaos violates the law of large numbers but not the central limit theorem. *Physics Review Letters*, **65**(12), 1391–1394.

Koen, B.V. (2003) On teaching engineering ethics: A challenge to the engineering professoriate. *Age*, **8**, 1.

Kramer, A.D.I., Guillory, J.E. and Hancock, J.T. (2014) Experimental evidence of massive-scale emotional contagion through social networks. *Proceedings of the National Academy of Sciences*, **111**(24), 8788–8790.

Pinelli, T.E. (2001) Distinguishing engineers from scientists – the case for an engineering knowledge community. *Science & Technology Libraries*, **21**(3&4), 131–163.

Schelling, T.C. (1978) *Micromotives and Macrobehavior*, Norton, New York, NY.

Simon, H.A. (1996) *The Sciences of the Artificial*, MIT Press, Cambridge, MA.

Wolfram, S. (1986) Random sequence generation by cellular automata. *Advances in Applied Mathematics*, **7**(2), 123–169.

Wolpert, D.H. (1996) The lack of a priori distinctions between learning algorithms. *Neural Computation*, **8**(7), 1341–1390.

Wolpert, D.H. and Macready, W.G. (1997) No free lunch theorems for optimization. *IEEE Transactions on Evolutionary Computation*, **1**(1), 67–82.

Wooldridge, M. (2000) The computational complexity of agent design problems, in *Proceedings of the 4th International Conference on Multi-Agent Systems*, IEEE Computer Society, New York, NY, pp. 341–348.

Xenitidou, M. and Edmonds, B. (2014) *The Complexity of Social Norms*, Springer-Verlag, New York, NY.

Zambonelli, Z. and Parunak, H.V.D. (2002) Signs of a revolution in computer science and software engineering, in *Proceedings of 3rd International Workshop on Engineering Societies in the Agents World*, Madrid, Spain. Available at: www.ai.univie.ac.at/~paolo/conf/ESAW02/.

4

Policy between Evolution and Engineering

Martin F.G. Schaffernicht

L'état, c'est moi

Louis XIV of France

What is good for Ford is good for America.

Henry Ford

The needs of the many outweigh the needs of the few. Or the one.

Mr. Spock

4.1 Introduction: Individual and Social System

We humans have a reflexive mind, giving meanings to what happens around us and making plans to adapt it to our needs and wants. Looked upon as a biological entity, each of us is a system of interdependent organs which do not have such a mind. Nor do the social systems we are subsystems of. We perceive ourselves as 'individuals', separated from what surrounds us and indivisible. However, without other individuals around us, none of us would walk or talk, and then there would be no other individuals from whom to learn how to do so. Many other species depend on their respective social systems to *stay* alive, but humans need it to *become* alive as such. There is no human *individual* without such a human *social system*: person and social system are indivisible, and therefore we are really 'indivi-duals', indivisible in two different and simultaneous ways.

For this reason, human social systems have a feature which distinguishes them from other social systems. Social systems can generally be defined by three aspects (Ackoff, 1994, p. 179): (1) they have purposes of their own; (2) they consist of subsystems with purposes of

Social Systems Engineering: The Design of Complexity, First Edition. Edited by César García-Díaz and Camilo Olaya.
© 2018 John Wiley & Sons Ltd. Published 2018 by John Wiley & Sons Ltd.

their own; and (3) they are contained in larger systems that have purposes of their own. Accordingly, a group of apes and a state of ants are social systems. But due to the specific human capabilities, this chapter distinguishes animals from humans and reserves the term 'social system' for human social systems (following Boulding, 1956, p. 205).

The human brain constantly abstracts patterns from perceptions and cannot not reason (Spitzer, 2006, p. 69). Since the wake of humanity, individuals have perceived the order which had somehow evolved in their societies and realized the need to bring order into their lives and relationships amongst one another. Over time, such order was brought by abstract 'tools', like *institutions* and *policies*, which have appeared, changed and disappeared. An *institution* involves shared concepts and consists of rules, norms and strategies (Nowlin, 2011, p. 43); it 'structures and regulates human behaviour by providing information about incentives for appropriate collective behaviour in different social situations' (Urpelainen, 2011, p. 217). Institutions like family, enterprise, school or hospital, market, government (democratic and otherwise), state and country are important parts of the lifeworld of humans. They provide the conceptual entities we deal with when we regulate our behaviour as individual members of social systems. This chapter refers to these regulations as our collective and individual *policies*.

The institutions listed above are also social systems. They have different sizes and therefore diverse levels of complexity, and they all intersect because at the base level they consist of the same individuals, with their needs and wants, their perceptive, cognitive and action processes. Over the past 200 years, social systems like companies and the global community of countries have grown, and therefore each individual is linked to many more other individuals and there are increasing overlaps between manifold social systems; consequently, the complexity of situations we need to deal with has increased. Whereas some needs are shared by all individuals, the resources to satisfy them are usually scarce, giving rise to competition. Other needs and wants are different across individuals, and sometimes they oppose one another. Conflicts of needs and wants are unavoidable, even more so in increasingly complex social systems. As a consequence, the need we detect to improve our social systems in general and policies in particular has risen.

Deliberate interventions into social systems ought to be carried out such as to increase the variety, that is, the number of options for the individuals living and working in the system (Ackoff, 1994). The individual freedom to do what one does best increases the likelihood of useful ideas and practices. At the same time, a richer pool of ideas and practices is important for developing useful new ideas and practices. The concern that deliberate intervention into the evolution of social systems might threaten exactly this variety has led to resolute critiques of the very idea that humans could indeed deliberately *improve* them (Hayek, 1988) – even though Hayek aimed his critique at socialism in the light of the regime of the former Soviet Union, which does not exist any longer. His argument was made using the concepts and analytical tools available more than 50 years ago, and what they could allow us to understand. However, his stance keeps being influential. As opposed to it, the present chapter argues that since humans are natural makers of artefacts, engineering can increase the pace at which cultural evolution shapes collective policies. Therefore, after a brief introduction of the term 'policy', it will propose a conceptual model of the human actor and then proceed to incorporate the artificial into the model. This prepares the scene for discussing the relationship between engineering and evolution proposed here, and it will become clear that policies can and should develop in an interplay between evolution and engineering.

4.2   Policy – Concept and Process

Generally speaking, a policy is 'a course or principle of action adopted or proposed by an organization or individual' (Oxford Dictionary 2015) and 'shared concepts used by humans in repetitive situations organized by rules, norms and strategies' (Ostrom, 2007, p. 23). We have inherited the work from the ancient Greek cities, where it had to do with πολιτεία (*politeía*: citizenship; polis, city state and government). Later it evolved into the Latin *politia* (citizenship; government).

During the epoch of Enlightenment, the idea of human beings living in society and the 'idea of developing social knowledge for the purpose of social betterment' (Wagner, 2006, p. 29) gave rise to different threads of social sciences: anthropology, economics are sociology. The so-called 'social question' led into a more empirically oriented direction, in which social knowledge soon appeared to be useful for planning interventions into the ongoing social processes. During this phase of development, specific actor-oriented areas profiled themselves, amongst others public administration and business administration (Wagner, 2006, p. 35). In the sphere of public administration, the policy orientation of social science has been connected to democracy and the self-determination of individuals, in the faith that human rationality would improve the quality of individuals' lives (Torgensen, 2006, p. 15). Inquiry in the tradition of John Dewey and the mapping of social processes and a reflexive and systemic stance of the mapper (the *self-in-context*) were thought to combine the creative power of imagination and the discipline of logic. In this process of inquiry, the activity of mapping articulates implicit knowledge of reality and makes it accessible to the reasoning power of the human mind (Torgensen, 2006, p. 12). It is worth noticing that while rationality and action are deemed to be features of the individual, policy concerns itself with the interactions amongst individuals, easily leading to the impression that policy is always collective. However, considering that interactions are actions, and that actions are carried out by a person, a policy must be or become personal in order to shape a collective pattern of behaviour; thus, this chapter is based on the idea that a policy is like a coin with two sides, one being personal and the other collective.

Originally, the concept of *policy* included social systems like firms, where the term 'business policy' was used. However, during the second half of the twentieth century, this term was gradually replaced by 'strategy' (Bracker, 1980). In turn, the term 'policy' is nowadays used as shorthand for 'public policy'. The Academy of Management has a specific division 'Business Policy and Strategy', but its topics are explicitly strategy-oriented: strategy formulation and implementation; strategic planning and decision processes. Consequently, much of the modern scholarly debate concerning policy is actually the discussion of *public policy*.

The enlightened faith in human reason and rationality has not been uncontested, and this in different ways. For Friedrich August von Hayek (1988), social systems and institutions develop 'between instinct and reason', and they are too complex for humans to safely intervene in without risking severe side-effects. Indeed, there have been deceptions and great difficulties with many public policies, and the policy sciences appear not to have a systematic influence on policy-makers' decisions (deLeon and Vogenbeck, 2006, p. 3).

Of course, the participation of a great number of actors, long timespans and multiple interacting government programmes lead to a great complexity of the topics to be resolved in designing policies for extensive social systems spanning cities or countries, or even the globe (Sabatier, 2007, p. 3). This complexity is further increased by the fact that the

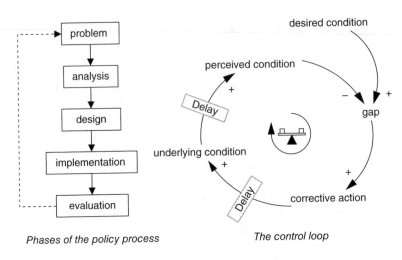

Figure 4.1   The policy process as control loop.

problems addressed are not obvious ones but *messes* in the sense of Ackoff (1994), and that many vested interests are at stake, thereby reducing the objectivity of the entire process. A number of theories concerning the 'policy process' and 'policy change' have been developed and continue to be developed to capture the phenomenon from different angles (Nowlin, 2011).

The overall policy process is always broken down into different steps, usually starting with the perception of an important problem, followed by policy analysis, policy design and formulation, implementation and evaluation, as shown in Figure 4.1.

The left side of Figure 4.1 represents the policy process, which deals with detecting and resolving problems. This reveals it as one form of closed control loop, shown on the right side: the problem is recognized as an increase in the gap between the desired and the perceived conditions of a particular issue. In response to this, corrective action is increased. After a delay it improves the underlying conditions and then, after an additional delay, evaluation will adjust the perception of the situation to the underlying conditions, which will close or at least diminish the gap. The '+' and '−' signs signal if the influence between two variables is reciprocal or inverse. Visibly, this chain of links constitutes a closed feedback loop; since the loop balances the initial disturbance indicated by the increasing gap, it is called a balancing loop. Balancing loops steer a variable towards its desired state or goal, and therefore the resulting behaviour is called goal-seeking. Such loops are used extensively in a policy design methodology called system dynamics, which has its roots in electrical engineering (Forrester, 1961; Richardson, 1999). In the logical shape of the policy process, no mention is made of whether it refers to individual or collective policies – indeed, there is no relevant difference between individual and collective policy as far as the fundamental logic of the policy process is concerned.

Of course, policy scientists have realized that by identifying such steps, one does not state how they are or should be carried out. In a quickly growing policy literature, different scholars have emphasized different aspects of the policy process. Figure 4.2 gives a brief overview of the threads, grouped by shared aspects.

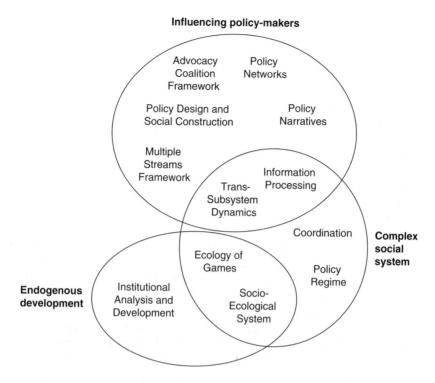

Figure 4.2 Different threads of policy research.

In Figure 4.2, eleven threads have been grouped together according to their respective emphasis on the process of influencing policy-makers, the endogenous character of development or the complexity of social systems. Each of these threads is developed by different groups of scholars, and the appendix offers a brief introduction as well as references.

Seven of the eleven threads emphasize the different ways of *influencing policy-makers*. They acknowledge that policies are shaped based on beliefs rather than objective truths. Actors build *coalitions* to exercise influence, and even wider *networks* of stakeholders and their topologies have their own influence. Politics, policy and problems are treated as *multiple streams*, and different political systems have an impact on the *social construction* of policies. *Narratives* are constructed and develop according to the influence of different parts of the policy system.

Other authors focus on the capability of social systems to *endogenously* develop solutions and policies. One can then conceptualize policy issue as a *game*, and as different games are played at the same time and interact with institutional arrangements a complex adaptive system arises, showing (amongst other phenomena) unintended consequences. This assigns the *ecology of games* thread to the intersection between *endogenous development* and *complex social system*. Another thread in this intersection emphasizes the interdependencies between *social and ecological* components of the overall system, using the concept of feedback loops to operationalize the complexity and the unintended consequences.

Still other scholars analyse how different *policy regimes* have an influence on the very policies which make up the regime. And of course, the high degree of complexity leads to challenges dealing with *coordination*, with *information processing* and with the dynamics of interaction between the different *subsystems*.

Across these threads of research, the vast majority of scholars acknowledge the *limited* nature of human *rationality*, the importance of *beliefs*, as well as the dynamic and *self-influencing nature* of the policies which a human society gives itself. Since at any given point in time the policies being applied have an influence on the future conditions under which they would be either maintained or modified, policies are recognized as components of *feedback systems* (Anderies and Janssen, 2013, p. 527). In this literature, policies translate information about a system with ecological and social components into actions influencing the system, thereby closing a feedback loop. In nature, closed control loops (or goal-seeking feedback) stabilize features of organisms, species and ecosystems (McCulloch, 1988) and it was only a question of time until humans would understand this principle and start to use it to adapt the state of the world to their needs and wants. Indeed, the structure and workings of feedback can be traced back in human history long before the modern state (Richardson, 1999). Of course, *endogenous development* also means that policies must be sufficiently stable and open to modification, which is where the *influencing policy-makers* thread, with its emphasis on *beliefs*, is relevant.

Beneath the diversity of topics investigated by these threads, there is a logical structure of interaction between individuals in a society. This chapter develops a conceptual model of the human actor and the social system, in which policies appear as a component in a system of feedback loops which can be developed by the process of engineering. The argument will be elaborated in three steps before coming to the central propositions. First, a conceptual model of the human actor is introduced. Second, the concept of 'artefact' is discussed and incorporated into the model. Third, the processes of evolution and engineering are inserted into the model. Overall, this conceptual model helps to see the fundaments common to the diverse research threads. Contemplating this diversity of threads in relationship with the model proposed here also shows that bringing an engineering approach to policy design implies transforming existing policies, and the intricacies of making policy have to be faced by those striving to make such changes.

4.3   Human Actors: Perception, Policy and Action

As humans, we interact with others. Interaction constitutes an enactive system in the sense of Varela (1995, p. 330): 'perception consists of perceptually guided action'. *Perceiving* is an internal activity of the individual, modulated by sensorial stimuli, and that which is perceived drives *decisions*, which drive *actions* on the environment, which then lead to new *perceptions*. Since our environment is made up by other, similar individuals, each actor is interconnected with others in a 'creative circle' (Varela, 1984), where one's *actions* trigger the other's *perceptions* and vice versa.

We also reflect upon what we perceive and what we decide to do. Consequently, we carry *beliefs* concerning what is going on around us, what we want and how we can act such as to obtain our desired results. These beliefs contain causal statements of what exists and how the existing entities affect one another, as well as values (what is good and desirable) and goals.

For instance, a mayor's beliefs will contain the entities voter, taxes and power; the relationships between them may be that voters give power and higher taxes reduce voters (in favour of the mayor). The mayor's values will contain his desire to stay in office, which leads to the goal of winning the next election. Our minds also contain individual *policies*, and these are driven by *belief* systems, which we can recognize. In the example of the mayor, the policy might be 'if voters seem unhappy, reduce taxes'. However, the *belief* systems also regulate what we *perceive* – which we usually do not notice. The exemplary mayor may fail to take into account that wisely spending the tax money increases the voters' well-being and therefore he may fail to pay attention to the deterioration of specific conditions in his city.

A *personal policy* is any rule, heuristic or strategy which allows the actor to decide a course of action which promises to achieve his goals, given his *beliefs*, which include his causal attributions, prejudices, values and goals. Since all actions are carried out by a person, all policies are *personal*. However, no person lives alone: there is no mayor without citizens (always in the plural) or voters. Therefore, there need to be *collective* policies – more or less internalized by the individuals – to bring orderly patterns into the interactions between individuals. Policy therefore always has a collective and a personal facet, just like a coin has two sides. A minimal social system would consist of two interconnected actors. This is shown in Figure 4.3.

In Figure 4.3, the large boxes show two interconnected individual actors, A and B; the smaller box in the middle represents social artefacts, which may be tools, machines, built objects, but also institutions and collective policies. The darker arrows correspond to links, which are usually most conscious to human actors. The perceptions of one actor lead to decisions, which translate into actions, which are then perceived by the other actor. Thus emerges a closed loop comparable to Bateson's 'learning 0' (1973), where one 'learns' what the conditions are. For example, the mayor of a city can 'learn' that voters are increasingly

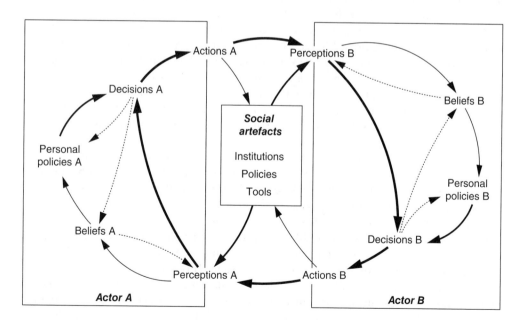

Figure 4.3 Human actors form a minimal social system.

complaining. The difference between the observed complaints and the number of complaints considered normal would be a first-order error – usually called a 'problem'. He may have a policy for dealing with this problem, and the resulting process would be exactly what the control loop in Figure 4.1 (right side) describes; it is single-loop learning in terms of Argyris (1976). This policy may already be an official and collective one – a social artefact – or the mayor may use his personal relationships. Of course, he must not act counter to existing collective policies.

There are other influences, though. They are less frequently taken into account, and therefore printed with an intermediate width. First, our perceptions are influenced by institutions and policies. Second, and at the same time, our actions also contribute to shaping the social artefacts. Third, the decisions we take are influenced by our personal policies, which tend to be unconscious under routine conditions. Still, these influences can be retrieved from the implicit into awareness with relative ease, for instance when something goes wrong. In the example, the mayor will wonder why voters are complaining increasingly. The results of this wondering may lead to changes in personal policies and then influence collective policies or other social artefacts, which will then have an influence on people's behaviours.

Other influences mostly remain implicit (lighter arrows): our personal policies are shaped by our beliefs (which are sometimes referred to as ideas or mental models). Our beliefs, in turn, are influenced by our perceptions, but the process of forming beliefs is usually unconscious. Additionally, our beliefs regulate what we perceive – and what we do not perceive (Senge, 1990) – and this influence is unconscious and very hard to bring into our awareness. Finally, the very fact of taking decisions reinforces our current policies and beliefs (Spitzer, 2006). In the example, maintaining taxes low or decreasing them may always have been accompanied by successful elections in the past, and each decision to cut taxes may have appeared to be a good one in this light; the belief that voters dislike taxes may therefore have become so second nature that it is not recognized as a belief any longer.

What would happen if the current policy of cutting taxes fails to correct the first-order error and complaints do not decrease? In the case that citizens appreciate roads without potholes and high-quality schools, and would be willing to pay even higher taxes for these things, tax cuts would still superficially be pleasing to them, but the ensuing deterioration of schools and streets would make them increasingly unhappy. This would be a second-order error, but the existing policy cannot detect it and therefore would keep trying to resolve the problem in the same way: 'more of the same'. This situation would require an instance of double-loop learning (Argyris, 1976).

If a policy changes decisions to correct first-order errors, what then can change a policy which has a second-order error? Since any decision to do something is driven by a policy, there must be a policy regulating how policies are changed (Schaffernicht, 2011). A prominent example of such second-order policies can be found in law: constitutions contain rules for how they can be changed. Such a policy has to answer two questions.

First, how can a second-order error (that is, the failure of a policy) be detected and the need to correct it be recognized? Given that the belief system frames policies, such a policy must also regulate the revision of the underlying belief system. This requires that those perceptions which could reveal flaws in our beliefs or policies are not unselected by our beliefs' influence on perception. How can the mayor realize that his loss of voters is not the fault of a political contender but of his own spending cuts, and that he should have school quality and the state of the roads on his scoreboard?

Second, what ought to be done upon detection of such an error? Which policies ensure adequate modifications of policies – including underlying beliefs? Human imagination can always generate new tentative ideas of what could be done – it generates *variation*. However, not all that would be logically plausible can really work out, and so we need a *selection* process to 'debug' such ideas. If someone suggests to the mayor that schools should raise tuition fees, so that quality can be funded without raising taxes, how can the mayor check if this would likely be a successful policy? There are two possible (second-order) policies for doing so. First, 'try in practice': actions which lead to undesired outcomes will trigger unexpected perceptions and, to the degree that these can be recognized, lead to 'surprises' which indicate the need to improve the belief system and/or the personal policy. So, if the mayor simply implements the suggested change in school funding policy, but if many families cannot pay the tuition fees and must stay in public schools, then the resulting inequality may turn away voters and he will lose the next election; in such a case, the price for finding out would have been high. The second possibility is to 'test drive': policy models can be developed to test possible policies without fully implementing them, reducing time and other resources spent in the process (Sterman, 2000). In the context of the mayor's example, this may avoid serious side-effects like school segregation and spare the mayor a lost election.

Figure 4.3 illustrates the complexity of the task. How can one person embody a personal policy which allows him or her to recognize the need to revise a policy and then develop a successful improvement, if humans tend to take their decisions with cognitive economy, settling for an idea as soon as cognitive ease is achieved (Kahnemann, 2011)? This is a fundamental question and it concerns every person, no matter which social systems he or she takes part in. However, the social systems we are actually part of are more complex than that in Figure 4.3: actors like colleagues, superiors, providers, customers, government representatives – all have their beliefs and policies. Since this has always been the case, manifold interactions have always occurred in parallel, and our beliefs and policies, as well as our social artefacts, have evolved: we are and always have been immersed in the process of *cultural evolution*, full of unanticipated turns. However, being the sense-making and planning beings we are, we are compelled to design explicit (in other words, artificial) aids that would help us to overcome those aspects of this evolution that we deem undesired. One of them is the mentioned, instinctive, self-concealing tendency.

This section closes with an important insight: policies do not exist without belief systems, which must be taken into account when analysing and designing policies.

4.4   Artefacts

What is made by humans is called 'artificial' as opposed to 'natural' (Remington *et al.*, 2012). As remarked by Hayek (1988, p. 143), nature changes due to evolutionary processes, and many if not most social institutions like 'family', 'enterprise' or 'market' are *natural* in this sense, rather than *artificial*. At the same time, creating artefacts is part of human nature, and has been elevated to the rank of a science (Simon, 1988). In the model presented in the previous sections, both policies and beliefs are originally tacit and therefore natural, but can be recognized: the actor can make them explicit and they can then be analysed and revised. In this sense, they are also artefacts.

We are used to calling 'artefact' the mass-produced devices of modern technology for two reasons:

1. they are *concrete* and we can physically touch them (as opposed to *abstract*);
2. they are *external* because individuals other than the user produce them (as opposed to *mental* or *personal*), which we will refer to as *social*.

Such tools and devices create a space for actions and the user develops a mental model of them; understanding this mental model is part of the process of designing these artefacts (Barfield, 1993).

However, beyond material tools, built objects and machines, there are other types of artefacts. In organizations, we are given manifold forms and standard documents to be used at work. As citizens, our life is regulated by many civic procedures. Forms may be just paper, but they are *concrete* and *social* too. Laws, official (stated) policies, procedures and official models are also defined by others and therefore *social*, but they cannot be touched: they are *abstract*. Even so, they regulate our actions, either by prescribing or by forbidding, because *concrete* artefacts are developed to implement them – and even so, they are doubtlessly produced by humans. Beneath these *external* artefacts, there are always personal processes like heuristics and cognitive structures like mental models and personal policies. They are *abstract* too, but additionally their locus of development is not *external*: they are mental objects, developed by the same actor who uses them. Since they are developed by a human being too, they are artefacts as much as the previous examples, though *personal abstract* ones. With these distinctions in mind, we can now inspect the area of social artefacts in more detail.

Compared to the way the human actor was represented above, beliefs are now separated into tacit and conscious ones, because only conscious beliefs can be deliberately worked with. Actions now appear only as the particular acts of articulating beliefs, policies and decisions. The area of personal artefacts is stacked to indicate that there are always many different individual human actors involved.

The human actor uses and thereby perceives *concrete social* artefacts, which are well described by the term 'appliance', which is 'an instrument, apparatus, or device for a particular purpose or use' (Oxford Dictionary 2015). Some appliances are machines or software converting a computer into a tool for a specific purpose, others are buildings and cities. Still others, like forms, standard documents and procedures, may not appear very engine-like; however, they are *concrete*, we use them and they have an influence on our actions. Such *concrete* artefacts are *social*, because the producer is usually not the user and vice versa; additionally, many users are taken into account in building such appliances, and at the same time these appliances are produced to be used by many.

By which kind of process do such *social* artefacts come into existence and evolve over time?

Organizational policies expressed in procedures, forms or software are in general developed by a process of hierarchical *authority* and *tradition*. In 'political' social systems, public policies are frequently shaped by *negotiation* and *voting*. In both types of cases, many different actors interact with different goals and expectations (also concerning the likely behaviour of other actors), and therefore beliefs (causal attributions as well as values) are rarely made explicit. Under such circumstances, mental models are left to themselves in the individual actors' minds, and therefore under the reign of the individual policies. Tradition, voting and hierarchical power do not use an explicit link from the beliefs of policies of individual actors to the shape of social artefacts. They depend on 'trying in practice' and will only be transformed when the external environment makes them fail.

This stands in remarkable contrast to *concrete, external* artefacts, which are usually developed by the process called 'engineering' (Kroes, 2012), where we are used to a high degree of *explicit* designing, testing and progressive implementation: engineered artefacts go through cycles of modelling from *abstract* to *concrete*, identifying and eliminating weaknesses and faults in the process: no person in their right mind would go aboard a plane if aircraft builders only improved their machines after crashes. The same care is frequently not given to our *abstract* artefacts, especially when dealing with collective policies. Think of schooling and health care in your country: have they been shaped by an engineering-like process?

Engineering is the practice of designing and developing engines. An *engine* is commonly understood as a machine or device that transforms one form of energy into another one; however, its origin refers to *genius*: Latin *ingenium* meant nature, innate quality, especially mental power. Engineers are individuals who can build devices that behave as if they had a mind of their own. Are policies like 'engines' and therefore amenable to engineering? The answer makes extensive use of the term 'model', because the deliberate construction, testing and improvement of external *models* (of increasing tangibility) is at the heart of engineering-like artefact development. As far as one can refer to *beliefs* and *policies* as *mental models*, can the disciplining power of testing be brought to bear on our *personal artefacts*?

A 'model' is generally defined as a representation, usually on a smaller scale, of a device or a structure. We can model for different reasons: understanding something or designing something (giving it a shape). There are many ways to distinguish different kinds of models (like Forrester, 1961). We focus exclusively on models developed as part of a design process (Schwaninger, 2010). The type of model one develops depends on the type of artefact to be developed. Such models are diagrams, equations and dictionaries of specifications, or any combination of the previous. They all conform to semantic rules (types of entities that can exist in a model) and syntactic rules (how they are represented), which reduce ambiguity.

Figure 4.4 is laid out so as to illustrate the engineering process as cyclical and linkable to the cyclical being of the human actor. The *concrete social* artefacts are the leftmost area, and bringing these into existence is the goal of all *abstract social* artefacts (in the neighbouring area), which are usually models in the sense of the above definition. Procedures as *social concrete* artefacts are implemented, for instance according to process models, which are frequently developed by management professionals rather than engineers. Business process modelling also establishes semantic and syntactic rules, conforming directed networks of activities (states) and transitions, which are hierarchically decomposed. Such models may be an input for engineering models (for information systems, for instance). They ought to be based on decision models, which in turn ought to draw on policy models, which ought to be developed out of articulated mental models or beliefs. At the same time, the development of decision models ought to be informed by decisions, policy models ought to be oriented by personal policies and, of course, articulated beliefs should articulate conscious beliefs. By crossing the border from the personal to the social (external) realm, articulating and modelling (producing abstract external artefacts) is an action that is perceived by the articulating and modelling actors, and therefore it has an influence on the actor's beliefs and policies. Figure 4.4 shows the interaction between a human actor – appearing as the system of personal artefacts – and the social artefacts.

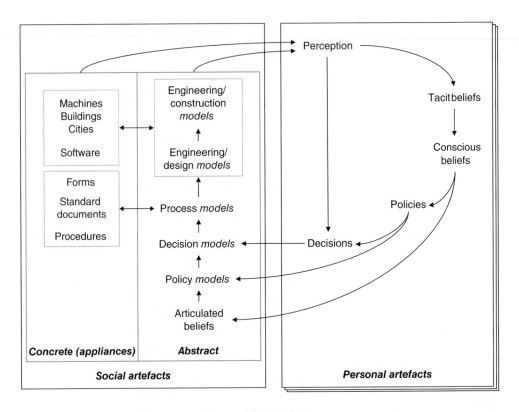

Figure 4.4   Artefacts.

The process description in the previous paragraphs contains many repetitions of 'ought to'. This may seem surprising when we think of the typical artefacts produced by engineering, like cars, household appliances, computers or smartphones. However, do the designers of such appliances take into account the ways in which their users' actions will influence ongoing institutions and policies? Do designers of business procedures take into account the ways in which the employees' actions will influence their own beliefs and policies? Do designers of public policy follow such a disciplined process, taking into account many different causal and other beliefs, systematically testing possible policies and systematically testing the likely outcomes of implementation, including side-effects? It is not a too far out guess to answer 'frequently not' to these questions – hence the importance of influencing policy-makers in the literature. And hence the 'ought'.

4.5   Engineering and Evolution: From External to Internal Selection

As stated above, personal beliefs and policies are usually shaped in a tacit as opposed to a deliberate way, and the same holds for many social institutions and policies. After all, the most fundamental institutions that human civilization has developed are 'human activity systems' (Checkland, 1981), influenced by the desires and actions of particular individuals but neither invented nor constructed by them. Arguably, our institutions are shaped by a process called *cultural evolution*.

In general terms, *evolution* has to do with 'adaptation to unforeseeable events' (Hayek, 1988, p. 25); it is a process operating on populations (not individuals) and proceeds in two interacting steps:

1. *variation*, which generates a pool of candidates from which to select;
2. *selection*, in the form of eliminating unsuccessful candidates, either by the environment or by internal selection processes.

This sequence leads to an increasing adjustment or fit of that which is evolving to its environment or selective systems (Campbell, 1965). It is an abstract process, which is not bound to the realm of biology but can take on many different concrete forms (Dennett, 1995).

Variation is a sub-process leading to differences between individuals in a population of some kind (Mayr, 2001). There are three different types of variation: random, blind, and biased. In *random* variation, all changes are equally likely due to random factors. In the case of *blind* variation, variations are independent of previous experiences and are produced without a priori knowledge of which ones can be successful, but not all variations are equally likely (Campbell, 1987); this type of variation is necessary for creativity to occur. Finally, *biased* variation is a corrective goal-seeking process in which the generation of variation is conditioned by previous experiences, like the control loop in Figure 4.1.

Selection is the process which favours the diffusion of successful candidates over the proliferation of less successful ones. It can occur in different ways and can be *natural* – resulting from interaction with the environment – or *artificial*, when carried out by a decision-maker (Mayr, 1991). Either way, selection is frequently *external* to the candidate population, but it can also be *internal* when there are requirements of organizational stability (Bickhard and Campbell, 2003). In particular, in engineering-like design processes, selection is mostly internal, whereas the 'try in practice' processes depend on external selection.

One may now infer that engineering and human attempts in general at maintaining something under control are essentially based on *biased* variation, in other words solving known problems. When an action fails to produce its expected outcome, and if this failure is perceived, then a second-order error is detected (see Section 4.3): the problem is not that the environment behaves other than desired, but that current actions driven by the policy do not yield the intended reactions in the environment. So, the problem is not the environment, but the policy. And assuming that the current policy is based on the current belief system, it follows that the belief system is part of the problem and must be revised. *Biased* variation of the policy would very likely be like trying to solve a problem with the same means used for creating it in the first place, like in the example of Section 4.3. But how do you search for better beliefs if you only know that your current ones have flaws? Acknowledging that the current beliefs are flawed opens the door to *blind* variation; that is, exploring novel alternatives no matter how unlikely they may look.

Different scholars have stressed that *blind variation*, combined with *selection*, drives technological evolution (Petroski, 1992, 1993; Ziman, 2000). Indeed, many technological artefacts that have changed our lives over the past 200 years were not piecemeal improvements of existing devices, but devices which had been deemed impossible until someone constructed them. Prominent examples are railroads (people believed the acceleration would press the air out and passengers would suffocate); aeroplanes (which would not fly because they were heavier than air); helicopters (which would not fly and would be useless); moving pictures with sound tracks (which were first seen as impossible because the brain does not process light the same way as sound, then believed to be commercially useless because the president of the Kodak Company did not believe that people would go to the movies to listen to a sound track) (von Foerster and von Glaserfeld, 2015). The washing machine is thought to be one of the major change agents in recent human history, because it freed women from washing laundry as a large part of their daily activities (Cardia and Gomme, 2014). The last examples illustrate that frequently the impact on our lives or social innovation was triggered by a new technological artefact, but the technological breakthrough was not necessarily made by seeking to achieve social change. In other words, the effects on collective and personal policies and beliefs were not anticipated – which does not mean that such effects are impossible to anticipate.

Engineering has had intended and unintended effects on natural as well as social systems, thereby influencing biological as well as cultural evolution – many times unintentionally. Suffice it to mention pharmaceutics and agriculture as domains where our *external concrete* artefacts modify the genetic base of organisms for the sake of health and food production. Even before the advances in genetic engineering, vaccination has long been a practice that some suspect to weaken the organism in the long term. And more generally, one could suspect modern medicine of weakening the human genome by artificially keeping alive individuals who, by the process of natural selection, would have been selected out of the population. By the same token, the possibility and practice of intervening in social institutions and policies has always raised the suspicion of critics. As one of the most outstanding critics, Hayek (1988) argued that human reason – itself a product of cultural evolution – is insufficiently apt to take into account the complexity of human society, and that therefore even well-intended interventions provoke too many unintended consequences. Specifically, no central authority should restrict individuals' freedom beyond warranting the equality of restrictions for all individuals, thereby protecting the potential for each individual to make important contributions in the form of success that would be imitated by others, and thus maintaining innovation and progress in human society.

With the sense-making and planning nature of human beings in mind, it can hardly be that we perceive undesirable aspects in the world and *do not* employ our creative powers to correct that which is unsatisfying. Since this is a natural way for humans to behave, Hayek's argument is itself an attempt to intervene in a spontaneous behaviour based on reason. Having developed our understanding of how evolution works, many of its principles have been successfully incorporated into engineering – to mention only two: genetic algorithms and complex adaptive systems (Holland, 1992). This has enabled human engineering to intervene in many aspects of our biological as well as our social life. The quick-paced advances in analytical methods and technological tools of the past decades can indeed encourage us to affirm that not only have the principles of evolution improved human engineering, but also that human engineering is capable of improving our biological and social systems, if not their evolution.

Of course, the quick scientific and technological advances that have enabled the policies of particular social systems – like commercial enterprises and secret services – to take advantage of a wealth of information about us as consumers and as citizens may have outpaced the speed of developing public policies ensuring that such organizations do not violate widely shared values and beliefs. However, this is not an argument against engineering in social systems;

rather, it is a call for improving the way engineering principles and methods are brought to bear on the process of public policy.

This requires recalling some of the less salient processes and links of the human actor and the inseparable social system he is embedded in. In particular, we must keep in mind that human actors will adapt their personal policies to new collective policies in their own way. Compared to the hierarchical command and control of enterprise and of the forces or order (police, armed forces), 'public' social systems and their collective policies leave more space for diversity of personal policies. This deserves special attention, because the resulting complexity makes it harder to anticipate the likely effects of intervention in policies. But rather than giving up on reason and yielding to instinct, the following section proposes some recommendations for influencing the evolution of policies by an engineering approach.

4.6   Policy between Cultural Evolution and Engineering

A society of human actors, as described in Section 4.2, necessarily generates artefacts, some of which are abstract. 'Policy' is one such type of artefact. While policies *evolve* over time, we cannot fail to reason about them and therefore would be ill-advised not to use our scientific understanding and engineering capabilities to influence their evolution. In order to reinforce this proposition, some other aspects ought to be noted.

Policies rather than policy. In a society consisting of individuals with their own needs, wants, reasoning power and roles, talking about 'policy' in the singular may be misleading. An 'engine' will have one design, defining a class of individual appliances that will have the same physical features upon production. The designers and builders will work towards a specific goal deemed to be meaningful assuming the interests of a particular actor (de Vries, 2010). Even so, they will be used in a variety of ways by many different individuals with their own interests and goals. The final meaning of the 'engine' comes into existence through the actions performed by the individual using it, rather than the designer or engineer: a knife or a gun is neither good nor bad, it can be used for peaceful, productive activities as well as for aggression and destruction. Analogously, a policy which has been devised as one to serve a concrete goal and the interests of one or several stakeholders will unfold in a range of ways, some of which may be unexpected for the policy designers. In this sense, terms like 'policy-maker' can be misleading too. As an example, think of tax policy in the context of the development of a city. The mayor needs financial resources to develop the city infrastructure, and therefore raises taxes. The tax policy is meant to support the city's development. However, firms striving to maximize their profits will use this tax policy in a way which maintains the tax burden as low as possible; this may even mean moving to a different city offering more attractive tax conditions. Such a move would arguably hinder the city's development. So, is there only one tax policy in this example? And is there only one policy-maker? If one acknowledges that the ultimate meaning of the policy is the level of tax revenues of the city, then the possible action outcome of the policy ought to be considered when designing the policy, so that the policy designer reduces the likelihood of other actors using the policy in ways which run counter to its intended purpose.

Versions rather than final designs. A second consideration is that any such policy would need to be constantly revised and adapted, in analogy with the versioning of usual 'engines'. When humans make 'engines', it is usually with a goal in mind – even though creativity (spontaneous generation of variety) plays a role – and a difference is made between design-time and use-time. It is interesting to remark that many 'engines' – like windows, cars, aircraft or software – are constantly changing through 'versions', where the experience of using the 'engine' is fed into a redesign process. Scholars in the realm of engineering have collected and analysed examples of the evolution of such 'engines' (Ferguson, 1992; Petroski, 2004, 2009). As far as engineering refers to the continuous process of adapting the current version of an 'engine' to the recent experience, it could be taken to be a process of variation and selection of policies, as will be discussed below.

Transformation rather than creation. A third consideration is the realization that human activity is always an ongoing process. Consider the case of transportation. Humans have always moved themselves and goods from one place to another; over time, quite different technologies have been used for this purpose. Originally only muscle power, then animal power for traction, then different kinds of engine – always converting one form of energy into a different one. The concrete, social artefacts produced and used have been carriages, ships, trains, cars and aircraft. Sometimes, a new type of artefact partially replaced an older one, like when the modern aeroplane replaced ships for intercontinental travel. However, such new artefacts did not really create a new domain of human activity; they simply intervened in existing ones. Of course, there are cases when that intervention had drastic consequences, like for instance the effect of washing machines on the role of women in society, a ubiquitous audio-visual medium on the reading-based culture, or Internet-based social-network software. Still, laundry was washed before, people communicated with one another before, and audio-visual transmission of messages existed before TV: theatre.

The same holds for organizational and public policies: whenever something new is planned or designed, it will never be put on a blank plate; people will always have been doing something already. Scholars have observed and analysed how firms and organizations deal with the unforeseeable features of what they occupy themselves with (Garud *et al.*, 2006, 2008). Policy designers should conceive of their designs as changing something which is already going on, and take the beliefs and policies of those involved into account. In the example of the mayor who wishes to stay in office and to develop the city, it may appear that some of the aspects of city development which have traditionally been looked after by the mayor could be transferred to a different actor, thereby reducing the need for financial resources coming from taxes.

Historical examples of policy reframing can be found in the influence of the 'urban dynamics' book and model (Forrester, 1969) on the way mayors thought about the role of social problems on the decay of inner cities in the United States; also the limits to growth model and book (Meadows *et al.*, 1972) appears to have had a profound influence on the belief systems of policy designers, even though the influence on concrete policies may be hard to trace. A more recent example is the role played by simulation-based role play in training nego-tiators and country representatives in global climate talks (ClimateInteractive, 2015; see also Wu and Lee, 2015).

The facts and relationships discussed in the previous section can be converted into a set of recommendations for continuous participatory modelling of collective policies:

1. Anchor *all the phases* of the policy process in the *belief systems* and current *policies* of the largest practically possible number of (groups of) actors. In *problem formulation*, partici-patory articulation of belief systems will raise individual consciousness and mutual under-standing of agreements and disaccords.

2. In *policy design*, the more actors can feed their personal policies into the process, the more diversity there is to select from. Such diversity calls for *internal* selection; indeed, possible policies must go through tests which do not depend on the stakeholder's opinions/interests. The interdependencies between policy subsystems make testing a demanding task. However, *external* selection in the form of implementation failure and unintended consequences will in general have a higher cost. Include indicators for goal achievement (indicating the policies' outcomes) and separate indicators for goal achievement changes, which reveal if policy performance increases over time.

3. In *implementation*, the delay between what actually happens and what is perceived decreases, making it easier to feed current developments into new decisions or even policy adjustments. Also, the actual variety of policies that individuals will have made from the initially designed policy will reveal useful information for deselecting some features and emphasizing others. This helps to take a dynamic stance, admitting that there is always already a policy in use, that policy diffusion takes time and that continuous versioning leads to an *evolution by design* of policies.

4. In *all phases*, both sets of indicators must be visible for the public at large. The process must include features which help actors to become aware of their beliefs and policies and their meaning in the wider social system, such as permanently linking them into the policy process.

These recommendations ensure that the almost invisible bond between perceiving, beliefs and policies is reinforced and becomes a regular part of the policy system. This in turn will help actors to become able to make sense of all the information, which is necessary to drive the iterations of variation – selection typical for engineering in complex adaptive systems, especially human social systems.

The systematic process of articulating and revising beliefs is called modelling, and some modelling methodologies are particularly useful for designing policies. System dynamics have been developed specifically for designing policies (Forrester, 1992), and explicitly stress the goal of improving mental models: they strive to facilitate learning (Sterman, 2000, chapter 2). Also, the concept of complex adaptive agents (Arthur, 2007) has diffused in economics, and agent-based modelling has become widely used to study how social patterns emerge from individuals' interactions.

Of course, models and belief systems driving policies have to be connected with selecting the types of information needed and making the information available, which is an area where computer science and informatics make contributions (Barrett *et al.*, 2011). Therefore, it is not surprising that over the past years, the emerging field of *policy informatics* has brought together information technology and systems with modelling and simulation (Johnston and Kim, 2011). 'Policy informatics is the transdisciplinary study of how computation and communication technology leverages information to better understand and address complex public policy and administration problems and realize innovations in governance processes and institutions' (ASU Center for Policy Informatics, 2015; eGovPolyNet, 2015). It is argued that the citizens can indeed be integrated into the policy process thanks to available technology (Dawoody, 2011). And informatics can help improve the integration between policy actors and across policy subsystems (Koliba *et al.*, 2011). By providing massive volumes of information, information technology increases the burden of selecting and processing described by

Workman *et al.* (2009), and by integrating the citizens – who have no specific preparation for processing this information and cannot call on specialists at their avail – into the policy process, it becomes ever more important to mind the role and the development of belief systems.

Climate change provides an example for the above recommendations. Since worries concerning the planetary climate became overwhelming and the International Panel on Climate Change (IPCC, 2015) started monitoring and investigating the climate, Earth's governments are meeting on a regular basis, trying to agree on goals and policies. The 1992 United Nations Framework Convention on Climate Change (UNFCCC, 2015a) was succeeded in 1997 by the Kyoto Protocol to the United Nations Framework Convention on Climate Change, but over the years the world's leading contributors to climate change did not commit to effective policies. For the 2009 Copenhagen conference (UNFCCC, 2015b), a team of system dynamicists at ClimateInteractive (2015) had developed the C-Roads simulation model based on and audited by the IPCC (recommendation 1) and a participative workshop using the model to train the negotiators (recommendation 2). In that 'game', negotiators can negotiate their pledges and then use the simulation model to discover the effects of the combined pledges on global warming and sea-level rise. Over the ensuing years, the model and the workshop materials have been expanded to be used in educational settings (C-Learn) and translated into other languages. The wider public of Internet users can also keep updated (recommendation 4), and even run their own simulations (recommendation 2) over the ClimateInteractive website. At the 2015 Paris conference (UNFCCC, 2015c), the C-Roads training had been in use for more than 5 years. It cannot be told how much influence the workshops and the simulation model have had on the process, and of course the global reorientation of national economic and industrial policies has long delays (recommendation 3), but the modelling work has been related to the – admittedly slow – progress of these negotiations (Mooney, 2015).

Climate change is one example which stands for many global challenges that confront us with the following question: Will we succeed in using an engineering approach to policy design and thus speed up the evolution of collective and personal policies by internal selection, or will we keep relying on external selection and risk failures with catastrophic consequences for humanity?

4.7   Conclusions and Outlook

Policies are driving our behaviours. As natural sense-makers, we can become aware of and intervene in them. We have to live together to be human, and therefore collective policies are of paramount importance – especially at the social level. We have built an advanced understanding of evolution in general, and have reached enormous power to analyse and synthesize, which has become manifest in countless products and services over the past decades: we have the methodological and technological means to apply an engineering approach to our collective policies.

Of course, a look at the sphere usually associated with policy reveals a less optimistic picture. The science and practice of public policy show us a discouraging picture of the policy process, and in many democratic industrialized countries the public appears to have a dire opinion of their politicians. At the same time, the breakdown of the former Soviet Union has been the reason to claim that decentralized political and economic orders are now proven to be superior to human planning and intervention. Global problems like climate change appear to thwart our capability to come up with solutions with sufficient speed, not to speak of peace between countries and ethnic groups. Scepticisms like those put forward by Hayek (1988) look difficult to counter in the face of such evidence.

However, a second look shows that processes like climate change-related policies, just like typical engineering endeavours, take iterative attempts. For instance, coming up with electric light bulbs took many years and countless trials (www.enchantedlearning.com/inventors/edison/lightbulb.shtml). This means that we need to be able to transform our policies with sufficient speed to keep up with the transformations our policies cause in our natural and social environment. Since engineering is the only known process able to use internal selection, this chapter comes to the conclusion that the design power of modern engineering methods and techniques – which has been able to transform the human society and the planetary eco-system in less than four generations by developing products and services at an astonishing pace – can and ought to be employed to speed up the evolution of policies.

While interventions like central planning may have been worth criticizing, modern governments are not pursuing such aims any longer. While, during the first half of the twentieth century, we may have lacked the methods and tools to intervene in society without provoking serious side-effects, it is now not a logical necessity that collective policies designed by human actors will lead to more damaging side-effects than positive impacts. Nor is success guaranteed, and such policies will not automatically bring about the best of worlds.

The pervasive availability of information makes it necessary to increase the ability of citizens, consumers and workers (in the broad sense) to know and develop their beliefs and policies, to select the relevant information and understand it in order to regulate their actions.

This, then, is a call to bring design and engineering to schools and universities. And since these institutions are run by individuals who have already left school and university, making the required transformation in educational policies happen is a huge engineering challenge. Even beyond school and university, the fact that humans learn as long as they live calls for evolving educational policies into learning policies, which will be even harder to achieve. It will serve us well to keep in mind the engineer's attitude that failure is not an option.

Policy engineering accepts the open-ended and decentralized nature of evolution without renouncing the power of human reason. In a way, it is our route to move from 'between instinct and reason' towards 'with reason-enhanced instinct'.

Appendix: Brief Overview of the Policy Literature

A summary of theoretical approaches to policy is presented in Table 4A.1. The *advocacy coalition framework (ACF)* (Sabatier, 2007; Weible *et al.*, 2009) puts emphasis on how groups of actors sharing beliefs and interests form *coalitions* and use information and strategies to influence policy-makers inside a policy subsystem, which deals with a wider system of relatively stable parameters, long-term coalition structures and emergent conditions. Beliefs (normative and causal) are given paramount importance and policies are considered as a translation of these beliefs. Actors are assumed to be boundedly rational. *Policy networks* (deLeon and Varda, 2009) stress that stakeholders interact and collaborate on policy issues, and expand the participating actors beyond those who usually participate in *coalitions*. The effects of

Table 4A.1 Threads of theoretical approaches to policy

Focus	Thread	Belief systems	Bounded rationality	Feedback loops / Unintended consequences	References
Influencing policy-makers	Advocacy coalition framework (ACF)	1	1		(Sabatier, 2007; Weible et al., 2009)
	Policy networks				(deLeon and Varda, 2009)
	Multiple streams framework				(Nowlin, 2011)
	Policy design and social construction				(Schneider and Sidney, 2009)
	Trans-subsystem dynamics	1			(Jones and Jenkins-Smith, 2009)
	Policy narratives	1			(Shanahan et al., 2013)
Endogenous development	Institutional analysis and development (IAD)				(Ostrom, 2007)
	Ecology of games		1	1	(Lubell, 2013)
Complex social system	Trans-subsystem dynamics				
	Ecology of games				(Feiock, 2013)
	Coordination	1			(Workman et al., 2009)
	Information processing	1			(May and Jochim, 2013)
	Policy regime	1			
	Socio-ecological system			1	(Anderies and Janssen, 2013)

different *network topologies* are studies with respect to hypothesis concerning diversity, reciprocity, horizontal power structure, embeddedness, trust, participatory decision-making and collaborative leadership.

The *multiple streams framework* (Nowlin, 2011) organizes the policy realm into three parallel streams: problems, politics and policies. Policy change occurs when *policy entrepreneurs* are able to influence actors in the politics stream with ideas from the policy stream to intervene in a theme of the problem stream.

Institutional analysis and development (IAD) (Ostrom, 2007) investigates the influence of institutions on human behaviour and states that individuals can create self-governing institutions – 'shared concepts used by humans in repetitive situations organized by rules, norms and strategies' (p. 23) – to mitigate conflicts. This framework has been influential for more recent developments, like the *ecology of games* (see below).

Policy design and social construction (Schneider and Sidney, 2009) asks how different political systems produce different designs and how these affect democracy. A policy is understood as consisting of a problem definition or goal, benefits and burdens to be distributed, target populations, rules, tools and implementation structures, social constructions, rationales and underlying assumptions. *Policy actors* are separated from the *public*, and it is essential that both groups accept particular concepts or constructs as 'real'. Target populations are organized into four ideal types: *advantaged* have political power and receive the benefits policy; *contenders* also have power but resist the policy. *Dependents* may benefit from the policy but lack power, and *deviants* have a negative image of the policy and also lack power. Policy actors are then related to possible policies in such a way that the advantaged receive the benefits and deviants the burden.

Jones and Jenkins-Smith (2009) study *trans-subsystem dynamics*, based on the fact that there are links between subsystems, and accordingly a meso-level analysis like the *ACF* needs to be complemented. In particular, they propose a system of linked subsystem clusters and public opinion. Whereas there is a tradition of being pessimistic concerning the cognitive capacities of the public, the so-called *revisionist* stance ascribes rational shortcuts and a reasonably well-defined belief system to the public. Public opinion is linked to the ideological tendency in the public, the policy issue's scope and the issue's salience. In this context, a particularly focused movement of public opinion can trigger policy change across multiple subsystems. However, the interactions between subsystems can also lead to so-called policy dimension shift.

Another approach stressing the multiplicity of subsystems and policy issues is the *ecology of games* approach (Lubell, 2013). A *policy game* is a set of policy actors taking part in a rule-governed collective decision-making process. Games and institutional arrangements form a *complex adaptive system*. The individuals are thought to be boundedly rational, a product of biological as well as cultural *evolution*. Multiple *games* are played at the same time and therefore the resulting policy decisions tend to have unintended consequences. Since there are multiple issues at stake, decision-makers have to face the dilemma between the short-term benefits of defection and the longer-term benefits of collaboration. This adds to the complexity of messy problems, for causes and solutions are not always known and there are multiple possible solutions. Therefore, learning is necessary.

Organizing our political system into many compartments of different scope – the world, country, region and so on – is in a way necessary for dealing appropriately with local issues. It also leads to dilemmas and conflicting mechanisms of *coordination* (Feiock, 2013). Coordination can come from embeddedness, contracts and delegated or imposed authority. According to the number of entities, coordination must be bilateral, multilateral or collective. As the number of entities increases, so do collaboration risk and transaction costs.

The role of *information processing* is essential (Workman *et al.*, 2009). Policy actors do not lack information on policy issues, quite the contrary: each interested actor, as well as the bureaucracy and research institutions, actively provide information. Therefore, the challenge is to prioritize which information to take into account. This is a decision in itself, and is framed by the belief system.

As far as the public is concerned, it has also been stressed that *policy narratives* have a great influence on sense-making (Shanahan *et al.*, 2013). A policy narrative consists of a judgement of policy-related behaviour plus the story character, potentially enriched by causal mechanisms, solutions and evidence. As far as influential policy narratives help to focus the public opinion on certain issues, the ongoing policy process may be disrupted and policy change occur.

There are two ways that policies have an influence on the conditions under which they have emerged. First, policies are part of a *policy regime*, including institutional arrangements, interest alignments and shared ideas (May and Jochim, 2013). Since policies will somehow redistribute benefits and burdens, the outcomes of any new policy would also affect the very policy regime, framed by the core ideas or beliefs behind the policies and how policy actors and coalition can influence them. Legitimacy, coherence and durability of policies, which depend on the policy regime, thereby have an influence on themselves.

Second, the social system, with its participating individuals, strives to regulate via policies that interact with the biophysical system we all are part of (Anderies and Janssen, 2013). Consequently, there is an interdependency between these systems. This is, in a way, an extension of the *IAD* framework, but it also proposes that policy subsystems have an influence on at least part of the parameters which the *ACF* assumes to be stable. Since our efforts to govern *social-ecologic systems* will always have to deal with imperfect information, causal ambiguity and external disturbances, an analogy is proposed with nature, where complex biological systems are composed of complex regulatory feedback networks. Indeed, the authors propose that 'SESs are feedback systems' (Anderies and Janssen, 2013, p. 527). Policies translate information about a system into actions influencing the system, and this system has ecological and social components, which can both be influenced by the actors, thereby closing a feedback loop.

These authors also note that the policy itself has to remain open to critique. Two essential questions in this context are which features transcend system scale and how complexity matters. Accordingly, system modules of polycentric governance units, ecosystems and infra-structure systems, combined with the feedback – in the form of rules, norms and shared strategies – should go through an iteration of studies of systems of different sizes and types in order to prove the policies' outcomes. Three design principles are recommended: redundancy, modularity and diversity. Redundancy makes a system failure-safe; upon failure of one sub-system, there is at least one backup system to fulfil the same function. Modularity avoids a failure extending over a specific module's scope. Diversity is a pool for creativity, required to refresh organizational mental models; in other words, belief systems.

References

Ackoff, R.A. (1994) Systems thinking and thinking systems. *System Dynamics Review*, **10**(2&3), 175–188.

Anderies, J.M. and Janssen, M.A. (2013) Robustness of social-ecological systems: Implications for public policy. *Policy Studies Journal*, **41**(3), 513–536.

Argyris, C. (1976) Single-loop and double-loop models in research on decision making. *Administrative Science Quarterly*, **21**(3), 363–375.

Arthur, W.B. (2007) Out-of-equilibrium economics and agent based modeling, in L. Tesfatsion and K.L. Judd (eds), *Handbook of Computational Economics, Vol. 2. Agent-Based Computational Economics*, North Holland, New York, NY.

ASU Center for Policy Informatics (2015) Arizona State University – Center for Policy Informatics. Available at: policyinformatics.asu.edu/content/about-0.

Barfield, L. (1993) *The User Interface: Concepts & design*, Addison-Wesley, Reading, MA.

Barrett, C.L., Eubank, S., Marathe, A., Marathe, M.V., Pan, A. and Swarup, A. (2011) Information integration to support model-based policy informatics. *Innovation Journal*, **16**(1), 1–19.

Bateson, G. (1973) *Steps to an Ecology of Mind: Collected essays in anthropology, psychiatry, evolution and epistemology*, Paladin, London.

Bickhard, M.H. and Campbell, D.T. (2003) Variations in variation and selection: The ubiquity of the variation-and-selective-retention ratchet in emergent organizational complexity. *Foundations of Science*, **8**, 215–282.

Boulding, K.E. (1956) General systems theory – the skeleton of science. *Management Science*, **2**, 197–208.

Bracker, J. (1980) The historical development of the strategic management concept. *Academy of Management Review*, **5**(2), 219–224.

Campbell, D.T. (1965) Variation and selective retention in socio-cultural evolution, in H.R. Barringer, G.I. Blanksten and R.W. Mack (eds), *Social Change in Developing Areas: A reinterpretation of evolutionary theory*, Schenkman, Cambridge, MA, pp. 19–49.

Campbell, D.T. (1987) Blind variation and selective retention in creative thought as in other knowledge processes, in G. Radnitzky, W.W. Bartley and K.R. Popper (eds), *Evolutionary Epistemology, Rationality, and the Sociology of Knowledge*, Open Court, La Salle, IL, pp. 91–114.

Cardia, E. and Gomme, P. (2014) The household revolution: Childcare, housework, and female labor force participation. Available at: www.emanuelacardia.com/wp-content/uploads/2014/05/household-revolution-2014-02-19.pdf.

Checkland, P.B. (1981) *Systems Thinking, Systems Practice*, John Wiley & Sons, Chichester, UK.

ClimateInteractive (2015) Available at: www.climateinteractive.org/.

Dawoody, A.R. (2011) The global participant-observer: Emergence, challenges and opportunities. *Innovation Journal*, **16**(1), 1–30.

de Vries, M.J. (2010) Engineering science as a 'discipline of the particular'? Types of generalization in engineering sciences, in I. Van de Poel and D.E. Goldberg (eds), *Philosophy and Engineering. An emerging agenda*, Springer-Verlag, Dordrecht, pp. 83–93.

deLeon, P. and Varda, D.M. (2009) Toward a theory of collaborative policy networks: Identifying structural tendencies, *Policy Studies Journal*, **37**(1), 59–74.

deLeon, P. and Vogenbeck, D.M. (2006) The policy sciences at the crossroads, in F. Fisher, G.J. Miller and M.S. Sidney (eds), *Handbook of Public Policy Analysis*, CRC Press, Boca Raton, FL.

Dennett, D.C. (1995) Darwin's dangerous idea. *The Sciences* (May/June), pp. 34–40.

eGovPolyNet (2015) eGovPoliNet – The policy community. Available at: www.policy-community.eu/results/glossary/policy-informatics.

Feiock, R.C. (2013) The institutional collective action framework. *Policy Studies Journal*, **41**(3), 397–425.

Ferguson, E. (1992) Designing the world we live in. *Research in Engineering Design*, **4**(1), 3–11.

Forrester, J.W. (1961) *Industrial Dynamics*, Productivity Press, Cambridge, MA.

Forrester, J.W. (1969) *Urban Dynamics*, MIT Press, Cambridge, MA.

Forrester, J.W. (1992) Policies, decisions and information-sources for modeling. *European Journal of Operational Research*, **59**(1), 42–63.

Garud, R., Kumaraswamy, A. and Sambamurthy, V. (2006) Emergent by design: Performance and transformation at Infosys Technologies. *Organization Science*, **17**(2), 277–286.

Garud, R., Jain, S. and Tuertscher, P. (2008) Incomplete by design and designing for incompleteness. *Organization Studies*, **29**(3), 351–371.

Hayek, F.A. (1988) *The Fatal Conceit: The errors of socialism*, Routledge, London.

Holland, J.H. (1992) *Adaptation in Natural and Artificial Systems* (2nd edn), MIT Press, Cambridge, MA.

IPCC (2015) International Panel on Climate Change. Available at: www.ipcc.ch.

Johnston, E. and Kim, Y. (2011) Introduction to the special issue on policy informatics. *Innovation Journal*, **16**(1), 1–4.

Jones, M.D. and Jenkins-Smith, H.C. (2009) Trans-subsystem dynamics: Policy topography, mass opinion, and policy change. *Policy Studies Journal*, **37**(1), 37–58.

Kahnemann, D. (2011) *Thinking – Fast and Slow*, Penguin Books, London.

Koliba, C., Zia, A. and Lee, B.H.Y. (2011) Governance informatics: Managing the performance of inter-organizational governance networks. *Innovation Journal*, **16**(1), 1–26.

Kroes, P. (2012) Technical artefacts: Creations of mind and matter – A philosophy of engineering design. *Philosophy of Engineering and Technology*, **6**, 127–161.

Lubell, M. (2013) Governing institutional complexity: The ecology of games framework. *Policy Studies Journal*, **41**(3), 537–559.

May, P.J. and Jochim, A.E. (2013) Policy regime perspectives: Policies, politics, and governing. *Policy Studies Journal*, **41**(3), 426–452.

Mayr, E. (1991) The ideological resistance to Darwin's theory of natural selection. *Proceedings of the American Philosophical Society*, **135**(2), 123–139.

Mayr, E. (2001) The philosophical foundations of Darwinism. *Proceedings of the American Philosophical Society*, **145**(4), 488–495.

McCulloch, W.C. (1988) *Embodiments of Mind*, MIT Press, Cambridge, MA.

Meadows, D.H., Meadows, D.L., Randers, J. and Behrens, W.W. (1972) *The Limits to Growth*, Universe Books, New York, NY.

Mooney, C. (2015) The world just adopted a tough new climate goal. Here's how hard it will be to meet. *The Washington Post*, 15 December 2015. Available at: www.washingtonpost.com/news/energy-environment/wp/2015/12/15/the-world-just-adopted-a-tough-new-climate-goal-heres-how-hard-it-will-be-to-meet/.

Nowlin, M.C. (2011) Theories of the policy process: State of the research. *Policy Studies Journal*, **39**(1), 41–60.

Ostrom, E. (2007) Institutional rational choice: An assessment of the institutional analysis and development framework, in P.A. Sabatier (ed.), *Theory of the Policy Process*, Westview Press, Boulder, CO, pp. 21–64.

Petroski, H. (1992) The evolution of artifacts. *American Scientist*, **80**, 416–420.

Petroski, H. (1993) How designs evolve. *Technological Review*, **96**(1), 50–57.

Petroski, H. (2004) Success and failure in engineering. *Journal of Failure Analysis and Prevention*, **1**(5), 8–15.

Petroski, H. (2009) Failing to succeed. *ASEE Prism*, **18**(9).

Remington, R., Boehm-Davies, D.A. and Folk, C.L. (2012) *Introduction to Humans in Engineered Systems*, John Wiley & Sons, Chichester.

Richardson, G.P. (1999) *Feedback Thought in Social Science and Systems Theory*, Pegasus Communications, Waltham, MA.

Sabatier, P.A. (2007) *Theories of the Policy Process*, Westview Press, Boulder, CO.

Schaffernicht, M. (2011) *Observer, modéliser, construire et agir comme un système autonome*, Éditions universitaires européennes, Paris.

Schneider, A. and Sidney, M. (2009) What is next for policy design and social construction theory? *Policy Studies Journal*, **37**(1), 103–119.

Schwaninger, M. (2010) Model-based management (MBM): A vital prerequisite for organizational viability. *Kybernetes*, **39**(9/10), 1419–1428.

Senge, P. (1990) *The Fifth Discipline*, Doubleday, New York, NY.

Shanahan, E.A., Jones, M.D., McBeth, M.K. and Lane, R.R. (2013) An angel on the wind: How heroic policy narratives shape policy realities. *Policy Studies Journal*, **41**(3), 453–483.

Simon, H.A. (1988) The science of design: Creating the artificial. *Design Issues*, **4**(1/2), 67–82.

Spitzer, M. (2006) *Lernen*, Spektrum Akademischer-Verlag, Heidelberg.

Sterman, J.D. (2000) *Business Dynamics – Systems thinking and modeling for a complex world*, John Wiley & Sons, New York, NY.

Torgensen, D. (2006) Promoting the policy orientation: Lasswell in context, in F. Fisher, G.J. Miller and M.S. Sidney (eds), *Handbook of Public Policy Analysis*, CRC Press, Boca Raton, FL.

UNFCCC (2015a) United Nations Framework Convention on Climate Change. Available at: newsroom.unfccc.int.

UNFCCC (2015b) United Nations Framework Convention on Climate Change. Available at: unfccc.int/2860.php.

UNFCCC (2015c) United Nations Framework Convention on Climate Change. Available at: www.cop21.gouv.fr/en.

Urpelainen, J. (2011) The origin of social institutions. *Journal of Theoretical Politics*, **23**(2), 215–240.

Varela, F.J. (1984) The creative circle, in P. Watzlawick (ed.), *The Invented Reality*, Norton Publishing, New York, NY, pp. 329–345.

Varela, F.J. (1995) The re-enchantment of the concrete. Some biological ingredients for a nouvelle cognitive science, in L. Steels and R. Brooks (eds), *The Artificial Life Route to Artificial Intelligence*, Lawrence Erlbaum Associates, Hove, pp. 320–338.

von Foerster, H. and von Glaserfeld, E. (2015) *Wie wir uns selber erfinden* (5th edn), Carl Auer-Verlag, Heidelberg.

Wagner, P. (2006) Public policy, social science and the state: An historical perspective, in F. Fisher, G.J. Miller and M.S. Sidney (eds), *Handbook of Public Policy Analysis*, CRC Press, Boca Raton, FL.

Weible, C.M., Sabatier, P.A. and McQueen, K. (2009) Themes and variations: Taking stock of the advocacy coalition framework. *Policy Studies Journal*, **37**(1), 121–140.

Workman, S., Jones, B.D. and Jochim, A.E. (2009) Information processing and policy dynamics. *Policy Studies Journal*, **37**(1), 75–92.

Wu, J.S. and Lee, J. (2015) Climate change games as tools for education and engagement. *Nature Climate Change*, **5**, 413–418.

Ziman, J. (2000) *Technological Innovation as an Evolutionary Process*, Cambridge University Press, Cambridge.

5

'Friend' versus 'Electronic Friend'

Joseph C. Pitt

When engineering a social system, much thought needs to be given to how these systems could disrupt other important current social systems, diminishing human value, and whether that disruption can be systematically anticipated. I view Facebook as an engineered (i.e., designed) social system. It was designed as a means to facilitate communication among a group. Determining who was to be included in this group resulted in the use of the term 'friend'. To become a member of the group, you 'friend' someone in the group – thereby creating, among other things, the verb 'to friend'. Little or no thought was given to the negative implications of this decision. Here I examine some of those implications. I will also look at the impact the electronic world of information has on our understanding of information and its impact on the way we live. But first some preliminaries.

Social systems are rarely engineered deliberately and even when they are, there is no guarantee they will work in the way they were intended to. Consider a housing development. Planned living environments can be traced back as far as Thomas Moore's *Utopia* (1516), but to refer to them as examples of social engineering is misleading. At best, a planned housing development creates an environment in which social systems can emerge. By 'social system' here I mean a set of relations among people with recognized rules of interaction, rarely set down on paper but understood well enough that we can teach newcomers how to behave in this environment.

Another social system is the law. Viewed as a social system, the law provides a mechanism to adjudicate disagreements and determine whether the properly enacted laws of the land have been violated. However, surrounding the formal machinery of police officers, judges, courts, etc. there is also an informal social system having primarily to do with how to behave when involved in the formal system. Thus, while there is no law requiring individuals in a courtroom to rise when the judge enters, it is expected that when the bailiff calls out '… all rise' everyone will stand up. In such a case we can view the law as a system composed of multiple systems. In fact, every system can be viewed as a system of systems, some formal, some based on tradition (Pitt, 2009). Social systems usually grow up either on their own or

Social Systems Engineering: The Design of Complexity, First Edition. Edited by César García-Díaz and Camilo Olaya.
© 2018 John Wiley & Sons Ltd. Published 2018 by John Wiley & Sons Ltd.

around some formal system. Thus we can look at a housing development as a formal system consisting of streets, houses, bus lines and people. How the people interact cannot be fully predicted ahead of time. Friendships, clubs and teams will emerge over time, but who becomes friends with whom and which clubs and what kinds of teams cannot be predicted with any degree of accuracy. For example, consider a planned retirement community for people aged over sixty-five. One of the teams you would not expect to find there is a women's basketball team for over eighties. But in a recent episode on the US TV programme '60 Minutes' (July 12, 2015), just such a team was highlighted, with its oldest member aged 90; the team competes in a league and wins!

An example of a failed social system is the American elementary and high-school educational system. Elementary and high-school requirements and structures are primarily controlled by the individual states. The number of schools at each level is determined by the localities. The individual systems are comprised of an infrastructure consisting of buildings, playing fields, a governing board of (usually) elected citizens, an administrative staff, a budget, secretarial and janitorial staff, sometimes school buses and bus drivers and maintenance personnel, supplies and finally teachers and students. The curricula are carefully scrutinized and argued over. The teachers are usually evaluated on some scale and the students pass through the system, graduate and go on to something else.

I claim this is a failed system. Its failure is multifaceted. Not all students graduate. Of those who do, many are so badly prepared that they fail at whatever next step they choose – job, college, etc. The cause of the failure is a function of unanticipated social interactions. Since schools are primarily locally funded, the source of that funding is contentious. It is primarily provided by the locality, usually through taxes. Thus, poor localities have less to fund their schools. That sets up obvious social implications. It creates a self-replicating mechanism where the rich districts have better-resourced schools and their students are better prepared, while the opposite is true of poor localities. Second, in the USA no one seems to want to pay taxes, even though they expect all the services taxes make possible – roads, schools, etc. So, schools are chronically underfunded. The ultimate victims here are the students, but there are other victims, such as the teachers. They are not paid well at all. In fact, it is rarely the case that a teacher makes enough to support a family. One of my former students went on to acquire a Master of Arts in Teaching. When he finished his education he was deeply in debt, an increasingly chronic American problem. He found a job teaching biology in a special school for exceptionally gifted students. He finally paid off his debts at age 40. He was unmarried and had no dependents. With low salaries you cannot expect to attract the most talented students to the teaching profession. My former student finally left high-school teaching to teach introductory biology at a nearby college. So, you get what you pay for, well-intentioned but woefully unprepared and/or incompetent teachers. It is a failed system because we haven't figured out how to pay for it. That is a social issue, not one derived from the basic structure of the design. Now it may be argued that it is in fact a function of the design, since part of the system is its economic underpinnings. By designing the system to be primarily in the hands of the localities, the economic failure is built in. But it need not fail if the individuals who want and need the system can be convinced to pay for it. However, telling people to raise taxes results in one system colliding with another – politics. The point here is that systems do not exist in isolation. Not only is a system a system of systems, it also interacts with other systems and the consequences are rarely fully predictable. With this understanding, let us turn to a more detailed examination of a designed system – Facebook.

On the assumption that thought requires language, where a language is broadly construed as a system composed of a grammar and a vocabulary, I want to examine how changes in the ordinary English terms deployed in the worlds of designed social media affect how we think and (more importantly) act. The main example I want to look at is the use of the word 'friend'. Consider the use of the word 'friend' on Facebook. In your email you receive a message informing you that a complete stranger wants to 'friend' you. Given our ordinary, pre-Internet understanding of what a friend is, and aside from the fact that as far as I know, prior to the creation of Facebook, 'to friend' was not a verb, this is a most unusual request. Can you seriously consider answering 'yes' or 'no' to a complete stranger who walks up to you in the grocery store and asks you to be their friend? The most probable reaction would be to ask the individual who they are and why you should even consider being their friend; do you, for instance, have friends in common, or interests? To be a friend in the non-social-media world entails certain responsibilities and obligations. For example, a real friend will drop what they are doing to come to your aid. No such obligation is entailed by being a social media friend. The question then becomes: How do we regard our real-time, real-world friends in light of our virtual personal relations? Or, to put it slightly differently: How has the use of 'friend' in the socially engineered world of Facebook changed its meaning and our understanding of friendship?

I am not a Luddite, but I do worry about the effect of the language we use in the virtual world on how we think in real time. More specifically, I worry about the speed with which these changes occur and what this rapidly changing way of communicating is doing to our ability to think and to our understanding of our personal relationships. One effect of this really new speak[1] is to limit options, or more precisely, to limit inventiveness and creativity. One of the beauties and frustrations of natural languages is the ambiguity of their vocabularies and the variety of ways they permit communication and thinking. This variety is one source of innovation, often by means of free association or analogical extension. If the meanings of words are limited, then the number of possible connections to other words is reduced. Employing a pragmatic theory of meaning, where the meaning of a word, or phrase, or sentence consists of the total number of inferences one can correctly draw from its use, consider what happens to the meanings of words when the number of potential inferences is restricted or radically changed. As we strip away meanings from common words or change the meanings, we seem to be reducing our capacity for thinking of new combinations and new ways of doing things. This is what happened with George Orwell's 'Newspeak' in *1984* (Orwell, 1948) when he introduced such phrases as 'war is peace' and 'love is hate'. Many of the things we tend to think regarding war are now blocked when it is equated with peace, and we also have new ways of thinking about both war and peace opened up for us. Likewise for love and hate. Friend?

Language is always changing – I am not concerned about change *per se*. It is the kind of change that has me concerned. Our new technologies open up new opportunities. I can imagine a burst of creativity unlike anything the world has seen in the coming century. But the paradox is that, if our capacity to think is restricted by reducing what our vocabulary can express and imply, then that very opportunity for creativity could be stymied. To avoid this situation requires that we think deeply about the consequences of embedding people in

---

[1] With apologies to George Orwell.

designed systems. The American educational system was not intended to fail. It was thought that by placing control in the hands of the localities, the particular needs of their students could best be attended to. Good thought – but not good enough. Let's start with Aristotle's account of friendship. For Aristotle there are three types of friend. The heart of the concept is based on loving another. One can love another for their usefulness, or for the sake of pleasure, or for a concern over their well-being. Aristotle says of friendship that:

> … It is a virtue or implies virtue, and is besides most necessary with a view to living. For without friends no one would choose to live, though he had all other goods. (McKeon, 1941)

> To be friends, then, they must be mutually recognized as bearing goodwill and wishing well to each other… (McKeon, 1941)

> Perfect friendship is the friendship of men who are good, and alike in virtue; for those wish well alike to each other *qua* good. And they are good in themselves. (McKeon, 1941)

If Aristotle is correct, and I have no reason to think he is not, then to be a friend you need to know the person whom you call 'friend'. What do I mean by 'know' here? To see what I am driving at, let me back off for a second and relay the circumstances of the origin of this discussion. I did my graduate work in Canada at the University of Western Ontario. In those brief 3 years we made many friends, some of whom we still stay connected with. A husband and wife with whom we remain very close came south to visit and see the hounds (we raise Irish Wolfhounds). Over drinks, I casually raised the issue of the use of the term 'friend' on Facebook, expressing some confusion as to how you could want to be a friend on Facebook with someone you had never met. This led to a general discussion of friendship, and I used our Canadian experiences as a touchstone. Canadians are known for being somewhat reserved and polite. That is until they decide you are okay and should be treated as a friend, then the barriers come down and you are welcomed into the family. Our visitors were very keen on distinguishing between being an *acquaintance* and being a *friend*. Fundamentally they held Aristotle's position.

An acquaintance is someone with whom you occasionally interact or share nothing more than, say, a work relationship. The Friend is someone you can count on and whose well-being concerns you. From here it is not hard to see that whatever 'friend' is supposed to mean on Facebook, it is not 'Friend' with a capital F. This is not to say that some of your real Friends aren't also friends on Facebook. But what does it mean to be a friend on Facebook *simpliciter*?

Being a Facebook friend is akin to participating in a Kevin Bacon six-degrees-of-separation game gone mad. When you receive a request from someone you have never heard of or met, all you have to do is browse their friends' list and eventually you will find someone you know for real. So the connection is that you are a friend of a friend of a friend of a friend of a … – well it is not quite *ad infinitum*, but close, or it can feel that way. In short, to be 'friended' on Facebook is to become part of someone's electronic network – nothing more is necessarily entailed. But what does that mean? It does not necessarily mean that every entry that person makes concerning every last detail of their utterly boring life is of interest to you – nor are you contacted by them through Facebook on a regular basis, both of which, however, may be the case. To be a Facebook friend means, first and foremost, that there is no obligation on your part to be concerned about their well-being.

The appropriation of the term 'friend' by Facebook has serious consequences. For one thing, it calls into question what we mean by 'friend' when used in a pre-Facebook sense; that is, when a friend was someone you cared about and who you could count on when you needed them. It used to be the case that when you met someone and they introduced you to the person they were with as their friend, you could make certain inferences: that, for example, they had known each other for a while; that they had a close relationship, etc. And depending on the intonation, you might even be able to infer with some hesitation that they were a couple. But one thing you didn't infer was that they were part of each other's social media network because it didn't exist back then, in the good old days. But now, one of the things you can infer is just that, and it says nothing in particular.

Now, consider a different scenario. You meet the same couple and the unknown person is not immediately identified as a friend. Later, you run into your friend and ask who that person was. If they respond with 'oh, she is just a friend', what can you conclude? That she is a Friend friend or a mere acquaintance or, perhaps, someone connected with on Facebook? In this sense being a Facebook friend – what I will refer to as an 'f-friend' – adds to the total possible set of inferences you can make, expanding the meaning of Friend. You might think of it as meaning number 5 in the *Oxford English Dictionary*.

But there is also a way in which using f-friend to talk about inclusion in a network also indicates a diminishing sense of friendship itself. As the quote from Aristotle above claims, 'To be friends, then, they must be mutually recognized as bearing goodwill and wishing well to each other'. To f-friend someone entails nothing about reciprocal good will. Rather, it indicates the kind of immersion in the digital world that increasingly characterizes the current generation of under-thirties, now coming to be called Millennials. That is a world in which having 2000 f-friends is some kind of status symbol. It entails nothing in the way of concern for the well-being of any of those individuals. It is, rather, an example of Aristotle's third reason for calling someone a friend – that is, for achieving some kind of advantage. You make 'friends' with someone because they can do something for you. This is not real friendship, for you care little about the well-being of the person, only to the extent that their staying healthy means they can provide you with access to X or help you acquire Y.

In an earlier piece (Pitt, 2008), I argued that the iPod was the single most pernicious piece of technology invented post-World War II. It is essentially a device that encourages people to be anti-social. You just plug in your ear buds and tune out the world. Well, I may just have to revise that opinion. The iPod is only one example of the way in which the world of electronic media, a world of intentionally designed means for connecting with someone for a specific purpose, is taking aim at the traditional social world. Connecting to your network via some devise – desktop, laptop or, increasingly, smartphone – *without personally (i.e., face-to-face in the same room) interacting with the person with whom you are communicating* is the latest form of what in an earlier age would be rudeness. In a recent article by Joel Stein in *Time Magazine* (20 May 2013), where he provides a penetrating analysis of the me-me-me generation, he conjures up an image of two 'friends' sitting at a bar next to each other, texting each other rather than talking – actually, not too far-fetched. The decrease in social skills is, however, matched by an increase in sophistication regarding network protocols and the development of new, if incomprehensible by us greybeards, languages. For what is emerging as the language by which the Millennial generation communicates has both a new vocabulary and a new grammar – witness Twitter. Language is always changing: just a short time ago 'bad' had a negative connotation; today, 'bad' means good – Orwell lives! It is not changing languages or even electronic communication that is the concern, it is the disintegration of non-electronic social skills. It is not just teenagers who can't communicate or won't, that has always been the case – but some of our most creative minds in developing new start-ups are socially inept when placed in a room with a real three-dimensional person or two. The very meaning of 'social' is being transformed. This suggests that referring to 'social media' is not referring to anything particularly social at all. It is more like putting crosses in a column. But if it is not social, then it is not a social system.

So that leads us to ask: What is so important about being able to chit-chat at a company party? What would be lost if we never interacted with another person in the flesh and just communicated electronically? Well, to begin with, it seems like a very effective form of population control. But, more to the point, we are playing with what it is to be human. It is a commonplace that humans are social animals. If person-to-person in social interaction is being devalued, then what are we doing to ourselves? The Millennial response is clear: we are simply redefining what it means to be social, no big deal. It happens all the time. We acted differently in a world defined by rigid social conventions (such as an aristocracy) than we do in a democratic world, where equality is the professed norm. What is wrong with being continually connected to your f-friends, even if you have never met them in person?

Well, that may be true, but it is also the case that in redefining 'social' you are fundamentally changing the meaning of what it is to be human. Even in a world where human relations were constrained by caste considerations, there were still some common fundamentals. For example, consider the role of the raised eyebrow in conversation or a shrug, or hand gesture (I have an Italian background) – in short, body language. Can all of that be replaced by emoticons? If it could, would they function the same way as looking into a person's eyes? There is much that body language contributes to the conversation – it permits whole ranges of inferences to be made or drawn from a single gesture. And it is also the case that it allows for incorrectly drawn inferences – the raised eyebrow may be seen as a warning about what you think that person understood to be a dangerous idea, or it could be a reaction such as 'there he goes again!'. And finally, there is the role of the social (i.e., in person) in understanding creativity.

In a recent article in *Wired*, Clive Thompson (2013) took a look at the ruckus created by Yahoo CEO Marissa Mayer when she stopped allowing employees to work at home. The earlier move to electronic commuting was itself a disruption to an established social system, the office environment. Mayer's claim was that 'speed and quality are often sacrificed when we work at home'. Thompson points out that reputable studies have shown that people are in fact more productive when working at home, for a variety of reasons. What is lost is the creativity that comes from the personal interactions at the office. Thompson asks that employers question what it is they wish to maximize, productivity or creativity. His discussion emphasizes, for me, the role of human personal interaction (by which I mean being in the same room together) in promoting human creativity. The typical picture of a Silicon Valley start-up, the epitome of creativity in today's digital world, is a group of very bright people in a room together bouncing ideas off each other. This is important – they are not at home communicating by computer. The face-to-face environment is crucial to the creative process.

Further, we don't need to think of creativity as something only businesses value. If we just stop and think about it, we value it in our everyday lives perhaps even more. What would your life be like if you never had a new and different thought every day? Is it really enough simply to do your job? Isn't part of the joy of living finding or thinking about something new every day? Without novelty, our lives would be comparable to Fritz Lang's portrayal of the workers in his 1927 masterpiece *Metropolis*,[2] drones marching in formation back and forth to work in a joyless world of endless similarity. It doesn't have to be much. It can be something as simple as being sparked to think about something you have never thought about before when a co-worker throws out an idea you had never considered before. Yes, this can take place over the Internet, but it is the give and take and the exploration of ideas and thoughts around the water cooler that is the real generator of new ideas – new thoughts.

Creativity is not the only aspect of our lives that living alone in the virtual world threatens. How about the sense of personal value? Eating dinner alone in front of a computer screen is just not quite the same as dining, even at home, with a loved one or a good friend. Socializing, in the form of face-to-face in the same place contact, brings a fullness to our lives that Facebook cannot. And that is the point – designed social systems like Facebook do not enhance the most important human qualities, they can only diminish them.

What worries me here is that it may already be too late. Here I am having to spell out what it means to be social in the 'old' sense, face-to-face in the same place. It seems that we may have already lost the war for humanity. It didn't take an alien invasion – it just took the Millennials, those kids born into the Internet age, coming of age. At one point in the introduction to *The Structure of Scientific Revolutions*, Thomas Kuhn (1964) noted that all it may take for a revolution to occur is for those in power to die off. Confirming that is a quote from one of the Millennials interviewed by Jerry Adler in another *Wired* article (Adler, 2013). Discussing what Millennials put on their Facebook pages and how it could possibly be used against them, 20-year-old Abigail Muir notes 'the people who care will all retire and the world will be run by my generation, which doesn't give a …'. Maybe I am just being an old reactionary – I still rage at my students about split infinitives, despite the fact that the *Oxford English Dictionary* has given in and said it is okay. Maybe the Millennials are evolving into a different form of humanity and I am just nostalgic for something that never really was.

Even if that is the case, I still remain convinced that there is something of immense value that we are losing by abandoning the old sense of 'social' – and that is spontaneity. Just as the idealized sense of 'family values' is hard to find real examples of – when examined closely, families can be fairly nasty and contentious groups of people – the social world of humans may be equally ephemeral. But just as there are those moments of tenderness when a new grandchild arrives or there is a wedding, there is also that dinner, that drink after work at the bar, rooting for your favourite football team at the stadium, etc. And tied to their computers as they may be, at the end of the semester – as they exit the exam – my students still feel the need to reach out and physically touch me.

So, just as Clive Thompson suggested that employers need to decide what they want from their employees, productivity or creativity, we might ask ourselves a similar question. What do we want from our social world, Friend or f-friend? In a materialistic world such as the one we live in, it is no surprise that Millennials – who value the accumulation of stuff – have come to see other people as part of that stuff. The world in which they live is an electronic world: their music is downloaded, their games are downloaded and their books are downloaded. They can look up anything on the Internet. They can order anything though the Internet. Is it any wonder that they can accumulate friends that way as well? Acknowledging this may force us to add a fourth kind of friend to Aristotle's three types. Friend as commodity. An f-friend is not

---

[2] *Metropolis*, motion picture, Kino International, Germany. Produced by Erich Pommer; directed by Fritz Lang.

a real person you care about and for whom you would drop what you are doing to assist if they asked. An f-friend is stuff in your electronic closet. A 'person' (and we may need to examine how the meaning of that term is changing, although the sociologists are ahead of us on that front – see the work of Sherry Turkle) may decide that they need to be socially schizophrenic – there is their electronic social world and then there is their physical social world. A problem with this view is the picture painted by Joel Stein when he describes that bar scene with two friends sitting side-by-side, texting each other.

But there may be a ray of hope – a slim ray, but a ray nonetheless. A year ago I ended my undergraduate philosophy of technology course by asking my students the following question: Which technology would you be willing to do without? The class clown suggested toasters, but the thoughtful engineering student in the front row slowly raised his hand and offered 'my cell phone'. His reasoning was that he lamented the loss of privacy that cell phones have forced his generation to accept. You can't get away from your friends or your mother. If you don't answer a text immediately, you get bombarded with ten more demanding to know if you are okay. And if you don't answer eventually, you are ostracized. 'Sometimes', he said, 'you just want some peace and quiet'. And that points to a final irony regarding f-friends. It isn't that they don't really care about you. Rather, it is that they really only care about themselves. It is all about self-gratification. It is all about me-me-me. Yes, kids are self-absorbed. We have always known that. But now they are so self-absorbed they can't recognize the harm they create. Maybe it is too early to say – it may be that the world of f-friends is just extending adolescence, extending this period of self-absorption. Maybe, after a travel adventure where there is no cell service, they will come to their social senses. Or maybe we will just have to wait and see what kind of a world they build – it is coming very quickly, so we may yet see what the postman brings. If we can still recognize it, and if there is still a postman...

When thinking further about the world of the Millennials, the crucial role played by smartphones needs to be examined as well. As a device, the smartphone has evolved over time from that large bulky bag phone you had for your car in case of emergencies to the centrepiece of your life. The telephone has always been a means to connect people. In that sense, the phone system is an engineered social system. But the world of the smartphone is also a world of unintended consequences. It is not just the world of f-friends – it is the world of information. As noted above, you can access information about virtually (no pun intended) anything on your cell phone. Furthermore, it has changed social arrangements in fundamental ways. If someone asks you a question, your first reaction, if you don't already know the answer, is to google it. Even my generation reacts that way. We no longer reach for a reference book – we no longer even use physical paper books. Everything is online, and if it isn't, it will be. The US Library of Congress is digitizing its entire collection. This has enormous implications for how we live our lives. For example, we no longer go to the library, we don't have to, we just do it online. So what will happen to physical libraries? If everything is online, that wonderful, serene environment for study and reflection – or simply reading – no longer serves that function. In addition, the world of electronic information has no filter. While we can answer just about any question by picking up our phones, how do know the answers are reliable?

The move to digitize all information has further implications for the future of information storage. For the electronic world is not static. It is undergoing constant change and many of these changes have not been thought through very well.

Consider the rapidity with which computer operating systems and word-processing systems are upgraded or even abandoned. How many of us can access data on a floppy disk today?

A number of years ago the Virginia Tech Graduate School decided to do away with hard copies of masters' theses and doctoral dissertations. Everything is submitted and stored electronically. But the word-processing systems on which these documents are created change. Some, like Word Perfect, have completely vanished. We have here another example of an exciting engineering development, electronic storage of data, being used in a manner probably not thought of by the designing engineers. In short, this is not a design problem. It is a social problem. The decision to store these scholarly documents was not made by engineers. It was made by academic bureaucrats. It is not clear that engineers should shoulder the responsibility for considering all possible uses of the technologies they develop, but someone should be charged with the task of considering the consequences of using a technology in such a sweeping manner. But while it is not the engineers' responsibility to think through every possible use of their creations, maybe the design team should include non-engineers whose expertise lies in that area: thinking about future possible uses and their impact on the social world that are not immediately evident. And even that will be very difficult, for when the cell phone was invented, Google didn't exist – who could have predicted that?

However, we cannot sit around and wait to see what the brave new world of the Millennials will be like. The need for greater understanding of the dynamics of social interactions in an increasingly technological world is upon us as we plan for, for example, a manned mission to Mars. The mission designers are keenly aware of the interpersonal problems presented by three people living together in a closed, cramped environment for six months. The thing is, they can't design the social system; they can only design the environment. The social system that emerges will evolve and change. No matter how careful the selection process for the astronauts is, how do you plan for the changes that will occur in the individuals selected for the mission while they are on the mission? How do you predict the personality evolution of three individuals (for argument purposes, think of it as three) who will not only be interacting with each other, but doing so within a very constricting environment?

If I am correct, social systems cannot be designed successfully ahead of time. Environments for social systems can be designed. What emerges when you pour in the people and further unanticipated technological developments is not predictable and hence, cannot be planned around (yet).

References

Adler, J. (2013) 1993. *Wired*, May, p. 192.

Kuhn, T. (1964) *The Structure of Scientific Revolutions*, University of Chicago Press, Chicago, IL.

McKeon, R. (1941) Nicomachean ethics, in *The Basic Works of Aristotle*, Random House, New York, NY, pp. 935–1126.

Orwell, G. (1948) *1984*, Secker & Warburg, London.

Pitt, J.C. (2008) Don't talk to me, in D.E. Whittkower (ed.), *The iPod*, Open Court, Chicago, IL, pp. 161–166.

Pitt, J.C. (2009) Technological explanation, in A. Meijers (ed.), *Philosophy of Technology and Engineering Sciences, Vol. 9, Handbook of the Philosophy of Science*, North-Holland, Amsterdam, pp. 861–879.

Stein, J. (2013) Millennials: The me me me generation. *Time Magazine*, May 20.

Thompson, C. (2013) How we work. *Wired*, June, pp. 107–108.

Part II

Methodologies and Tools

6

Interactive Visualizations for Supporting Decision-Making in Complex Socio-technical Systems

Zhongyuan Yu, Mehrnoosh Oghbaie, Chen Liu, William B. Rouse and Michael J. Pennock

6.1 Introduction

Complex socio-technical systems are typified by behavioural and social phenomena that significantly impact the functioning and performance of complicated engineered or designed systems. Hospitals are good examples, as are transportation systems, financial ecosystems, and many other types of enterprises. The key stakeholders in such systems are often important and influential policy- and decision-makers. They need means to understand and manage the complexity of their systems, yet seldom do the systems have science and engineering skills to develop or perhaps even understand the models of their enterprises. Models and simulations allow the inclusion of organizational details yet afford a systematic analysis (Siggelkow, 2011), as well as bound outcomes to plausible ranges (Epstein, 2008).

We have many experiences of key stakeholders referring to our mathematical models or simulations as 'magic boxes' that produce predictions of how their enterprises will likely respond to a range of scenarios. These experiences have led us to realize that such stakeholders need much more hands-on experience with models and simulations. They need to take the controls (sliders, radio buttons, etc.) and explore the nature of these creations. Such experiments can confirm or challenge their intuitions. In either case, over time, they will gain a sense of ownership of these models and simulations and develop a level of trust in the insights they gain from their explorations.

Social Systems Engineering: The Design of Complexity, First Edition. Edited by César García-Díaz and Camilo Olaya.
© 2018 John Wiley & Sons Ltd. Published 2018 by John Wiley & Sons Ltd.

This possibility can be greatly enhanced via compelling interactive visualizations. Such visualizations go far beyond Excel spreadsheets and charts. Indeed, we have been most successful with large-scale interactive visualizations that enable users to immerse themselves in the complexity of their enterprises. This enables them, usually working as groups, to share ideas and insights as they explore. They can move controls and observe responses, and then discuss and debate relationships among phenomena and the implications of these relationships for policies and other decisions.

This chapter discusses and illustrates how these benefits can be achieved. The next section is entitled 'policy flight simulators', a name for these interactive visualizations that many stakeholders have found compelling. This leads to in-depth discussion of two applications: the first example addresses hospital consolidation via mergers and acquisitions in New York City; the second example focuses on enterprise diagnostics, the process of determining whether one's enterprise is performing well and, if not, identifying the causes of poor performance. This example addresses enterprises in the automobile industry. A final section of the chapter brings together the many results and insights from previous sections.

6.2 Policy Flight Simulators

As we have worked with groups of senior decision-makers and thought leaders using interactive visualizations of their complex domains, they have often asked: 'What do you call this thing?' I suggested 'multi-level simulations', but I could tell from the polite responses that this did not really work. At some point, I responded 'policy flight simulator' and immediately knew that this was the right answer. Numerous people said, 'Ok, now I get it'. This led to a tagline that was also well received. The purpose of a policy flight simulator is to enable decision-makers to 'fly the future before they write the check'.

Policy flight simulators are designed for the purpose of exploring alternative management policies at levels ranging from individual organizations to national strategy (Rouse, 2014). This section focuses on how such simulators are developed, and on the nature of how people interact with these simulators. These interactions almost always involve groups of people rather than individuals, often with different stakeholders in conflict about priorities and courses of action. The ways in which these interactions are framed and conducted are discussed, as well as the nature of typical results.

6.2.1 Background

The human factors and ergonomics of flight simulators have long been studied in terms of the impacts of simulator fidelity, simulator sickness, and so on. Much has been learned about humans' visual and vestibular systems, leading to basic insights into human behaviour and performance. This research has also led to simulator design improvements.

More recently, the flight simulator concept has been invoked to capture the essence of how interactive simulations can enable leaders to interactively explore alternative organizational designs computationally rather than physically. Such explorations allow rapid consideration of many alternatives, perhaps as a key step in developing a vision for transforming an enterprise.

Computational modelling of organizations has a rich history in terms of both research and practice (Prietula *et al.*, 1998; Rouse and Boff, 2005). This approach has achieved credibility in organization science (Burton, 2003; Burton and Obel, 2011). It is also commonly used by the military.

Simulation of physics-based systems has long been in common use, but the simulation of behavioural and social phenomena has only matured in the past decade or so. The latter involves much higher complexity, but nevertheless allows exploration of a wider range of 'what if' questions (Levitt, 2012). Simulations that involve social phenomena are of particular value for exploring alternative organizational concepts that do not yet exist and, hence, cannot be explored empirically. The transformation of health delivery is, for example, a prime candidate for exploration via organizational simulation (Basole *et al.*, 2013).

This section focuses on the nature of how people interact with policy flight simulators that are designed for the purpose of exploring alternative management policies at levels ranging from individual organizations to national strategy. Often, the organizations of interest are best modelled using multi-level representations. The interactions with simulators of such complex systems almost always involve groups of people rather than individuals, often with different stakeholders in conflict about priorities and courses of action.

6.2.2 Multi-level Modelling

To develop policy flight simulators, we need to computationally model the functioning of the complex organizational system of interest to enable decision-makers, as well as other significant stakeholders, to explore the possibilities and implications of transforming these enterprise systems in fundamental ways. The goal is to create organizational simulations that will serve as policy flight simulators for interactive exploration by teams of often disparate stakeholders who have inherent conflicts, but need and desire an agreed-upon way forward (Rouse and Boff, 2005).

Consider the architecture of public–private enterprises shown in Figure 6.1 (Grossmann *et al.*, 2011; Rouse, 2009, 2010a). The efficiencies that can be gained at the lowest level (work practices) are limited by the nature of the next level (delivery operations). Work can only be accomplished within the capacities provided by available processes. Further, delivery organized around processes tends to result in much more efficient work practices than are typical for functionally organized business operations.

However, the efficiencies that can be gained from improved operations are limited by the nature of the level above (i.e., the system structure). Functional operations are often driven by organizations structured around these functions (e.g., manufacturing and services). Each of these organizations may be a different business, with independent economic objectives. This may significantly hinder process-oriented thinking.

And, of course, potential efficiencies in system structure are limited by the ecosystem in which these organizations operate. Market maturity, economic conditions and government regulations will affect the capacities (processes) that businesses (organizations) are willing to invest in to enable work practices (people), whether these people be employees, customers or constituencies in general. Economic considerations play a major role at this level (Rouse, 2010b, 2011).

These organizational realities have long been recognized by researchers in socio-technical systems (Emery and Trist, 1973), as well as work design and system ergonomics (Hendrick

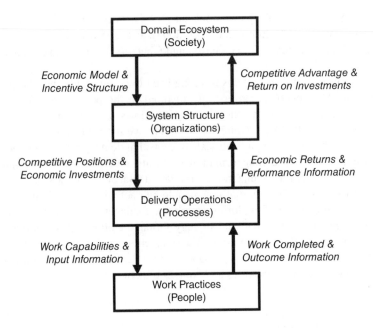

Figure 6.1 Architecture of public–private enterprises.

and Kleiner, 1999). The contribution of the concept of policy flight simulators is the enablement of computational explorations of these realities, especially by stakeholders without deep disciplinary expertise in these phenomena.

6.2.3   People's Use of Simulators

There are eight tasks associated with creating and using policy flight simulators:

- Agreeing on objectives (the questions) for which the simulator will be constructed.
- Formulating the multi-level model (the engine for the simulator), including alternative representations and approaches to parameterization.
- Designing a human–computer interface that includes rich visualizations and associated controls for specifying scenarios.
- Iteratively developing, testing and debugging, including identifying faulty thinking in formulating the model.
- Interactively exploring the impacts of ranges of parameters and consequences of various scenarios.
- Agreeing on rules for eliminating solutions that do not make sense for one or more stakeholders.
- Defining the parameter surfaces of interest and 'production' runs to map these surfaces.
- Agreeing on feasible solutions and the relative merits and benefits of each feasible solution.

The discussions associated with performing the above tasks tend to be quite rich. Initial interactions focus on agreeing on objectives, which includes output measures of interest, including units of measure. This often unearths differing perspectives among stakeholders.

Attention then moves to discussions of the phenomena affecting the measures of interest, including relationships among phenomena. Component models are needed for these phenomena and agreeing on suitable vetted, and hopefully off-the-shelf, models occurs at this time. Also of great importance are uncertainties associated with these phenomena, including both structural and parametric uncertainties.

As computational versions of models are developed and demonstrated, discussions centre on the extent to which model responses are aligned with expectations. The overall goal is to computationally redesign the enterprise. However, the initial goal is usually to replicate the existing organization to see if the model predicts the results actually being currently achieved.

Once attention shifts to redesign, discussion inevitably shifts to the question of how to validate the model's predictions. Decision-makers usually ask about predictive validity: Will the model's predictions turn out to be what actually happens? However, as these predictions inherently concern organizational systems that do not yet exist, we shift such discussions to construct validity: Do the structures embodied in the models and the trends exhibited by the prediction make sense? Thus, validation is limited to discussing the believability of the insights emerging from debates about the nature and causes of model outputs. In some cases, deficiencies of the models will be uncovered, but occasionally unexpected higher-order and unintended consequences make complete sense and become issues of serious discussion.

In general, explorations are scenario-based. What would happen if certain sets of decisions were made, or a particular set of economic circumstances emerged, or government statues and regulations changed? Socio-technical systems often involve deep uncertainties about behavioural and social reactions to such changes (Lempert *et al.*, 2013). Consequently, 'narrative scenarios' (Lempert *et al.*, 2006) may be useful to formulate alternative futures.

Model-based policy flight simulators are often used to explore a wide range of ideas. It is quite common for one or more stakeholders to have bright ideas that have substantially negative consequences. People typically tee up many different organizational designs, interactively explore their consequences and develop criteria for the goodness of an idea. A common criterion is that no major stakeholder can lose in a substantial way. This rule can pare the feasible set from hundreds of thousands of configurations to a few hundred.

Quite often, people discover the key variables most affecting the measures of primary interest. They can then use the simulator in 'production mode', without the graphical user interface, to rapidly simulate ranges of variables to produce surface plots. The simulator runs to create these plots are done without the user interface, which would very much slow down the simulation process.

Discussions of such surface plots, as well as other results, provide the basis for agreeing on pilot tests of apparently good ideas. Such tests are used to empirically confirm the simulator's predictions, much as flight tests are used to confirm that an aircraft's performance is similar to that predicted when the plane was designed 'in silico'.

Policy flight simulators serve as boundary-spanning mechanisms, across domains, disciplines and beyond initial problem formulations, which are all too often more tightly bounded than warranted. Such boundary spanning results in arguments among stakeholders being externalized. The alternative perspectives are represented by the assumptions underlying – and the elements that compose – the graphically depicted model projected on the large screen.

The debate then focuses on the screen rather than being an argument between two or more people across a table.

The observations in this section are well aligned with the findings of Rouse (1998) concerning what teams seek from computer-based tools for planning and design:

- Teams want a clear and straightforward process to guide their decisions and discussions, with a clear mandate to depart from this process whenever they choose.
- Teams want capture of information compiled, decisions made and linkages between these inputs and outputs so that they can communicate and justify their decisions, as well as reconstruct decision processes.
- Teams want computer-aided facilitation of group processes via management of the nominal decision-making process using computer-based tools and large-screen displays.
- Teams want tools that digest the information they input, see patterns or trends, and then provide advice or guidance that the group perceives they would not have thought of without the tools.

Policy flight simulators do not yet fully satisfy all these objectives, but they are heading in this direction.

It is useful to note that the process outlined in this section is inherently a participatory design process (Schuler and Namioka, 1993). Participatory modelling employs a wide variety of analytic tools (Voinov and Bousquet, 2010), for example, agent-based modelling and social-based system dynamics (Vriens and Achterbergh, 2006) to model different levels of stakeholder participation (Pretty, 1995). Such a human-centred process considers and balances all stakeholders' concerns, values and perceptions (Rouse, 2007). The result is a better solution and, just as important, an acceptable solution.

6.3   Application 1 – Hospital Consolidation

Skyrocketing costs, long waiting times, an aging population, declining reimbursements and a fragmented system are the words people use when talking about healthcare. This industry also involves complicated relationships, as revealed in the systemigram of Figure 6.2, which applies systems thinking methods such as causal loop diagrams and soft systems methodology to transform 'rich text' into a structured diagram, in which the principal concepts and relationships are identifiable and the sentence structures recoverable (Boardman and Sauser, 2008). The system is so complex and interconnected, it suggests that stakeholders' strategic decisions – for example, healthcare providers' mergers and acquisitions (M&A) plans, health insurance's reimbursement schemes and preventive healthcare initiatives – will have both positive and negative (and often, unforeseen) consequences. Because of this complexity, even proactive decision-makers who thoroughly game out what might happen can easily fail to anticipate how various actors or competitors will respond. Methods drawing upon significant insights into decision-making processes are needed (Rouse and Serban, 2014).

Among the stakeholders in the complex U.S. healthcare system, hospitals are dominant players, with significant contributions in the overall economy (Boccia, 2014). In many cities, hospitals are leading employers (Chandler, 2014). For instance, 7 out of the 25 largest employers in New York City are hospital systems, including New York Presbyterian Hospital

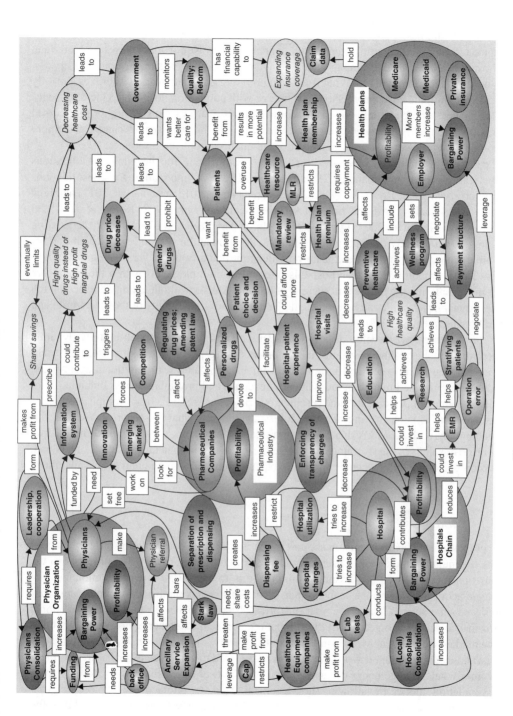

Figure 6.2 Healthcare systemigram.

and Mount Sinai Health System. Because of the complexity, importance and uncertainty that lies in the U.S. healthcare system, it is quite understandable that hospitals are uncertain about how they should best respond to pressures and opportunities. This is particularly relevant for hospitals located in competitive metropolitan areas such as New York City, where more than 50 hospitals are competing, among which many are the nation's best. Questions that arise in this uncertain environment are: What if we wait until the healthcare market stabilizes and only invest in operational efficiency? Should we merge with competing hospitals to increase negotiation power? Shall we only focus on acquiring physician practices in highly reimbursed diagnostic groups? Aggregated decisions from numerous hospitals could change the future hospital market, potentially affecting the cost and quality of delivered services.

Extensive research has described providers' enthusiasm for hospital consolidations (American Hospital Association, 2013; Yanci et al., 2013), motivation for consolidations (Casalino et al., 2003; Rosenthal et al., 2001; Saxena et al., 2013; Yanci et al., 2013) and how M&As affect quality, cost and price of services (Connor et al., 1998; Keeler et al., 1999; Krishnan and Krishnan, 2003; Lynk, 1995; Melnick and Keeler, 2007). Most of the existing research does not study this substantive strategic decision at the system level, accounting for dynamic interactions among healthcare providers, particularly under various regional structures. This is important, because hospitals interact on multiple levels within the healthcare system; considering the failure and success of hospital M&As without taking into account the system dynamics provides an incomplete understanding of the viability of the managerial and operational strategies.

In this research, we develop a data-rich agent-based simulation model to study dynamic interactions among healthcare systems in the context of M&A decision-making, where by 'rich' we mean extensive rule sets and information sources, compared with traditional agent-based models. The proposed model includes agents' revenues and profitability (i.e., financial statements), operational performance and resource utilization, as well as a more complicated set of objectives and decision-making rules to address a variety of 'what if' scenarios. We pilot our modelling approach on M&A dynamics of hospitals in New York City, informed by in-depth data on 66 hospitals of the Hospital Referral Region in Bronx, Manhattan and East Long Island. The proposed methodology can be applied to other regions, depending upon data availability.

6.3.1 Model Overview

The proposed agent-based decision support model for hospital consolidation has two components: a simulation model and a user interactive system (see Figure 6.3). The objective of the simulation model is to assist hospital administrators to assess the impact of implementing strategic acquisition decisions at the system level. This is accomplished by simulating strategies and interactions based on real historical hospital balance sheet and operational performance data. The outcomes of the simulation include the number of hospitals remaining in the market and frequent M&A pairs of hospitals under various settings. By varying strategy inputs and relevant parameters, the simulation can be used to generate insights as to how these outcomes would change under different scenarios. The user interactive system complements the simulation model by allowing non-technical users to interactively explore relevant information, input parameter values for different scenarios, as well as view and validate the results of the simulation model.

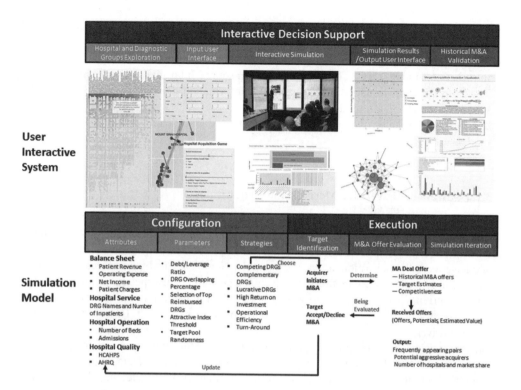

Figure 6.3 Two components in the agent-based decision support model for hospital consolidation: simulation model and user interactive system.

This section presents the hospital consolidation simulation model and an overview of its implementation, including details on: (1) data sources; (2) strategic drivers for M&As; (3) the simulation process; and (4) the user interface.

6.3.1.1 Data Sources

The validity of models relies partially on the selection and interpretation of data. We piloted our modelling approach with an in-depth 'story' of the hospitals in the New York area to identify the information that is essential for making M&A decisions. The key characteristics of M&A targets as sought by acquirers and bankers include sector, customers, geography, profitability, growth profile and return on investment. To obtain such data, we used historical data from the Center for Medicare and Medicaid Services (CMS), the American Hospital Association (AHA) and Medicare.gov for the Medicare-involved hospitals in New York City, as well as some overarching characteristics for the 5000+ hospitals across the United States.

Specifically, data sources include:

- Provider utilization and payment data consisting of both inpatient and outpatient data. This provides hospital payments and charges for predefined diagnostic-related groups (DRGs).

Within our modelling framework, these data are used for defining M&A motivations based on competing or complementary diagnostic groups.

- The hospital cost report, including hospital revenue, costs, assets, discharges and bed information, among others. This serves as a major source of hospital attributes in our model.
- The hospital compare data, describing quality information. Quality is used as one of the hospital attributes in our model.
- The M&A transaction database, providing historical transaction details to validate our model.

6.3.1.2 Strategic Drivers for Mergers and Acquisitions

Several key parameters in this hospital consolidation model are the motivations for initiating the hospital M&A activities; such motivations are also called strategic drivers for M&A. A strategic driver could be a combination of different drivers, which means each hospital could have more than one motivation. For example, New York Presbyterian Hospital may be interested in acquiring its competitors, acquiring top performers in a specific diagnostic group, as well as turning around poorly performing hospitals. In this section, we introduce six strategic drivers for hospital acquisitions, while our model maintains high flexibility for adding additional drivers.

Strategic driver 1. Acquiring hospitals that are competing with each other to increase negotiation power with insurance companies and medical suppliers. Data employed include hospitals' similarity in terms of diagnostic group offerings. For example, the NYU Hospitals Center may regard Mount Sinai Hospital as a competing hospital, because Mount Sinai has 100% of what NYU offers and the NYU Hospitals Center has 98% of DRGs offered by Mount Sinai.

Strategic driver 2. Acquiring hospitals that serve complementary DRGs to get access to more patients and provide access to comprehensive diagnosis and treatment services. This includes consideration of the percentage of DRG overlaps among hospitals. Two hospitals with low DRG overlap suggests that they offer complementary services.

Strategic driver 3. Acquiring hospitals with high market shares or high-paying DRG services. This requires identifying hospitals with high market share in terms of disease-specific total charges. For example, one of the target hospitals could be the Hospital for Special Surgery – although it is a highly specialized hospital, its offerings, including spinal fusion and major joint replacement services, have high market shares and high reimbursement.

Strategic driver 4. Acquiring hospitals with high return on investment (ROI). To calculate ROI, the benefit (return) of an investment is divided by the cost of the investment. In our case, the benefit includes the target hospital's current assets and estimated future profitability. The cost of the investment is the acquisition cost.

Strategic driver 5. Acquiring hospitals whose operation and management could be easily turned around. If hospitals choose to invest in 'turning around a hospital', the targets are hospitals with consecutive negative profits and low acquisition costs. One scenario is that the acquirer has good profit margin and shares most DRGs with the target. Such targets need to be relatively smaller than the acquirer. This scenario suggests that the acquirer has the capability to turn around the target and make it profitable. Another scenario is that the target has attractive assets, which the acquirer could use or sell.

Strategic driver 6. Acquiring hospitals with operational efficiency to learn from best practices. For this strategy, we need to go beyond simple summaries of hospital service delivery and financial statements to incorporate informed decisions within our hospital consolidation model. Particularly, we apply data envelopment analysis (DEA) to derive the operational efficiency index for all hospitals considered in our study. DEA is one of the most widely used and proven effective tools for healthcare efficiency measurement (Banker *et al.*, 1984; Charnes *et al.*, 1978; Hollingsworth, 2008; O'Neill *et al.*, 2008). The DEA model uses bed capacity, number of employees and operational expense as inputs to the transformation process (i.e., patient treatment), and the outputs are represented by the number of patients served (inpatients, outpatients and births), as well as the perceived quality (Hospital Consumer Assessment of Healthcare Providers and Systems, or HCAHPS) of these services. Another aspect of Strategy 6 is that M&A activities often lead to overhead reduction. The acquirer has a better chance of making targets more efficient and achieving higher returns if the target hospital has lower labour costs but higher overhead costs compared with the acquirer. Linear regression was used to fit overheads to total salary.

6.3.1.3 Merger and Acquisition Process

In this section, we describe the mergers and acquisition process (see Figure 6.4). For each simulation period (year), M&A targets were identified for each acquiring hospital based on the strategic drivers, then acquisition offers were estimated based on similar historical transactions. Throughout the process, teaching affiliation compatibilities between the two sides and regulations were checked for all deals. After a transaction was finalized, hospitals' attributes were updated in the model. We assume each transaction involves only two hospitals, acquiring hospitals can only submit one target offer per period, and acquired hospitals cannot acquire any hospital. M&A prices in this model are determined using historical data. The simulation stops after period 10. This is because realized M&A activities decrease significantly during the first four periods. A successful M&A is very rare after period 8. Fifty simulation replications were run across different sets of analysis to generate stable results.

Decision-makers vary the input parameters to create and test different scenarios. Besides the six strategic drivers discussed above, other input parameters include: M&A transaction-related parameters and market predictions. Simulation output includes: (1) number and identities of hospitals that remain in the market at the end of each stage, along with their financials and patient counts; (2) a detailed list of M&A activities that happened throughout the simulation process. From that, we are able to determine the most frequently appearing M&A pairs, and identify hospitals that have greater interest in and capability of acquiring other hospitals. While there are other considerations that are relevant to hospital mergers and acquisitions that are not included in the simulation (e.g., payment structure), the simulation is flexible enough to incorporate many of these for future studies.

6.3.1.4 User Interactive System

The purpose of this simulation model is to serve as a means to facilitate strategic decision-making processes. To this end, we have developed an interactive visualization environment, where market dynamics can be simulated and decision-makers can interact with different

For each acquiring agent, evaluate each target agent in the hospital database:

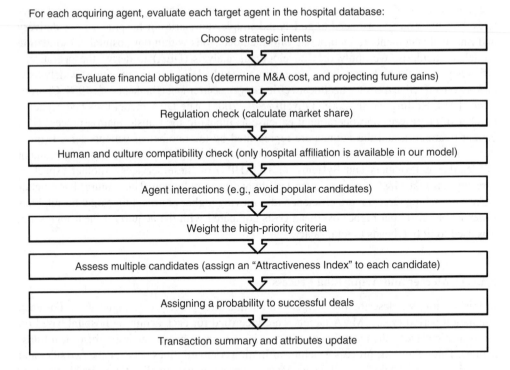

| Choose strategic intents |
| Evaluate financial obligations (determine M&A cost, and projecting future gains) |
| Regulation check (calculate market share) |
| Human and culture compatibility check (only hospital affiliation is available in our model) |
| Agent interactions (e.g., avoid popular candidates) |
| Weight the high-priority criteria |
| Assess multiple candidates (assign an "Attractiveness Index" to each candidate) |
| Assigning a probability to successful deals |
| Transaction summary and attributes update |

Figure 6.4 Mergers and acquisitions process.

settings to address 'what if' scenarios. Users interact with several interactive visualizations concurrently in a large-scale interactive environment that includes an array of seven 8 ft by 20 ft, 180-degree touchscreen monitors installed in the CCSE *Immersion Lab* (Center for Complex Systems and Enterprises) at Stevens Institute of Technology. This approach enables non-technically oriented stakeholders to have an immersive experience that greatly increases their comfort levels with model-driven decision-making.

Typically, there are multiple models running on each of the seven screens, with outputs from one model feeding another, and results fed back both computationally to other models and visually to participants. In Figure 6.5, one of the authors is leading and facilitating the discussion. The seated participants will typically ask initial questions that help them explore the linked models. Once comfortable, they will propose 'what if' scenarios for exploration. When really engaged, participants will leave their sets and want to make their own choices of slider positions and radio buttons. Participants, many of them now standing in the 180-degree presentation, will start to point out to each other insights about what affects what – or perhaps does not. The facilitator now plays the role of a guide, to help them embody their scenarios.

We used the R statistical software to implement the hospital consolidation model and used Shiny for the interactive web interface. We employed d3.js, which is a JavaScript library for producing dynamic, interactive data visualizations in web browsers, to visualize and explore strategic motivations and study historical M&A events. These capabilities are intended to lead to meaningful evaluations and better decisions.

Figure 6.5   CCSE *Immersion Lab* at Stevens Institute of Technology.

It is difficult to understand the 100 DRG offerings and overlapping relationships among the 66 hospitals in a static display. The interactive heat-map visualization (shown in Figure 6.6) overcomes this difficulty, and allows users to conduct in-depth studies of hospitals and their offerings. In Figure 6.6(left), when a hospital is selected, its corresponding column will be highlighted and sorted according to different diagnostic groups. Clicking the label in each row enables sorting by diagnostic groups. Such functionality helps decision-makers to identify their competitive diagnostic groups and assists policy-makers with assessing hospital market share and preventing regional hospital monopolies.

Figure 6.6(right) shows the interactive visualization of the M&A transaction database, which allows users to explore and understand past M&A activities involving hospitals as well as to compare actual transactions with M&A results from the model. The visualization consists of four parts:

- A zoomable and draggable M&A transaction time series from 1990 to 2014, where each bubble represents one transaction and the size of the bubble indicates total transaction value.
- Top acquirers' market share, acquisition volume and lists of activities.
- Acquirers' financial highlights captured in over 30 key statistics, including revenues, profits and returns.
- Categorized news articles from various sources, including new product releases, executive changes, business expansions and lawsuits, to name but a few (Yu *et al.*, 2013).

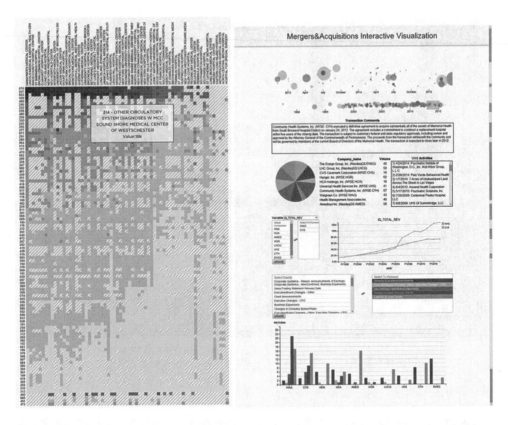

Figure 6.6 Interactive visualization using d3.js: left figure shows payment, charges and market share information for each hospital (rows) and diagnostic group (columns), with sorting enabled for both rows and columns; right figure explores M&A transactions database, from M&A activities, to top acquirers' financials, to news reports.

6.3.1.5 Model Validation

Validation of agent-based simulations is challenging, especially for high-level strategic decision simulations. We compared our simulation results with Capital IQ's hospital mergers and acquisitions transaction dataset. Although there is a limited number of cases under our regional constraint in Capital IQ's database, the realized M&A transactions do appear in our results. For validation purposes, we created various interactive visualizations (Figure 6.6) that provide a means for decision-makers to evaluate the validity and representativeness of model outputs. This not only serves as a way to alleviate validation concerns, but also aligns with the purpose of this work – providing tools to facilitate gaining insights important to decision-making processes.

Another approach to validation is feedback from users. There were many, roughly 30, demonstrations to hospital decision-makers and healthcare consultants as well as senior executives from insurance, government, foundations, etc. In total, perhaps 200 people participated in the demos and many took the controls and tried various options. They made many suggestions, and the number of types of interactive visualizations increased iteratively.

6.3.2   Results and Conclusions

Some of the key outputs of the simulation are the number and identities of hospitals that remain in the market at the end of each simulation run, along with the financials and patient count for each. Recall that acquisitions are initiated based on the six strategies described in Section 6.3.1.2. Results show that the more diversified the strategies are, the greater the number of M&A deals. The initial pool of 66 hospitals was reduced to an average of 31 hospitals under varying combinations of all strategies (all hospitals randomly choose one of the six M&A drivers), 45 with the turnaround strategy (all hospitals adopt a turnaround strategy) and 50 when all hospitals adopt the competing strategy.

As expected, hospital market share depends on the choice of strategies. Since the goal of the simulation is to help decision-makers compare strategies, for each simulation run we record the top five hospitals according to net patient revenue under different strategies. Figure 6.7 shows how often each hospital belongs to the top five market leaders based on net patient revenue. New York Presbyterian Hospital is always among the top hospitals; it could gain as much as 25% of the market share if it chooses a good strategy. But there is still a possibility

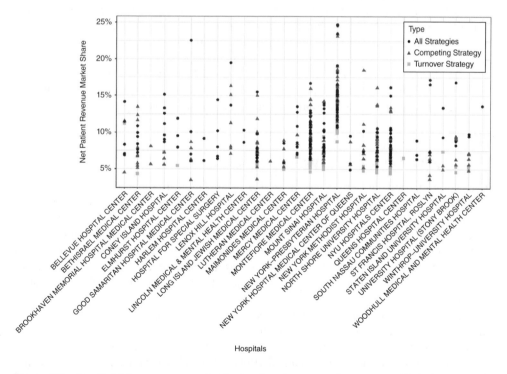

Figure 6.7   Hospitals' frequency as market leaders based on net patient revenue. Each dot represents one instance of a particular hospital entering the top tier.

that New York Presbyterian will drop out of the top three if other hospitals pursue better strategies. Another example is the Coney Island Hospital. Note that all circles are above the triangles, which suggests that the competing strategy is not a good one for the Coney Island Hospital. Some hospitals have very limited dots (for example, the Queens Hospital Center), which has little chance of achieving a high market share, but if it chooses a good strategy, while others are still struggling to make decisions, it could also succeed.

Bidding and deals evolve at each stage. As time goes on, there are less successful bids, as the good targets have already been taken, which suggests that 'do nothing' is a safer strategy when the market is noisy (Yu *et al.*, 2011), but may result in a 'too-late-to-do-anything' situation when the market has matured. Decision-makers have to stay alert, because the market and the competitors change continuously. A hospital which is not an M&A target previously can become one in the future when it acquires other hospitals and becomes competitive.

The results from the simulation model facilitate M&A decision-making, particularly in identifying desirable acquisition targets, aggressive and capable acquirers, and frequent acquirer–target pairs. The frequency level is a relative value, in that it depends on the number of strategies included and hospitals involved. A high frequency of acquirer–target pairs suggests a better fit and also repeated attraction. A high frequency of successfully being an acquirer suggests a greater interest and capability level to acquire under a given strategy. Hospitals with frequency greater than one are likely to make more than one acquisition during the simulation time period.

The key value of the overall model and set of visualizations is, of course, the insights gained by the human users of this environment. For example, they may determine the conditions under which certain outcomes are likely. They can then monitor developments to see if such conditions are emerging. Thus, they know what might happen, even though they cannot be assured what will happen. The greatest insights are gained not only from simulation, but also from interactive visualizations that enable massive data exploration, which moves from a 'one-size-fits-all' static report to a more adaptable and useful decision process.

6.4   Application 2 – Enterprise Diagnostics

When designing and developing the interactive visualizations to support a policy flight simulator, a logical question is what principles should one follow in order to aid decision-makers in drawing inferences from the simulator. As a first step to identifying those principles, we developed an interactive visualization that leveraged two approaches to thinking about complex problems: Rasmussen's abstraction–aggregation hierarchy and Pirolli and Card's intelligence analysis model (Pirolli and Card, 2005).To provide context, we applied the visualization to the diagnosis of an enterprise's decision to withdraw brands from the automobile market.

This application focused on using Rasmussen's abstraction–aggregation hierarchy as the foundation of a policy flight simulator for enterprise diagnostics. Rasmussen's early research focused on troubleshooting of electronics (Rasmussen and Jensen, 1974). His research on this topic aligned closely with one of the author's (Rouse) research at the time, prompting in-depth discussions around the swimming pool in Mati, Greece at the NATO Conference on Mental Workload in 1977. The result was a plan for a NATO Conference on Human Detection and

Diagnosis of System Failures held in Roskilde, Denmark in 1980, as well as a subsequent book (Rasmussen and Rouse, 1981).

By that time, Rasmussen and Rouse's research had moved on to decision support for detection, diagnosis and compensation of system failures in power and process plants, as well as complex vehicle systems like aeroplanes, supertankers and space shuttles. Subsequently, our focus shifted to enterprises as systems (Rouse, 2005a,b, 2006). This has led recently to the concept of enterprise diagnostics. The idea is simple – create an interactive visualization that enables executives and senior managers to assess whether or not their enterprise is functioning correctly. If it is not functioning correctly, enable these decision-makers to determine why their enterprise is malfunctioning.

While Rasmussen's abstraction–aggregation hierarchy (Rasmussen, 1983, 1986) provides a framework for organizing evidence, we still need to support the user's dynamic process of navigating the abstraction–aggregation levels. To provide this support, we applied the learning loop developed by Russell et al. (1993) to study the inner representation of the sense-making process. Their model was designed to adjust the representation to better portray the information in support of two main tasks: searching for a good representation of the problem and excluding representations that do not fit.

Pirolli and Card (2005) subsequently developed an intelligence analysis model based on expert behaviours and the set of patterns that experts develop from past experiences. The model consists of two major loops of foraging and sense-making. The foraging loop includes searching for information, moving the desired information items to a shoebox, examining the information items in the shoebox and then deciding whether or not to move them into the evidence file. The sense-making loop involves iteratively developing a mental model. Pirolli and Card's model includes multiple feedback loops. An aiding interface should support the user in the execution of both the forward flows and the feedback loops.

6.4.1 Automobile Industry Application

A recently published study (Liu et al., 2015) addressed the withdrawal of 12 car brands from the market during the 1930s, 1960s and 2000s, including the following cars:

- Cord, Duesenberg, LaSalle, Pierce Arrow (1930s);
- DeSoto, Packard, Rambler, Studebaker (1960s);
- Mercury, Oldsmobile, Plymouth, Pontiac (2000s).

The study focused on why these cars were removed. Explanations were derived at four levels: automobile, company, industry and economy. Interestingly, only one of the twelve decisions was driven primarily by the nature of the car. Other forces usually dominated.

Data sources included quantitative data such as production levels for each car, as well as text sources such as the *New York Times* archive, which contributed almost 100 articles published over the past 100 years on these vehicles. Using this data, we created an interactive visualization for enterprise diagnostics in the automobile industry context. The user's task is to determine why brand X failed. To support users in performing this task, we leveraged Rasmussen's abstraction–aggregation hierarchy (Rasmussen, 1983, 1986), as well as Pirolli and Card's (2005) intelligence analysis model.

6.4.1.1   Abstraction–Aggregation Space

To illustrate how Rasmussen's abstraction–aggregation hierarchy would apply to an enterprise diagnostic problem, we consider an idealized use case for the automobile example where the decision-maker has access to any relevant piece of data that he or she would desire. Table 6.1 provides an abstraction–aggregation hierarchy with associated example data sources. Table 6.2 provides mnemonics for each point in this hierarchy. The use case is expressed using these mnemonics.

Each point in the abstraction–aggregation hierarchy may include:

- visualizations of quantitative data (e.g., production, economic projections);
- visualizations of geographic data (e.g., geographic markets);
- visualizations of temporal data (e.g., company timelines);
- visualizations of structural relationships (e.g., product assembly);
- newspaper and magazine articles (e.g., interviews, product announcements);
- financial statements (e.g., Q-1s and K-1s);
- company product and process descriptions;
- company strategic, tactical and operational plans;
- photographs of management, factories, dealers, cars, etc.

Table 6.1   Abstraction–aggregation hierarchy

Level of abstraction	Level of aggregation		
	High	Medium	Low
Economy	GDP, inflation, unemployment, interest rates	Availability and prices of raw materials, labour and investment	Costs of acquisition and use of product/service
Industry	Current and projected sales of products/services across companies	Published or purloined strategies and plans of competing companies	Alternative available products/ services and consumer ratings of offerings
Company	Current and projected market share, revenues, profits, share price	Points of view of company leadership, overall strategy and plans	Strategy and plans for evolution and management of brand
Products/ services	Current and projected revenues and costs across brands	Current and projected production levels across brands	Current and projected production level within brand

Table 6.2   Mnemonics for abstraction–aggregation hierarchy

Level of abstraction	Level of aggregation		
	High	Medium	Low
Economy	E-H	E-M	E-L
Industry	I-H	I-M	I-L
Company	C-H	C-M	C-L
Products/services	P-H	P-M	P-L

6.4.1.2 Example Use Case

Problem statement. Brand X was removed from the market. Why did the company make this decision? Provide evidence to support your answer.

There are many ways to approach this question. One could search for articles at the C-M and C-L levels, looking for direct answers to the question from company press releases, industry publications or the *Wall Street Journal*. Such articles might, for example, attribute the withdrawal of brand X to decreasing sales. An article in *Fortune* on General Motors stated that the company's problem was loss of market share. Not much of an insight!

Liu *et al.* (2015) have shown that the causal chain can be traced back from the symptoms (withdrawal) to earlier decisions (investments, acquisitions, etc.). Thus, deeper answers are needed than 'brand X was withdrawn because it was not selling'. The goal is to support users to identify the reasons it was not selling. They also should be able to determine the source(s) of the reasons. How did the company get into this situation?

To find deeper answers, one might start at the P-H, P-M and P-L levels. How was brand X doing relative to competing brands? Were sales in this market increasing, flat or decreasing across all companies? Causes of brand X sales decreasing in a decreasing market are likely very different from causes of decreasing sales in an increasing market. Is brand X losing and others winning, or is everybody losing?

If everybody is losing, one might then move from P-H to C-H, I-H and E-H to determine why. If only brand X is losing, one would likely move from C-L to C-M to C-H to see whether brand X is really the problem rather than the rest of the company. To determine if brand X is the problem in itself, one might move from C-L to I-L to see how it competes with other brands in the market.

If brand X is not the source of its own problems, one would dig more deeply, likely into C-M and C-H. One would look into company leadership and financial situations as potential reasons that brand X was sacrificed. It could be that product lines had to be trimmed and brand X was selected as the least painful alternative.

Looking at C-H could lead to M-H to determine if the company is having difficulties competing in general. Perhaps the bigger players with more resources are quickly absorbing every good idea, cloning them and rolling them out in a big way. Thus, brand X may have been highly competitive until the bigger competitors swamped it.

Another path would arise from discovering that everyone is losing, but other companies are not withdrawing brands. They may be better managed and have deeper pockets, or they may have a strategy that requires sustaining all of their brands. Comparing I-H with C-H, as well as I-M with C-M, would enable identification of such differences.

In summary, the use case is as follows.

- User would start at P-L, then move to P-M and P-H.
- If everybody is losing, user would move to C-H, then I-H and E-H.
  - o If others are not withdrawing brands, user would compare C-H with I-H and then C-M with I-M to determine why and *report that conclusion.*
- If only brand X is losing, user would move to C-L, then C-M and C-H.
  - o If comparison of C-L with I-L leads the user to conclude that brand X is the problem, user would *report that conclusion.*

- If brand X is not the problem, user would move to C-M and C-H.
  - ○ If the company management or financial situation is the problem, user would *report that conclusion.*
  - ○ If the company is not the problem, user would compare C-H with I-H and then C-M with I-M to determine whether the competition is the problem and *report that conclusion.*

6.4.2   Interactive Visualization

In order to evaluate whether or not decision-makers would actually leverage available data in this manner, we developed an interactive visualization that allowed users to explore a subset of the data types discussed in the idealized use case. Again the objective is to diagnose why a given car was withdrawn from the marketplace (Figure 6.8). Quantitative and textual data were organized by the four levels of abstraction described above. The visualization allows the user to explore the data by these levels of abstraction if so desired.

A screen shot of the interface is shown in Figure 6.9. The information sources include production data (for the car of interest and the industry as a whole), U.S. economic data such as GDP and CPI, and published news articles. Information sources were identified as belonging to one or more layers of abstraction. A filter allowed the subject to reduce the data available in the display to one layer of abstraction at a time if so desired.

Subjects move information sources from the left of Figure 6.9 to the right and bottom, where they accumulate evidence to subsequently review. They can remove evidence that does not prove useful. This process of selection, review and retention or discarding of information provides data on where subjects move in the abstraction–aggregation space, as illustrated in the above use case.

The interface design was strongly influenced by the work of Pirolli and Card (2005). They developed an analysis model that is intended to support an analyst as he or she forages for information and then makes sense of it. In the spirit of their model, the interface consists of three key components.

- *Available data window* – implements Pirolli and Card's 'Shoebox' of potentially relevant data.
- *Article evidence and chart evidence windows* – implements Pirolli and Card's 'Evidence File' that allows users to separate out the evidence that supports reaching a conclusion.
- *Data filter* – allows the user to filter the available data by the level of abstraction.

Leveraging the interactive visualization, a user should be able to identify which of a set of potential factors determined by Liu *et al.* (2015) contributed to the outcome for a given car. We were interested in how subjects' use of displays at different levels of abstraction and aggregation affects the speed and accuracy of their decisions. However, we did not expect subjects to necessarily think in terms of abstraction and aggregation. Instead, we expected them to simply use the information sources they find most useful for informing their decisions. Their focus would be on what went wrong rather than the fundamental nature of the information sources.

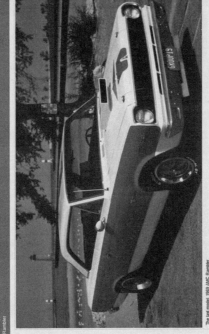

Auto Data Decision Dashboard

What Happened to This Car?

Rambler

The last model: 1969 AMC Rambler

The most popular model: 1961 Rambler American

Rambler entered the automobile business in 1900 and ceased production in 1969.

The Thomas B. Jeffery Company produced the Jeffery automobile in 1900. Nash Motors Company bought Jeffery in 1917. Nash later dropped the Jeffery brand. In 1937, Nash and Kelvinator merged with each other.

In 1950, the first Nash Rambler was designed as a compact two-door sedan. The motivation for producing the Rambler was the post World War II economy and the Korean War's steel quota policy.

Similar to other automakers' strategies under pressure from the Big Three, Nash-Kelvinator and Hudson Motor Car Company merged in 1954. The new American Motors Corporation (AMC) discontinued the Nash Rambler in 1955. In 1958, AMC reintroduced the Nash Rambler and renamed it Rambler American.

After reentering the compact car market, the Rambler American as an American compact-car helped AMC into the black for the first time since the merger of the Nash-Kelvinator and Hudson. Rambler built strong market perceptions of high quality, lower gas consumption and, in general, a high-value economy car. After 1960, the Big Three entered the compact car market. President George W. Romney led Rambler towards diversification by sharing production lines, chassses, and parts to broaden its market offerings. AMC decided to broaden its market offerings and entered the standard and full-size automobile market.

In 1963 George Romney left AMC to become Michigan governor. Roy Abernethy took the helm and switched strategy to compete against the Big Three head to head.

In 1967, Roy D. Chapin replaced Roy Abernethy. By 1968, Rambler only had the Rambler American in the market. After that, Mr. Chapin began to broaden AMC's compact car offerings. For competing with Ford's Maverick, AMC produced a new compact model, Hornet. For competing against imported cars and exporting create a new niche market, AMC created other sub-compact car, Gremlin. The Hornet replaced Rambler in 1969 even though it still had reasonable sales volumes.

Figure 6.8 Introduction screen.

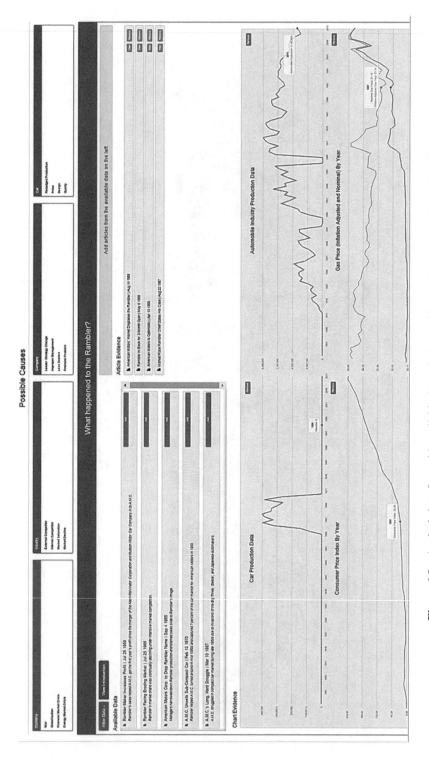

Figure 6.9 Analysis interface with available data window, evidence windows and abstraction filter.

6.4.3   Experimental Evaluation

In order to understand how the interactive visualization would actually be used, we conducted an experiment that involved ten faculty members and graduate students in science and engineering. [The full results of this experiment are discussed in Rouse (2015) and Rouse *et al.* (2017).] Five of these subjects had a high level of expertise, for some of them due to having participated in the study of these cars. Five of the subjects had a low level of expertise in the topic.

Each of them solved the twelve problems in a semi-random order – four subjects started with the 1930s, four with the 1960s and two with the 2000s. They each addressed the four cars in the assigned period before moving on to the next period. Thus, they learned much about each era as they addressed the four cars from that era. The cars were addressed in alphabetical order, with two subjects starting with, for example, Cord, two with Duesenberg, two with LaSalle and two with Pierce Arrow. After they completed their initial car, they moved to the next one in alphabetical order. Subjects that started with Pierce Arrow, for example, next addressed Cord. Subjects who first addressed the 1930s, then moved to the 1960s. In contrast, those who started with the 2000s then moved to the 1930s, and so on.

Each subject's choices of information sources were captured, as were their final decisions. Speed was simply the time from selecting the assigned car until the decision was entered, although time was also partitioned into two segments, as discussed below. Accuracy was measured by comparing subjects' decisions to 'ground truth' from Liu *et al.* (2015). The measure of accuracy was fraction correct.

We also measured the use of the predefined levels of abstraction and aggregation of the information sources chosen. We hypothesized that these average levels would correlate with speed and accuracy, but also vary by car. More specifically, we expected that diagnosis of the reasons for some cars failing would be more difficult than others. Consequently, these more difficult diagnoses would require accessing more information sources and/or result in slower, less accurate diagnoses.

6.4.4   Results and Discussion

Figures 6.10 and 6.11 show the accuracy results for each era and each car, respectively. Subjects found the more recent cars more difficult to diagnose, particularly those from the 2000s. Figures 6.12 and 6.13 show the accuracy for each era and car, for each level of expertise. Subjects with high expertise are more accurate than those with low expertise, and the difference is more pronounced for the more recent cars, again particularly those from the 2000s.

MANOVA results showed that the accuracy varied by car ($p < 0.01$) and era ($p < 0.01$). There was a significant interaction of car or era with expertise ($p < 0.01$). In other words, the superior accuracy of subjects with high expertise was only the case for some cars or eras. Subjects with high expertise read more articles ($p < 0.01$).

Accuracy differed significantly as a main effect for the levels of abstraction ($p < 0.01$) and as an interaction of cars or eras with levels of abstraction ($p < 0.01$). Thus, displays with various levels of abstraction were used differentially, depending on the car or era.

The time required for each subject to complete each car was tracked, but none of the differences were statistically significant.

In additional to the statistical results, the analysis identified several questions that could not be answered with these results and will need to be explored in future experiments. First, we

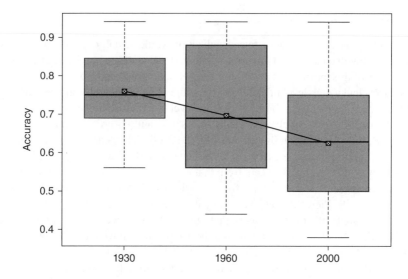

Figure 6.10   Accuracy results for each era.

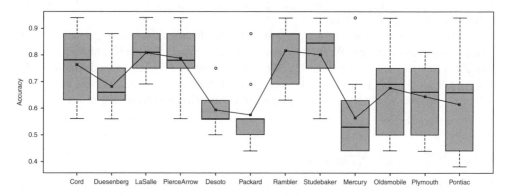

Figure 6.11   Accuracy results for each car.

could not rule out the possibility that differences in the level of accuracy between high-expertise subjects and low-expertise subjects were due to motivation effects. Second, several subjects reported that they spent their first few cases learning the interface and the problem-solving approach. This could be managed through additional subject training in future experiments. Third, limitations in how the experimental setup captured which data objects the subject accessed resulted in data that were too coarse to disentangle when subjects were accessing data at different levels of aggregation within each level of abstraction. As a result, abstraction and aggregation were confounded in this experiment. Future experiments will need a more capable mechanism for capturing evidence use.

In summary, subjects with high expertise were more accurate and sought more information to support their decisions, but were not faster. They inherently employed information from higher levels of abstraction, because they sought much more information in general.

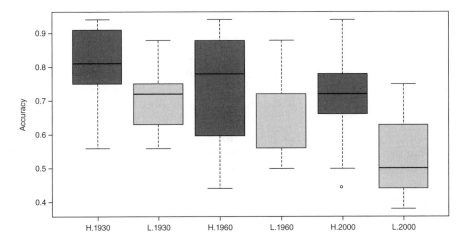

Figure 6.12   Accuracy results for each era for each level of expertise.

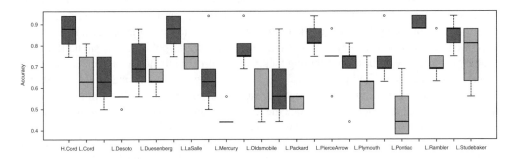

Figure 6.13   Accuracy results for each car for each level of expertise.

This result is aligned with other findings for Rasmussen's abstraction–aggregation hierarchy (Frey *et al.*, 1992).

Subjects with less expertise sought less information. They had particularly low accuracy for the 2000s era. This may be due to the complexity of the economic situation during this period. The Great Recession dominated this period. However, the relentless globalization of the automobile industry was also pervasive. Consequently, the four cars removed from the market in this era were not just the victims of the Great Recession.

Ironically, this may have confused the low-expertise subjects. They had all personally experienced the 2000s era and knew, obviously idiosyncratically, much more about the era than was presented in the experiment. For example, they knew the four cars removed from the market, may have owned one or more of these cars, and may have read various articles concerning their demise.

In general, the better performers have quite different information-seeking and utilization behaviours. This raises an interesting question that will be addressed in a future experiment. Can we provide training and/or aiding that will enable poor information-seekers and utilizers to match the results of the top performers?

6.4.5   Implications

As we have learned from psychology, human beings often give more weight to the coherence of an explanation than its consistency with data or theory. This tendency can be particularly problematic when we consider complex socio-technical systems. In such systems, events are not the product of simple cause and effect relationships, but rather result from a confluence of circumstances. Consequently, there may be no coherent story to explain why something happened.

While it might seem more satisfying to say that a company failed because of a recession in the economy, that assertion, by itself, cannot possibly be true because there were other companies that did not fail during the same recession. In the car experiment, it certainly looks like what differentiated the experts from the non-experts is that they knew to consider more than just the most immediate contributing factors. While the results of the car experiment do not establish this definitively, they do raise an important question.

The ultimate motivation behind a policy flight simulator is to facilitate the 'engineering' of a socio-technical system. While engineering is too strong a word in the context of something as complex as a socio-technical system, it does capture the intent of the policy-maker to design or redesign the systems and policies that he or she can control to influence outcomes in the larger socio-technical system. If one misdiagnoses the contributing factors to an existing or potential state of the system, it is unlikely that one is going to make the policy and technical design choices that will achieve the desired outcomes. While policy flight simulators may enable the discovery of these contributing factors, there is certainly no guarantee that decision-makers will leverage the full potential of these tools as opposed to latching onto the first correlation that appears in the visualized data and results. It is possible that policy flight simulators, through their flexibility, actually facilitate the identification of perceived but spurious correlations that the policy-maker might not have conceived of without them.

Consequently, the question is: How does one go about designing a policy flight simulator so that it aids users in both fully exploring the space of possibilities but also discarding spurious relationships? Answering this question is critical to realizing the full potential of this method to positively influence organizational and social outcomes.

6.5   Conclusions

This chapter has focused on the use of interactive visualization – policy flight simulators – to support diverse groups of stakeholders wrestling with issues associated with complex socio-technical systems. Such support systems are undoubtedly beneficial if designed and evaluated appropriately. A methodology was presented for accomplishing this end.

The two applications discussed were quite different. The hospital consolidation application focused on informing key stakeholders of the complexity of the healthcare-delivery ecosystem and the implications of competitors' strategies and decisions for their decisions. Succinctly, pursuing what seems best for your organization, independent of what strategies your competitors are pursuing, is a recipe for significant negative surprises.

The enterprise diagnostics application was less concerned with broad strategy than with supporting decision-makers to figure out what went wrong and resulted in an esteemed automobile brand being withdrawn from the market. The task in this case is troubleshooting in a complex socio-technical system composed of automobiles, companies, markets and economies. Why is the vehicle and in some cases the company failing? The answer is some mixture of bad economic situations, poorly run companies and occasionally unfortunate design choices for the car.

These two applications provide an interesting contrast. This contrast is between broad strategies in response to tectonic shifts of the healthcare market and tactical mistakes in a highly competitive automobile market. We think that policy flight simulators can help decision-makers faced with this wide spectrum of problems. The key is creating the capabilities that enable them to drive the future before they write the check.

References

American Hospital Association (2013) *How Hospital Mergers and Acquisitions Benefit Communities*, AHA, Chicago, IL.

Banker, R.D., Charnes, A. and Cooper, W.W. (1984) Some models for estimating technical and scale inefficiencies in data envelopment analysis. *Management Science*, **30**, 1078–1092.

Basole, R.C., Bodner, D.A. and Rouse, W.B. (2013) Healthcare management through organizational simulation. *Decision Support Systems*, **55**, 552–563.

Boardman, J. and Sauser, B. (2008) *Systems Thinking: Coping with 21st century problems*, CRC Press, Boca Raton, FL.

Boccia, R. (2014) *Federal Spending by the Numbers, 2014: Government spending trends in graphics, tables, and key points (including 51 examples of government waste)*, The Heritage Foundation, Washington, D.C.

Burton, R.M. (2003) Computational laboratories for organization science: Questions, validity and docking. *Computational & Mathematical Organization Theory*, **9**(2), 91–108.

Burton, R.M. and Obel, B. (2011) Computational modeling for what-is, what-might-be, and what-should-be studies – and triangulation. *Organization Science*, **22**(5), 1195–1202.

Casalino, L.P., Devers, K.J., Lake, T.K., Reed, M. and Stoddard, J.J. (2003) Benefits of and barriers to large medical group practice in the United States. *Archives of Internal Medicine*, **163**(16), 1958–1965.

Chandler, M. (2014) *Great Graphic: Leading industry employer by state 1990 and 2013*. Available at: www.marctomarket.com/2014/08/great-graphic-leading-industy-employer.html.

Charnes, A., Cooper, W.W. and Rhodes, E. (1978) Measuring the efficiency of decision making units. *European Journal of Operational Research*, **2**(6), 429–444.

Connor, R.A., Feldman, R.D. and Dowd, B.E. (1998) The effects of market concentration and horizontal mergers on hospital costs and prices. *International Journal of the Economics of Business*, **5**(2), 159–180.

Emery, F.E. and Trist, E.L. (1973) *Toward a social ecology: Contextual appreciation of the future in the present*, Plenum Press, New York, NY.

Epstein, J.M. (2008) Why model? *Journal of Artificial Societies and Social Simulation*, **11**(4), 12.

Frey, P.R., Rouse, W.B. and Garris, R.D. (1992) *Big Graphics and Little Screens: Model-based design of large scale information displays*, DTIC, Fort Belvoir, VA.

Grossmann, C., Goolsby, W.A., Olsen, L. and McGinnis, J.M. (2011) *Engineering a Learning Healthcare System*, National Academy Press, Washington, D.C.

Hendrick, H.W. and Kleiner, B.M. (1999) *Macroergonomics: An introduction to work system design*, Human Factors & Ergonomics Society, Santa Monica, CA.

Hollingsworth, B. (2008) The measurement of efficiency and productivity of health care delivery. *Health Economics*, **17**(10), 1107–1128.

Keeler, E.B., Melnick, G. and Zwanziger, J. (1999) The changing effects of competition on non-profit and for-profit hospital pricing behavior. *Journal of Health Economics*, **18**(1), 69–86.

Krishnan, R.A. and Krishnan, H. (2003) Effects of hospital mergers and acquisitions on prices. *Journal of Business Research*, **56**(8), 647–656.

Lempert, R.J., Groves, D.G., Popper, S.W. and Bankes, S.C. (2006) A general, analytic method for generating robust strategies and narrative scenarios. *Management Science*, **52**(4), 514–528.

Lempert, R., Popper, S., Groves, D., Kalra, N., Fischbach, J., Bankes, S. and McInerney, D. (2013) *Making Good Decisions Without Predictions: Robust decision making for planning under deep uncertainty*, RAND Corporation, Santa Monica, CA.

Levitt, R.E. (2012) The virtual design team: Designing project organizations as engineers design bridges. *Journal of Organization Design*, **1**(2), 14–41.

Liu, C., Rouse, W.B. and Yu, Z. (2015) When transformation fails: Twelve case studies in the American automobile industry. *Journal of Enterprise Transformation*, **5**(2), 71–112.

Lynk, W.J. (1995) The creation of economic efficiencies in hospital mergers. *Journal of Health Economics*, **14**(5), 507–530.

Melnick, G. and Keeler, E. (2007) The effects of multi-hospital systems on hospital prices. *Journal of Health Economics*, **26**(2), 400–413.

O'Neill, L., Rauner, M., Heidenberger, K. and Kraus, M. (2008) A cross-national comparison and taxonomy of DEA-based hospital efficiency studies. *Socio-Economic Planning Sciences*, **42**(3), 158–189.

Pirolli, P. and Card, S. (2005) The sensemaking process and leverage points for analyst technology as identified through cognitive task analysis, *Proceedings of International Conference on Intelligence Analysis*, McLean, VA, pp. 2–4.

Pretty, J.N. (1995) Participatory learning for sustainable agriculture. *World Development*, **23**(8), 1247–1263.

Prietula, M., Carley, K. and Gasser, L. (1998) *Simulating Organizations: Computational models of institutions and groups*, MIT Press, Cambridge, MA.

Rasmussen, J. (1983) Skills, rules, and knowledge; signals, signs, and symbols, and other distinctions in human performance models. *IEEE Transactions on Systems, Man and Cybernetics*, **13**(3), 257–266.

Rasmussen, J. (1986) *Information Processing and Human–Machine Interaction. An approach to cognitive engineering*, North-Holland, New York, NY.

Rasmussen, J. and Jensen, A. (1974) Mental procedures in real-life tasks: A case study of electronic trouble shooting. *Ergonomics*, **17**(3), 293–307.

Rasmussen, J. and Rouse, W.B. (1981) *Human Detection and Diagnosis of System Failures*, Plenum Press, New York, NY.

Rosenthal, M.B., Landon, B.E. and Huskamp, H.A. (2001) Managed care and market power: Physician organizations in four markets. *Health Affairs*, **20**(5), 187–193.

Rouse, W.B. (1998) Computer support of collaborative planning: An applications report. *Journal of the American Society for Information Science*, **49**(9), 832–839.

Rouse, W.B. (2005a) Enterprises as systems: Essential challenges and approaches to transformation. *Systems Engineering*, **8**(2), 138–150.

Rouse, W.B. (2005b) A theory of enterprise transformation. *IEEE International Conference on Systems, Man and Cybernetics*, Waikoloa, HI, pp. 966–972.

Rouse, W.B. (2006) *Enterprise Transformation: Understanding and enabling fundamental change*, John Wiley & Sons, Hoboken, NJ.

Rouse, W.B. (2007) *People and Organizations: Explorations of human-centered design*, John Wiley & Sons, Hoboken, NJ.

Rouse, W.B. (2009) Engineering perspectives on healthcare delivery: Can we afford technological innovation in healthcare? *Systems Research and Behavioral Science*, **26**(5), 573–582.

Rouse, W.B. (2010a) *Engineering the System of Healthcare Delivery*, IOS Press, Amsterdam.

Rouse, W.B. (2010b) Impacts of healthcare price controls: Potential unintended consequences of firms' responses to price policies. *IEEE Systems Journal*, **4**(1), 34–38.

Rouse, W.B. (2011) *The Economics of Human Systems Integration: Valuation of investments in people's training and education, safety and health, and work productivity*, John Wiley & Sons, Hoboken, NJ.

Rouse, W.B. (2014) Human interaction with policy flight simulators. *Applied Ergonomics*, **45**(1), 72–77.

Rouse, W.B. (2015) *Modeling and Visualization of Complex Systems and Enterprises: Explorations of physical, human, economic, and social phenomena*, John Wiley & Sons, Hoboken, NJ.

Rouse, W.B. and Boff, K.R. (2005) *Organizational Simulation: From modeling and simulation to games and entertainment*, John Wiley & Sons, Hoboken, NJ.

Rouse, W.B. and Serban, N. (2014) *Understanding and Managing the Complexity of Healthcare*, MIT Press, Bostom, MA.

Rouse, W.B., Pennock, M., Oghbaie, M. and Liu, C. (2017) Interactive visualizations for decision support: Application of Rasmussen's abstraction–aggregation hierarchy. *Journal of Applied Ergonomics*, **59**(B), 541–553.

Russell, D.M., Stefik, M.J., Pirolli, P. and Card, S.K. (1993) The cost structure of sensemaking. *Proceedings of the INTERACT'93 and CHI'93 Conference on Human Factors in Computing Systems*, ACM, New York, NY, pp. 269–276.

Saxena, S.B., Sharma, A. and Wong, A. (2013) *Succeeding in Hospital & Health Systems M&A: Why so many deals have failed, and how to succeed in the future*, Booz & Co., New York, NY.

Schuler, D. and Namioka, A. (1993) *Participatory Design: Principles and practices*, CRC Press, Boca Raton, FL.

Siggelkow, N. (2011) Firms as systems of interdependent choices. *Journal of Management Studies*, **48**(5), 1126–1140.

Voinov, A. and Bousquet, F. (2010) Modelling with stakeholders. *Environmental Modelling & Software*, **25**(11), 1268–1281.

Vriens, D. and Achterbergh, J. (2006) The social dimension of system dynamics-based modelling. *Systems Research and Behavioral Science*, **23**(4), 553–563.

Yanci, J., Wolford, M. and Young, P. (2013) *What Hospital Executives Should be Considering in Hospital Mergers and Acquisitions*, DHG Healthcare, New York, NY.

Yu, Z., Rouse, W.B. and Serban, N. (2011) A computational theory of enterprise transformation. *Systems Engineering*, **14**(4), 441–454.

Yu, Z., Serban, N. and Rouse, W.B. (2013) The demographics of change: Enterprise characteristics and behaviors that influence transformation. *Journal of Enterprise Transformation*, **3**(4), 285–306.

7

Developing Agent-Based Simulation Models for Social Systems Engineering Studies: A Novel Framework and its Application to Modelling Peacebuilding Activities

Peer-Olaf Siebers, Grazziela P. Figueredo, Miwa Hirono and Anya Skatova

7.1 Introduction

Agent-based modelling and simulation (ABMS) is a well-established method for studying human-centric systems and is therefore well suited for supporting the investigation of different scenarios related to social systems engineering. An example of social systems engineering opportunities where ABMS could be applied to evaluate different strategies is that of peacebuilding efforts to bring peace and stability to conflict-affected countries. But how do we develop such a model? What are the steps required for model conceptualization, design and implementation? We found ourselves left alone with these kinds of questions when we aimed to build a simulation model to study the impact of current international peacebuilding efforts in conflict areas such as South Sudan. After many focus group meetings and discussions within our multidisciplinary team, we derived a framework to develop such models in a more formal way.

In this chapter we introduce the framework as a tool that employs the possibilities of computer modelling and simulation for advancing the way in which peacebuilding processes can be better understood and improved, given the complexity that social systems represent. The aim is to assist advancing prospects towards social systems engineering in two ways: (1) by presenting a

Social Systems Engineering: The Design of Complexity, First Edition. Edited by César García-Díaz and Camilo Olaya.
© 2018 John Wiley & Sons Ltd. Published 2018 by John Wiley & Sons Ltd.

guide to model, reuse and extend ABMS systems and (2) by employing the obtained simulation models as decision-support tools to investigate the result of applying public policies, actions and interventions to promote societal changes.

Our work represents the outcome of a study bringing together computer scientists, psychologists, political and sociological researchers into the development of a framework to facilitate model design, social engineering and decision-making for peacebuilding. The approach focuses on applying software engineering techniques in agent-based social simulation to describe factors that would trigger individuals' actions/responses to their environment. We establish a framework that integrates individual stakeholders within the social scenario, with their physical and psychological properties embedded within agents. The initial goals are (1) to advance modelling and understanding of real-world peacebuilding scenarios using a graphical notation for agent-based simulation modelling and (2) to produce a reusable toolkit to assist in peacebuilding and other social simulation exercises. In addition, we establish a less-restrained, intuitive manner of incorporating theories of human behaviour into social simulation models. We therefore introduce our framework as a new intervention tool to assist researchers from different research communities who are interested in social systems engineering, to investigate 'what if' scenarios, as well as assess the plausibility of activities and solutions in social simulation contexts.

Software engineering techniques can enable stakeholders from different communities (within research and society) to communicate their ideas in a structured manner. Using a graphical notation, such as the unified modelling language (UML) or system modelling language (SysML) has several advantages within a multidisciplinary context: (1) it is easy to communicate and unambiguous; (2) it is robust as the complexity of the system being modelled increases; (3) it is easily adaptable/extensible to new system requirements; (4) it allows reusability and is easy to maintain; (5) it allows hierarchical modelling; (6) the final implementation decisions of the system can be delayed, as the refinement of functionalities is defined during the design phases; (7) it allows for automated implementation (at least to a certain degree); and (8) it is ideal for modelling within groups of people, as it requires several iterations for the characterization of system requisites. In addition, the iterative process by which the documentation is produced allows social scientists to better understand the model, as well as the final simulation system. This framework aids researchers to scrutinize the real world and subsequently pinpoint useful inputs and abstractions for the simulation system.

In the remainder of this chapter, we provide some background information regarding simulation and peacebuilding, before we explain the steps of the framework and exemplify how to employ it for social systems engineering studies through an illustrative example. This example investigates the perceptions and behaviours of the South Sudanese people towards peacebuilding efforts. Through simulation experiments with the model, we show how changes in internal and external peacebuilding policies affect the behaviour of individuals and impact the dynamics of the population.

7.2   Background

7.2.1   Simulation

A computational simulation of a dynamic system can be defined as an 'imitation (on a computer) of a system as it progresses through time' (Robinson, 2004). The model user determines the possible scenarios to be investigated and the simulation predicts the outcomes.

Simulation can, therefore, also be seen as a decision-support tool. The purpose of simulation is to understand, change, manage and control reality (Pidd, 1992). Moreover, simulation can be used to obtain a better understanding and/or to identify improvements to a system (Robinson, 2004). Simulation models are focused on the main aspects of the real system; they are therefore a simplified version that excludes unnecessary details of the original system.

ABMS is a modelling and simulation approach largely employed in the field of social simulation due to its characteristics and capabilities. It is a technique that employs autonomous agents that interact with each other (Macal and North, 2005). The agents' behaviour is described by rules that determine how they learn, respond to the environment, collaborate with each other and adapt. The overall system behaviour arises from the agents' individual dynamics and their interactions (Siebers and Aickelin, 2008). For social simulation, it can amalgamate real-world data on distinct interactions between individuals in society to give an impression of the system as a whole.

There is no consensus about a definition of an agent among the ABMS community. Macal and North (2005) define some characteristics for an agent:

- A self-contained, modular and uniquely identifiable individual. An agent has a set of attributes, the values of which will define it as a unique individual in the system.
- They are situated in an environment where interactions with other agents occur. Agents are capable of responding to the environment and have protocols to communicate with other agents. Their responses to environmental stimuli and interactions are defined by rules that determine their reactive and proactive behaviours. Apart from behavioural rules, agents communicate with each other through message exchange.
- Agents are autonomous and self-directed.
- Agents are flexible, with the ability to learn and adapt their behaviours according to the environment and past experiences, and they also can have memory.
- They are goal-directed, having objectives to achieve determined by their behaviour.
- Agents constitute a construct with states that vary over time.
- Agents are social, having dynamic interactions with other agents that impact on their behaviour.

Part of the process of developing the social simulation within our framework comprises determining the groups of individuals, their attributes, behaviour, interactions and network memberships, which are all relevant for the social simulation scenario studied. Once these elements are defined, we incorporate them into the design of the agents that will mostly represent these societal stakeholders in the simulation environment.

7.2.2   Peacebuilding

Peacebuilding can be defined in a variety of ways, but there seems to be a consensus that peacebuilding is a multidimensional process of transformation from war to peace, aiming 'to identify and support structures that tend to strengthen and solidify peace in order to avoid a relapse into conflict' (United Nations, 1992, para. 21). The 'multidimensional process' encompasses a wide range of political, economic, developmental, humanitarian and human rights programmes, in order to address both the root causes and immediate consequences of a conflict.

Since 1989, international peacebuilding missions have been deployed to a number of conflict-affected countries such as Angola, Mozambique, Rwanda, Cambodia and East Timor. Under the name of peacebuilding, the United Nations (UN) and other international organizations have undertaken programmes to prepare for, oversee or administer elections in conflict-affected countries, to reform the security sector, strengthen the rule of law, assist the process of disarmament, demobilization and reintegration, facilitate political dialogue and national reconciliation and promote democratic governance.

Current peacebuilding research faces four major shortcomings. First, it adopts a top-down approach by focusing on formulating neoliberal policy 'solutions' to fix the 'problems' of conflict-affected countries (Richmond, 2004). However, as has been widely noted in relation to cases of peacebuilding in Iraq and Afghanistan (e.g., Berdal, 2009), such a solutions-oriented peacebuilding policy has not been effective, because those states that received the neoliberal 'solutions' are left with high levels of insurgency and instability after these peacebuilding efforts. We believe that peacebuilding 'solutions' have faced problems because top-down policy solutions neglect the perspectives of the people in conflict-affected countries. The second shortcoming is Western-centrism. This is manifest in neoliberal policy 'solutions', which are formulated largely by the West (the United States and Europe) to 'fix the problems' of non-Western states, irrespective of local cultures and traditions. In addition, the role of non-Western emerging powers (such as those called the BRICS – Brazil, Russia, India, China and South Africa – countries) in peacebuilding is understudied. These countries are increasingly more powerful and influential in conflict-affected countries, but the effect of their interventions cannot currently be predicted. The third shortcoming is the static nature of research. 'Conflict' is often regarded as a bounded and one-time event, often predicated in descriptions of conflict with particular starting and end dates. However, 'conflict is a social process in which the original structural tensions are themselves profoundly reshaped by the massive disruption of complex political emergencies' (Goodhand and Hulme, 1999, p. 18). It is therefore important to examine how people in conflict-affected countries interact and how such interactions affect peacebuilding efforts. The fourth shortcoming is that current peacebuilding research is fragmented. A small number of advanced studies take a bottom-up, local-culture-oriented and/or dynamic approach to peacebuilding research (Hirono, 2011), but none of them offers a robust quantitative framework that systematically analyses the available data.

To address all the above problems and gain a more comprehensive understanding of the dynamics of perceptions and behaviours in conflict-affected countries, one needs to overcome the fundamental challenge that has contributed to the above problems – a lack of appropriate methodological tools. Our work attempts to overcome this methodological challenge by bringing computer science technology and social science studies into peacebuilding research. We offer a framework (from planning to implementation) with which peacebuilding researchers and other international relations entities can make qualitative and quantitative analyses simultaneously, and gain a comprehensive understanding of the dynamics within the systems they observe. In short, we provide scholars and international policymakers with a novel research framework, which will guide them step by step from creating stereotypes that represent the characteristics of the people (or groups of people) involved in conflicts and conflict resolution to implementing a multi-agent simulation model that can be used as an artificial laboratory to test the potential impact of different peacebuilding activities.

7.3   Framework

Our framework (depicted in Figure 7.1) captures two main activities: 'toolkit design' and 'application design'. While in the toolkit design generic reusable components are created, in the application design we develop simulation models for specific purposes. A toolkit is designed once (and perhaps improved if required) and its components (agent templates and stereotypes) can be used for many applications (simulation models) that somehow relate to the general objective set for developing the toolkit.

The purpose of our framework is to define a more formal approach to allow for the creation of generic templates to be reused and adapted within the modelling and development of social simulations. In addition, as contrast to employing existing, predefined social simulation tools, this framework seeks to empower researchers to design and implement their own entities. As a consequence, they have total control over the factors that might influence the dynamics of the system under investigation over time. This enables researchers to conduct artificial longitudinal studies of the social systems they are aiming to engineer. One can thereby test the short- and long-term impact of different interventions in the social system modelled.

The methodological basis for the framework is ABMS, which is particularly well suited to developing simulations of human-centric systems. ABMS became widespread in the early 1990s and is now well established in fields like economics, sociology and political science (Squazzoni, 2010). It is a bottom-up approach, where real-world actors are represented as intelligent entities (agents) that have a memory, can make decisions, interact with other entities and the environment, and be proactive (Macal and North, 2005). This simulation paradigm is chosen as it allows one to simulate the system from a position that values individual perceptions and behaviours (and their changes over time), and therefore allows the study of the aggregate impact of all the individuals (i.e., the entire population) on social scenarios and, in particular, in situations in conflict-affected countries. Our framework, however, goes beyond

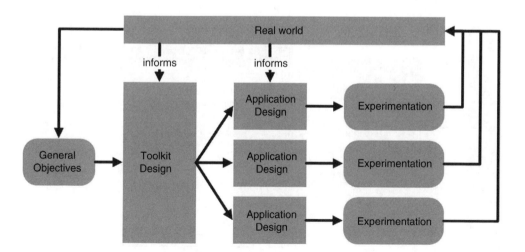

Figure 7.1   Overview of the framework.

simply using ABMS and delivering models of peacebuilding activities in conflict-affected countries, which provides a one-off analysis of a particular region. Rather, it offers an innovative template-driven method for repeated model construction.

7.3.1 Toolkit Design

The toolkit design part of the framework consists of three main activities, once the objectives of the social simulation model are defined: (1) knowledge gathering; (2) defining the stereotypes of relevant actors; and (3) using software engineering tools (UML/SysML) to create generic components in the form of agent templates that are capable of representing the different types of actors with regard to their attributes, behaviours and interactions in an abstract way. Figure 7.2 shows the general structure of this part of the framework.

In these three steps, knowledge about patterns, concepts, existing entities and the relationships among them are described, abstracted and finally represented in graphical notation (toolkit design). The description of external factors – such as the environment, events, networks, etc. – can also be considered. Our focus in this chapter is on producing templates that do not rely on specific simulation tools and provide a generic approach for reusability in social simulation engineering.

7.3.1.1 Knowledge Gathering

The initial step related to the toolkit design is to collect knowledge regarding the model/ scenario to be simulated. In social simulation exercises it has become a norm to build hypotheses and justify the model construction by offering some empirical and theoretical facts. Diverse sets of data can be used based on the objectives established – from surveys and reports to theoretical models and observations in real-world settings (Moss and Edmonds, 2005). In our framework, the knowledge gathered is employed: (1) to describe the system

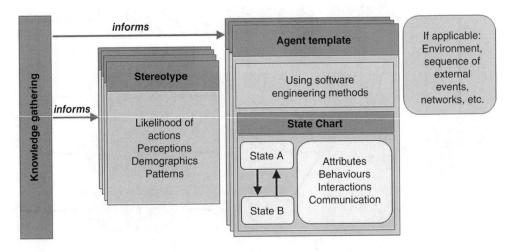

Figure 7.2 Toolkit design part of the framework.

studied; (2) to inform the design of stereotypes and agent templates; (3) to specify the agents and the simulation building; and (4) to validate the results. Ultimately, we hope that the simulation results can also contribute to further knowledge about the system by providing insights into possible scenario outcomes that might occur in the real world. In addition, this output might be useful to update or adapt the existing templates if necessary.

The process of extracting knowledge from the real world to build a toolkit can be done in different ways. Robinson *et al.* (2007) review and exemplify five approaches to empirically inform agent-based modelling: sample surveys, participant observation, field and laboratory experiments, companion modelling and remotely sensed data. The authors describe each approach and discuss their suitability for the modelling process. According to Robinson *et al.*, surveys, as part of the qualitative data pool, are useful: to provide information on the distributions of characteristics, beliefs and preferences within a population of agents; to estimate behavioural models based on economic theory; to provide rough estimates of local-level change variables; and to identify constraints on decision-making. Field experiments address questions regarding resource use, whether or not it is possible to forecast subject behaviour, how changes in environment rules affect resource use and which competing theories best explain behaviour. Companion modelling helps designers: to analyse the interactions among actors, their institutions and the natural environment; to evaluate the process of collective decision-making as observed within the role-playing game context; and to improve the stakeholder's knowledge of the diversity of perceptions and beliefs held in the community. Remotely sensed data answers questions concerning: the relative influence of biophysical factors that an agent will convert from one land use (i.e., human modification of the Earth's terrestrial surface) to another; how biophysical factors interact to affect particular decisions; how neighbourhood characteristics affect decision-making; and how spatial relationships vary over time and space.

The type of method to be applied to the modelling depends very much on the problem domain, objectives, research questions, resources available, personal preferences and how familiar the modellers are with the approach. Furthermore, how the information gathered is incorporated into the model is determined by the level of detail required (Figueredo *et al.*, 2014). For instance, most market dynamic models are based on theoretical assumptions and rely on empirical results to validate the theory (Parker *et al.*, 2003). Other models, such as urban systems, employ mass data and produce large-scale simulations. Schenk (2014) propose model development based on stakeholders' descriptions and observation results in a political process scenario. Yang and Gilbert (2008) explore some of the methodological and practical problems involved in basing an agent-based model on qualitative participant observation (ethnographical data) and suggest ways of converting this information into rules that can be adopted by the simulation system model. Empirical material combined from different sources and mixed methods has also been adopted in modelling. A practical example is an agricultural land-use model, which was initially based on a number of theoretical assumptions combined with earlier empirical findings and further extended by results from qualitative interviews (Bharwani, 2004). It is not our purpose in this chapter to perform an extensive review of methods of knowledge gathering performed for modelling. Instead, we just want to present some alternative routes that modellers can take when they reach this stage of our framework. For the large projects we work on, information comes from different sources and it is important to enhance the integration between different types of data (e.g., quantitative and qualitative) by performing information triangulation. In this manner the qualitative, quantitative and theoretical information can be checked against each other, thereby assisting to establish a more accurate picture of the system modelled. Therefore, in our practical exercises employing the framework, in order to achieve the design objectives, we organize focus groups – internal ones (with the development team) and external ones (with the different stakeholders). The objective is to bring together a team of experts that provide theories, data (quantitative and qualitative) and knowledge regarding the real-world system. The team is responsible for describing the system and also assisting in validating and translating this knowledge into the abstract templates in an interactive manner.

This process involved several meetings with the team, in which we clarified the toolkit design objectives, gathered ideas for relevant actors (a type of role played by an entity – e.g., peacekeepers, citizens), stereotypes and use cases and developed some initial state machine diagrams to define actors' behaviours. Once we had gathered all the relevant information, we worked on the main design, which is discussed further in the next sections. During this phase we conducted further internal focus groups to discuss improvements of the initial design and how best to translate the knowledge gathered into mechanisms or factors that we can model. We also considered the level of abstraction at this phase. As a final template design was achieved, we organized a final external focus group for its evaluation.

The last step of the knowledge-gathering phase consists of defining the scope for the toolkit design. Here we look at the elements we want to include in the toolkit, considering agent types, environment and psychological factors. We provide a table (Table 7.3) in which we list the potential elements and factors, provide the decision for either including or excluding them, and provide a justification for this decision. This table was created during internal focus group discussions.

7.3.1.2   Stereotype Design

We employ the concept of stereotypes to establish patterns of behaviour, perceptions, habits, demographics, personality factors and emotional reactions, which agents copy or emulate. A stereotype is a thought that can be adopted about specific types of individuals or certain ways of doing things (McGarty et al., 2002). The usage of stereotypes allows for a holistic approach, where the behaviour and characteristics defined are universally accepted within a domain area. The use of stereotypes in the design process, rather than solely extracting statistics from personal and/or demographics information, avoids discrepancies during the simulation validation process. Such disparities occur due to incorrect assumptions regarding groups of individuals in the population. For instance, if the designer specifies that in a model, single male individuals aged between 25 and 30 years with a certain level of education behave or respond to the environment in a particular way (for instance, working-class single males tend to adhere to violent acts as a response to high rates of unemployment in their country), this might not reflect the real-world scenario, as an individual's characteristics do not always have direct implication in their behaviour. Stereotypes, in contrast, allow for a more flexible and accurate modelling of individual actions and responses.

The identification of stereotypes (i.e., the names of the different stereotypes), as well as the defining categorization criteria (i.e., the factors that identify stereotypical behaviour), is an iterative and sometimes also long and laborious process. The modellers need to rely on focus groups, expert input and data analysis (data visualization, feature selection, identification of clusters with patterns of behaviour, etc.) to determine what are the important patterns that emerge during the knowledge gathering that should be included in the design. For complex problems, involving multiple elements to be investigated/simulated, an agent can assume one or more stereotypes. In addition, over the course of simulations, agents can change from one stereotype to another.

7.3.1.3 Agent Template Design

Agent templates define all relevant states, and all transitions between these states for a specific type of actor. These are organized in the form of state charts and can be used as blueprints. An agent template can have multiple state charts, representing physical or mental states. The agent template acts as a pattern for creating virtual representations of typical real-world actors (e.g., drivers, citizens, peacekeepers).

To define our agent templates, we employ state charts from the standard UML. State charts are a graphical notation commonly used in software engineering for the purpose of software design. We advocate the use of this tool as it describes the agents in an intelligible, intuitive manner, suitable for cross-disciplinary tasks, facilitating therefore the communication between the different stakeholders during the modelling process. The state charts represent the behaviour of an agent at discrete points in time. Agents assume an initial state in the start of the simulation and transit between states as the simulation proceeds. During the transitions between the states, actions may be executed. The main graphical elements of a state chart employed in template definitions are shown in Table 7.1.

In our toolkit design part of the framework there is no link between stereotypes and agent templates, and therefore both development processes are not directly related. However, the stereotypes should be developed before the agent templates to inform their design. Obviously, as this is an iterative process, amendments in both stereotypes and templates might occur until the

Table 7.1 The main elements that constitute a state chart

Graphical element	Description
	State chart entry point: Indicates the initial state an agent is in once created.
State	State: The particular condition that someone or something is in at a specific time.
	Composite state: A state that has substates (nested states).
	Initial state pointer: Points to the initial state within a composite state.
	Final state: Indicates the termination of a state chart.
	Transitions: These arrows indicate the movement between the states. They can be triggered by events (first arrow), timeouts (second arrow) or a conditional (third arrow).
	Branch: Represents the branching of transitions and/or connection points.

final design is achieved. The stereotypes are going to be linked to the agent objects in the *application design* phase, as explained further in the next section. This happens after the *tool-kit design phase* is complete.

One question that might arise during creating an agent template is when a template should be considered 'finished', accurate or ready, so that agents can be derived from it. There is no obvious answer for this question, as in software engineering templates can be reused, modified and/or extended to adapt to new requirements. As a rule of thumb, we consider that a template is ready to be employed in a simulation problem when it is able to capture (or represent) the behaviours of the stereotypes defined within the context of the domain model (or conceptual model). Defining a template is an iterative process, where several prototypes are produced and refined.

7.3.2 Application Design

The application design part of the framework consists of: (1) the development of a problem-specific conceptual model and (2) the development of the simulation model defining agent objects (which are instantiations of the agent templates) and their interactions (communication and network of contacts) and behaviours (by linking them to a specific stereotype). The proportions within the stereotypes to be adopted in the simulation are informed by the real-world knowledge, conceptual model and research questions regarding the scenarios simulated. Figure 7.3 provides an overview of the application design part of the framework

A simulation model should always be built for a specific purpose or set of objectives. For the application design we have to create a problem-specific conceptual model, which will lead to a problem-specific application. Here we use the conceptual modelling framework of Robinson (2004), which consists of describing the following.

- Objectives: The purpose of the model (e.g., the hypotheses to be tested).
- Inputs: Elements of the model that can be altered.

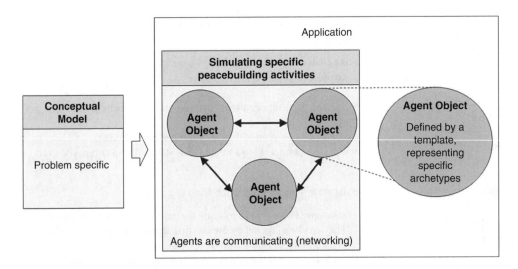

Figure 7.3 Application design part of the framework.

- Outputs: Measures to report the results from the simulation runs.
- Content: Components represented in the model and their interconnections.
- Assumptions: Uncertainties and beliefs about the real world to be incorporated into the model.
- Simplifications: Reduction of the complexity of the model.

For the content definition, the model scope (what to include in the model) and level of detail (how to include it in the model) should be defined. For the implementation, the knowledge gathered also specifies the simulation parameters, population size, rates, etc. and assists in the simulation verification and output validation processes.

As the purpose of the toolkit design is to provide a complete, reusable, extensible set of templates in a social simulation context, different application designs might employ distinct subsets of the available templates, depending on the problem to be simulated. The ultimate aim of using this framework for social simulation is to bring together a set of experts that will enable the creation of libraries (toolkits) of generic, reusable templates of different social actors (individuals, organizations, government, peacekeepers, etc.) that can be combined together in a simulation design.

7.4 Illustrative Example of Applying the Framework

In order to demonstrate how the framework can be employed to create simulations that assist researchers in social systems engineering studies, we apply it to develop a toolkit for studying peacebuilding activities in general. Our application is developed for the specific case of studying international peacekeeping activities in South Sudan. Our objective is to employ the developed toolkit to design a decision-tool application that allows us to investigate, through 'what if' scenarios, how the Sudanese people respond to different governmental actions and societal interventions. In this tool, the user acts as 'the government', taking decisions and applying several public policies to different regions of the country and observing the population's responses to these changes. The purpose of the tool is to allow policymakers and other interested parties to investigate peacekeeping efforts in South Sudan and to evaluate how actions interfering in security, governance, economy and wellbeing might influence South Sudanese individuals towards violent acts, which are triggered by their levels of anger and fear.

7.4.1 Peacebuilding Toolkit Design

In this section we explain how the elements of the framework are employed to design the toolkit for peacebuilding simulations. We show how the knowledge was gathered and converted into stereotypes and agent templates. Subsequently, we introduce how the toolkit was employed to develop the decision tool.

7.4.1.1 Peacebuilding Knowledge Gathering

Within a peacebuilding context, the actors to be incorporated in the simulation model could be citizens and other peacekeeping stakeholders, for example government officials, organizations, business people, industry and the army. Moreover, we consider their actions (and perceptions) towards and in response to peacebuilding activities. To keep things simple, we focus solely on citizens (using the South Sudan citizens as a representative example) and therefore the knowledge-gathering activity sought to collect information regarding citizens' characteristics, habits and responses to conflict and peacekeeping actions. In addition, we look at individual behavioural changes, which are governed by their inner perceptions and emotions regarding their micro and macro environment. Consequently, we adopt three foundational dimensions that influence individual behaviour: rational, emotional and social (Epstein, 2014). The rational and social aspects are modelled using knowledge obtained: (1) from a report by the Center for Nation Reconstruction and Capacity Development (CNRCD) on post-conflict indicators in South Sudan (CNRCD, 2011); (2) from studies including Hirono (2011); and (3) through expert input from political scientists. An expert in psychology provided the information regarding hypothetical individual actions and associated changes as a consequence of emotional reactions.

According to the findings of the CNRCD report, the four main indicators affecting violence in South Sudan are security, governance, economy and social wellbeing. Each of these indicators is subsequently divided into several categories. For our tool design, during our focus group meetings, we chose those more relevant in our context, as shown in Table 7.2. In addition, information such as demographics per region, ratio between male and female citizens, and number of violent acts and regions with more propensity of violence was collected.

Expert knowledge regarding individual citizens was provided to us during our focus meetings with political scientists. According to these scientists, apart from the factors described by the CNRCD report, the employment status also influences people's likelihood of committing acts of violence. Furthermore, in South Sudan, citizens from different categories have different perceptions of peacebuilding activities. Senior citizens (over 65 and retired) are likely to be more concerned about their welfare and safety rather than unemployment, for instance. Adults (between 18 and 65), as the workforce, are concerned with employment and the means

Table 7.2 Violence factors affecting South Sudanese people (derived from CNRCD, 2011)

Factor	Category	Description
Security	Security forces	Effort to provide security to the population
	Attacks by rebel groups	
	Public perception of security	Individual fear of violence
Governance	Govern representation	Individual perception of being represented by the government
	Fairness in justice	Individual perception of fairness in the justice system
	Citizen participation	Citizen participation in local government
Economy	Inflation rate	
	Per capita consumption	
	Unemployment	
Wellbeing	Water sources	Access to improved water source
	Human rights index	A-E rating (for more, see uhri.ohchr.org/en)
	Religious oppression	
	Tribal discord	Hostility between tribes increasing anger and violence

to support and guarantee their own safety and that of their families. Without a job and/or dissatisfied with their current environment, adults are more likely to engage in violent acts. As a consequence, they can become part of riot groups, be arrested or face premature death. Environments with a high likelihood of violence also stimulate the migration of citizens in fear to safer areas. There are also children or junior citizens (under 18), and their perception of the peacebuilding activities is unspecified.

Individual perceptions, such as a feeling that the world is unfair, or perceiving the uncertainty of the situation, are shown to evoke emotional reactions, specifically anger and fear (Kuppens *et al.*, 2003; Smith and Ellsworth, 1985). In return, the emotional reaction biases people's future decisions: for example, individuals experiencing fear tend to be more risk-averse, while those who are angry tend to be risk-seeking (Lerner and Keltner, 2001). We incorporate into our knowledge pool, therefore, psychological mechanisms such as anger and fear, allowing individuals to change between neutral and emotionally loaded states of anger and fear. High levels of anger can trigger highly risky, impulsive behaviour; in our case it could be violence or participation in riots. High levels of fear increase risk-averse behaviours (i.e., migration), as individuals want to move to safer areas with better opportunities. The values for individual anger and fear in our context are determined by citizens' overall satisfaction with the region they live in, how they perceive external events and their individual proneness to fear or anger. The satisfaction/perception of the external events is a result of government/ peacebuilding policies applied in that region.

This information was obtained during the course of four months, when we met with experts and discussed the population we wanted to model, their habits and the likelihood of their actions in conflict areas. The focus groups helped filter the information to obtain what seemed to be more relevant to the toolkit design next steps. During this time, we also defined the scope of the toolkit, through several internal focus group discussions. In order to keep the model transparent, we decided to only model the citizens as agents. As stated before, the whole development is an iterative process and we could always come back to this point to extend the scope of the toolkit. Table 7.3 shows the details related to the current scope of the toolkit.

7.4.1.2 Peacebuilding Stereotype Design

In our example, we have citizens that respond to peacekeeping efforts based on what they perceive as satisfactory in terms of security, governance, economy and wellbeing. If individuals do not feel well represented according to these four factors, which determine their levels of anger and fear, they respond either with violence (when there is high anger) and/or migration (when there is high fear). In addition, peacekeeping activities aim to reduce violence and the number of refugees in conflict areas. Having this information and the objectives in mind, we defined three citizen stereotypes. The first stereotype is the 'fighter'. This is someone with high anger and low fear, having a high likelihood of participating in outbursts of violence in response to changes in environment. The second stereotype is the 'conformer'. This is someone with moderate anger and fear, who tends to be less likely to migrate or participate in violent acts. The third stereotype is the 'refugee'. This is someone with high fear and low anger, who has a high likelihood of migrating to areas of less conflict. Table 7.4 shows a summary of the definition of these stereotypes.

Table 7.3   Scope to be considered for the toolkit design

Type	Subtype	Decision	Justification
Actors (these will be the agent objects in the simulation)			
Government		Exclude	Impact will be modelled through 'indicators affecting violence'
Oil company		Exclude	Impact will be modelled through 'indicators affecting violence'
Citizens	Workers	Include	Key 'role' to be considered in our model (employed/ unemployed)
	Families	Exclude	Not considered to keep the model transparent (perhaps consider in future)
	Rebels	Exclude	To be defined as a 'role' within the worker actor
	Migrants	Exclude	To be defined as a 'role' within the worker actor
Peacekeepers		Exclude	Impact will be modelled through 'indicators affecting violence'
Environment (here we consider the environment that influences the actors; usually in the form of physical representations or resources)			
Weather		Exclude	Not directly relevant
Location		Include	For being able to provide graphical representations of segregation
Networks		Exclude	Not considered to keep the model transparent (perhaps consider in future)
Peacekeepers		Exclude	Impact will be modelled through 'indicators affecting violence'
UNO		Exclude	Impact will be modelled through 'indicators affecting violence'
Local police		Exclude	Impact will be modelled through 'indicators affecting violence'
Businesses		Exclude	Not considered to keep the model transparent (perhaps consider in future)
Job vacancies		Include	Will be a resource (number of vacancies)
Money		Exclude	Not considered to keep the model transparent (perhaps consider in future)
Weapon delivery		Exclude	Not considered to keep the model transparent (perhaps consider in future)
Interventions by	Homeland	Exclude	Can be set by 'indicators affecting violence' (experimental factors)
	UNO	Exclude	Can be set by 'indicators affecting violence' (experimental factors)
Indicators affecting violence		Include	Experimental factors (economy, security, governance, wellbeing)
Psychological factors			
Anger		Include	Anger/fear model considered in form of state variables
Fear		Include	Anger/fear model considered in form of state variables
Hate		Exclude	Expressed by consideration of anger/fear
Trust		Exclude	Exclude due to time constraints; to be added in future

Table 7.3 (Continued)

Type	Subtype	Decision	Justification
Fairness		Exclude	Exclude due to time constraints; to be added in future
Violence		Exclude	Defined by the 'indicators affecting violence'
Representativeness		Exclude	Represented by the 'indicators affecting violence'

Table 7.4 Citizen stereotypes

Stereotype	Likelihood of violence	Likelihood of migration	Anger level	Fear level
Fighter	High	Low	High	Low
Conformer	Moderate	Moderate	Moderate	Moderate
Refugee	Low	High	Low	Medium

This is only one possible group of stereotypes suitable for this problem, which we employ in this example for simplicity of presentation. Other stereotypes can be defined if there is the need. In addition, when different populations are considered in the simulation (for instance, government, peacekeepers, industries, etc.), these should also be included in the stereotype group, with their own set of patterns of behaviour.

7.4.1.3 Peacebuilding Agent Template Design

When developing agent templates, it is good to start with a very simple prototype and then work towards the final template by adding more details. One thing that is quite difficult, in particular for modelling novices, is to find the right level of abstraction (a model is always a restricted copy of the real world, and is created through abstraction). There are no strict rules for this process, and it needs some experience to work it out. A good approach is to follow the KISS principle (keep it simple, stupid), which was introduced to ABMS (Axelrod, 1997). This states that only essential elements and relationships should be considered within the model. It does not state how to know what is essential. A good way to gain experience in this process is to look at model examples, which are usually shipped with the software tools.

Our population is composed of junior, adult and senior citizens. With time, junior citizens become adults, adults become seniors and seniors die (for simplification, let us not consider that there is the possibility of juniors or adults dying). Therefore, our actor (citizen) can be represented by a state chart with junior, adult and senior states (Figure 7.4). The transitions between these states occur as the individuals reach age thresholds (i.e., 18 to transit from junior to adult and 65 from adult to senior). As these transitions are triggered by individual age, which increases with time, the 'death' state is also included to finalize the lifecycle of the agent.

To start building a general template for representing citizens in conflict areas, it is necessary to know who the actors in our model are. Our actors, as mentioned previously, are the citizens. We decided to start by modelling citizens' life stages: junior, adult and senior citizens. These are the first three states we add to the state chart, which will become the basis for our template.

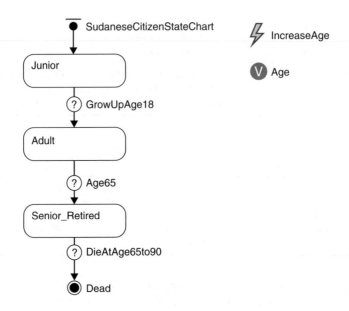

Figure 7.4 Citizen in conflict areas (early prototype).

Figure 7.4 shows the early prototype state chart. This prototype could be implemented, and we could create an application with an aging population. Having an application at this stage allows us to test the basis of our future template.

As we have already seen in Figure 7.1, the process of stereotyping and creating the templates is informed by data, theories, expertise, knowledge, etc. Apart from the theory employed, in order to complete our template, we had the assistance of one of our team members, who is a political scientist, who provided us with further information, as described below and in Hirono (2011). Therefore, we had further guidance regarding the states and attributes found necessary in our diagram, in a peacebuilding context.

The states are defined as follows. Citizens could be in the 'unemployed' state. In this state they can either be in a (nested) state 'looking for a job' or in a state 'loafing around'. If a long time elapses while in the state unemployed, dissatisfaction and therefore anger rises, which might trigger violent behaviour. If individuals become part of a rebel movement, they can move to the state 'arrested', the state 'recruit' (where they enrol other individuals to help with acts of violence) or the state 'dead'.

Conversely, if they are successful in getting a job, they assume the 'employed' state. While employed, they can be in the states 'at home' or 'at work'. This location differentiation can be useful when/if individual emotions modify or are influenced by the environment or the person's network of contacts, for more complex simulation models. When citizens are located in violent areas, their fear increases and they migrate.

Furthermore, we added some more avenues towards death. If citizens suffer premature death, the 'dead' state assists us to identify what triggered the transition that led to this state (e.g., cases of violent acts). As noted in the system description, employment appears to be an important factor for adults, as they are more likely to get involved in violent activities when there are no jobs available.

We also add four attributes to the template: the parameter *Male*, to indicate whether the individual is male or female; the individual thresholds of anger and fear that would prompt a transition to being violent or migrating, respectively (this provides diversity to the population, as each citizen will have a different threshold); and the attribute *ProneToRebel*, which determines whether a violent individual is likely to join rebel groups based on their anger level. All these parameters allow for variability and flexibility of the model. In addition, the parameters and rates at which transitions and events occur are to be determined by data and/or information about the population. Figure 7.5 shows the final template for citizens in conflict areas.

The template is ready to be employed in other peacebuilding modelling and simulation activities. It is generic enough so that other researchers can reuse it in their peacebuilding simulation studies. It is our intention in the future to extend our framework and create a library of templates with other actors involved in peacekeeping activities (organizations, foreign peacekeepers, government, business people, army, etc.).

Additionally, the framework can be extended by other users. Therefore, if one feels that the template provided is not suitable for a particular problem, further knowledge can be employed to modify the existing templates. In the case of the template in Figure 7.5, for instance, one could reuse the outer shell of the template (i.e., the states *Junior*, *Adult* and *Senior_Retired*) and modify the states inside to better suit the problem being addressed.

7.4.2   Peacebuilding Application Design

Now we use the newly developed toolkit to develop a specific application: a decision-support tool for studying the impact of different peacebuilding activities for the case of South Sudan. The tool was developed mainly for the purpose of testing the usability of the toolkit. But it is also part of a proof-of-concept study, where we were aiming to better understand the principle dynamics and opportunities of social systems engineering within the context of South Sudan. We have used the decision-support tool during our multidisciplinary focus group discussions while we were writing a larger funding proposal, related to peacekeeping activities in South Sudan. For the development of this specific application we used knowledge about the situation in South Sudan, on the one hand gained from the literature and on the other hand gained from the expertise of the multidisciplinary focus group members.

The initial step in the application design is the development of a conceptual model. Here we should look back at the objectives and scope definitions of the toolkit design (if available) and adapt these to fit the purpose of the application design (which is usually a special case of the more general one for the toolkit design). In our case, rather than focusing on peacebuilding activities in general (toolkit objective), we focus on peacebuilding activities in a specific location – South Sudan (application objective). For the conceptual modelling we follow broadly the ideas of Robinson (2004), defining objectives, inputs, outputs, content, assumptions and simplifications.

- *Objectives*
  - To better understand the impact of different peacebuilding activities (defined through indicators affecting violence) for the case of South Sudan.
- *Inputs (things we can vary during runtime)*
  - Security: Security forces.
  - Governance: Government representation; fairness in justice; citizen participation.

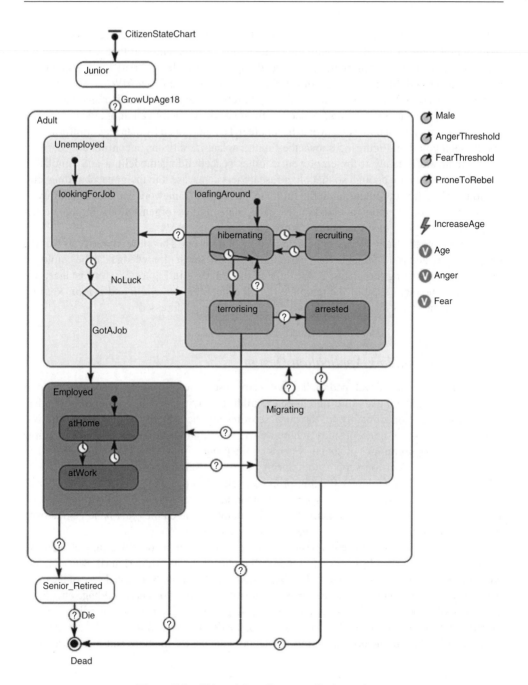

Figure 7.5 Citizen in conflict areas (final template).

o Economy: Inflation rate.
o Wellbeing: Water source; human rights index; religious oppression; tribal discord.
- *Outputs (responses from the model that help us with achieving the objectives)*
  o States of the adult population (unemployed, employed, migrating) [percentage of population].
  o Population anger level [percentage of population].
- *Content*
  o Scope is identical with that defined for the toolkit (see Table 7.3); the only difference is that we are focusing on a specific conflict region – South Sudan.
- *Assumptions*
  o Changes happen 24/7 (we do not consider time of day/day of week).
  o Initial values and relationship dynamics are based on best guesses.
- *Simplifications*
  o These are listed in the scope definition of the toolkit (see Table 7.3).

For the application implementation we use AnyLogic University 7.0 (AnyLogic Company n.d.), a multi-method simulation IDE that allows the automated translation of state charts into Java code. Our decision-support tool consists of a graphical user interface (GUI) to make it easy to use and informative. Underneath the hood of this GUI we have created an agent population representing South Sudanese citizens, by linking each agent to a specific stereotype.

The variables shown in Figure 7.6 are those employed to run simulations to study the dynamics of anger and fear, and therefore the occurrences of violence and migration. In the environment, everything besides the South Sudanese people (i.e., the peacekeeping factors of Table 7.2) is modelled using variables that can be changed (by researchers) before and during the simulation. The diagram of Figure 7.6 shows all the variables that affect individual emotions in the tool. The levels of anger and fear depend on the outcome of the weighted sum of all these variables. In the diagram, the circles with a 'V' inside represent the variable numbers that will emerge (and change) as the simulation is run. For instance, the number of rebels will increase if the policies of the government generate anger. The dark circles are the main factors of the sum. Each of these factors is a combination of other variables. For instance, the economy factor is a combination of the inflation rate (which is a parameter modifiable by the user), the *per capita* consumption (which is calculated based on the inflation rate of the country) and the number of job vacancies (which is variable). It is up to the user of the decision tool to define these values and the weights of the sum based on real-world information about Sudan and analyse how changing the policies (sliders in Figure 7.7) affects the behaviour of the agents (citizens). As we do not have information regarding job creation, we have defined it as a constant rate in the system (in the top rectangle of Figure 7.6 the cloud named *JobCreation* represents a constant input of new job vacancies).

The GUI is designed to allow easy access to the parameters, which can be changed even during runtime. Figure 7.7 shows a screenshot of the main screen of the tool. The user can modify the values of the security, governance, economy and wellbeing variables (even during runtime) and observe how the behaviours and numbers change over time for several scenarios.

The left panel shows the state of the individuals (at home, at work, looking for a job and rebellious) and the number of citizens (agents) in each state. On the right side of the screen the

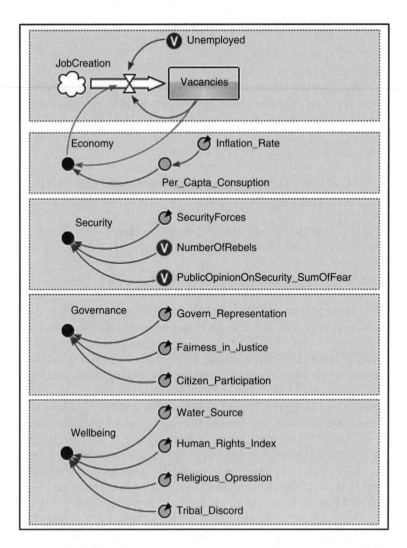

Figure 7.6 Variables included in the decision to study the dynamics of South Sudanese citizens.

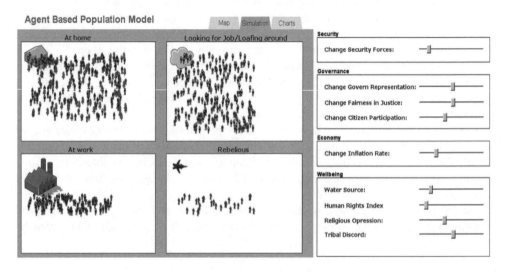

Figure 7.7 Screenshot of the decision tool main screen.

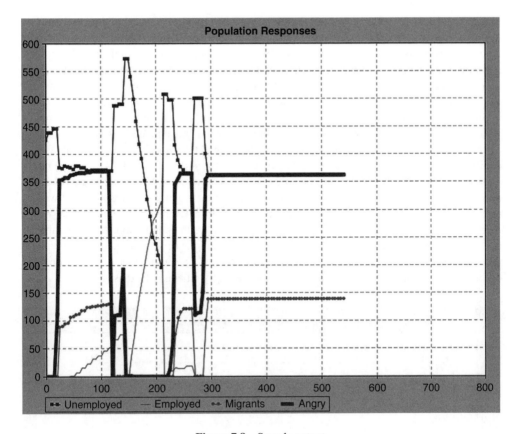

Figure 7.8   Sample output.

user has the possibility to change the peacekeeping variables by using the sliders and observing the impacts of these changes. For example, if citizen participation is reduced, this will generate more anger and as a consequence more citizens become rebellious and the number of agents in the 'rebellious' box will increase.

The chart tab provides access to population statistics. Figure 7.8 shows an example of statistics that can be collected as output. The *x*-axis is the number of periods of the simulation, which can be days, months, etc. The *y*-axis represents the number of agents in each category. The user of the decision tool can observe when the outbursts of anger and therefore violence occur and the associated reasons (e.g., the growth in unemployment). The user can also observe the numbers of migrants and how they arise due to increased violence.

7.4.3   *Engineering Actions and Interventions in a Peacebuilding Context*

As this tool is an illustrative example only, further data is necessary to validate the outcomes produced. However, once the parameters of the tool are adjusted to assume those values close to the current context of a country (in our case, South Sudan), several scenarios can be validated, studied and evaluated by peacekeeping policymakers. Experimental results are therefore useful to better understand social phenomena or, in our case, the impact of peacebuilding activities on the psychological and physical state of the individual citizens (micro level). For instance, an investigation on which factors are primarily responsible for migration or outbursts of violence, and how these numbers alter as a result of changes in the current policies. Furthermore, the scenario outcomes also indicate the impacts of global measures (macro level) such as, for example, employment rate, financial measures, level of unrest in the country, etc.

Such decision tools can also be employed as an educational or communication interactive instrument by demonstrating to politicians and citizens the consequences of behaviours such as radicalization. For example, in some public engagement settings such as a science fair or exhibition at a museum, public members could change the sliders themselves to see the micro and macro impacts of different options and compare alternatives. Through this exercise, the public will learn the implications of various actions on the population in general, as well as individual groups of citizens. For instance, Figure 7.9 shows the results of experiments where measures are taken to secure the safety and employability of the population. As the number of unemployed people decreases, together with their perceptions of security and representativeness adjusted in the sliders of the tool (shown in Figure 7.7), the number of migrations and outbursts of violence (represented by the number of angry citizens) decreases.

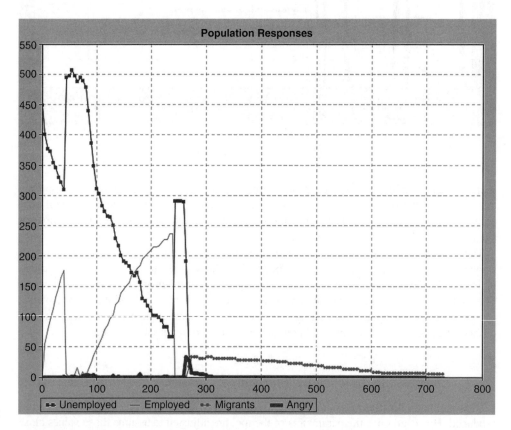

Figure 7.9 Example of the use of the tool to test actions towards creating jobs and increasing safety and representativeness.

7.5   Conclusions

In this chapter we have presented a novel framework that employs software engineering methods to build agent-based social simulation models. The objective for developing this framework was to enable researchers to conduct artificial longitudinal social systems engineering studies in which they have control over factors influencing the development of the system under investigation over time. Therefore, one can test the short- and long-term implications of different interventions on the evolution of social systems. We have illustrated the application of this novel framework through the development of a decision-support tool for studying the impact of different peacebuilding activities in South Sudan. We believe that the framework will be easy to use by a multidisciplinary team for the development of models supporting social systems engineering studies, as it employs clearly defined steps and a simple graphical notation for the conceptual modelling, which can then be implemented using simulation software packages that support automatic translation of state charts, such as, for example, Repast Simphony (Argonne National Laboratory, n.d.) or AnyLogic.

A possible next step in this project would be to turn the illustrative example into a case study. This would require some additional data collection, with a focus on South Sudan and a better understanding of the dynamics of anger and fear, and therefore the occurrences of violence and migration. The latter we could get from the literature and by talking to domain experts. It would also be interesting to apply the framework presented here to multidisciplinary projects in other domains. A potential application area would be the sustainability and resilience of cities, which is one of our current research priority areas. We could use this framework to build models that allow testing the impact of social systems engineering on the sustainability of urban habitats.

References

AnyLogic Co. (n.d.) AnyLogic University 7.0. Available at: www.anylogic.com/ (retrieved 23 March 2016).

Argonne National Laboratory (n.d.) Repast Suit, Available at: repast.sourceforge.net/ (retrieved 23 March 2016).

Axelrod, R. (1997) *The Complexity of Cooperation: Agent-based models of competition and collaboration*, Princeton University Press, Princeton, NJ.

Berdal, M.R. (2009) *Building Peace After War*, Routledge, Abingdon.

Bharwani, S. (2004) Adaptive knowledge dynamics and emergent artificial societies: Ethnographically based multi-agent simulations of behavioural adaptation in agro-climatic systems. PhD thesis, University of Kent, Canterbury.

CNRCD (2011) The Sudan and development of post-conflict indicators. Center for Nation Reconstruction and Capacity Development. Available at: www.usma.edu/cnrcd/cnrcd_library/conflict.pdf (retrieved 23 March 2016).

Epstein, J.M. (2014) *Agent_zero: Toward neurocognitive foundations for generative social science*, Princeton University Press, Princeton, NJ.

Figueredo, G., Siebers, P.O., Owen, M., Reps, J. and Aickelin, U. (2014) Comparing stochastic differential equations and agent-based modelling and simulation for early-stage cancer. *PLoS ONE*, **9**(4), 2.

Goodhand, J. and Hulme, D. (1999) From wars to complex political emergencies: Understanding conflict and peacebuilding in the new world disorder. *Third World Quarterly*, **20**(1), 13–26.

Hirono, M. (2011) China's charm offensive and peacekeeping: The lessons of Cambodia – what now for Sudan? *International Peacekeeping*, **18**(3), 328–343.

Kuppens, P., Van Mechelen, I., Smits, D.J. and De Boeck, P. (2003) The appraisal basis of anger: Specificity, necessity and sufficiency of components. *Emotion*, **3**(3), 254.

Lerner, J.S. and Keltner, D. (2001) Fear, anger, and risk. *Journal of Personality and Social Psychology*, **81**(1), 146.

Macal, C.M. and North, M.J. (2005) Tutorial on agent-based modeling and simulation, in M.E. Kuhl, N.M. Steiger, F.B. Armstrong and J.A. Joines (eds), *Proceedings of the 2005 Winter Simulation Conference*, Orlando, FL.

McGarty, C., Yzerbyt, V.Y. and Spears, R. (2002) Social, cultural and cognitive factors in stereotype formation, in C. McGarty, V.Y. Yzerbyt and R. Spears (eds), *Stereotypes as Explanations: The formation of meaningful beliefs about social groups*, Cambridge University Press, Cambridge.

Moss, S. and Edmonds, B. (2005) Sociology and simulation: Statistical and qualitative cross-validation. *American Journal of Sociology*, **110**(4), 1095–1131.

Parker, D.C., Manson, S.M., Janssen, M.A., Hoffmann, M.J. and Deadman, P. (2003) Multi-agent systems for the simulation of land-use and land-cover change: A review. *Annals of the Association of American Geographers*, **93**(2), 314–337.

Pidd, M. (1992) *Computer Simulation in Management Science* (3rd edn), John Wiley & Sons, New York, NY.

Richmond, O.P. (2004) UN peace operations and the dilemmas of the peacebuilding consensus. *International Peacekeeping*, **11**(1), 83–101.

Robinson, D.T., Brown, D.G., Parker, D.C., Schreinemachers, P., Janssen, M.A., Huigen, M. et al. (2007) Comparison of empirical methods for building agent-based models in land use science. *Journal of Land Use Science*, **2**(1), 31–55.

Robinson, S. (2004) *Simulation: The practice of model development and use*, John Wiley & Sons, Chichester.

Schenk, T. (2014) Using stakeholders' narratives to build an agent-based simulation of a political process. *Transactions of the Society for Modeling and Simulation International*, **90**(1), 85–102.

Siebers, P.O. and Aickelin, U. (2008) Introduction to multi-agent simulation, in F. Adam (ed.), *Encyclopaedia of Decision Making and Decision Support Technologies, Vol. 2*, IGI Global, Hershey, NY.

Smith, C.A. and Ellsworth, P.C. (1985) Patterns of cognitive appraisal in emotion. *Journal of Personality and Social Psychology*, **48**(4), 813.

Squazzoni, F. (2010) The impact of agent-based models in the social sciences after 15 years of incursions. *History of Economic Ideas*, **18**(2), 197–233.

United Nations (1992) *An Agenda for Peace. Preventive diplomacy, peacemaking and peace-keeping*. UN Document A/47/277. Available at: www.undocuments.net/a47-277.htm (retrieved 23 March 2016).

Yang, L. and Gilbert, N. (2008) Getting away from numbers: Using qualitative observation for agent-based modelling. *Advances in Complex Systems*, **11**(2), 175–185.

8

Using Actor-Network Theory in Agent-Based Modelling

Sandra Méndez-Fajardo, Rafael A. Gonzalez
and Ricardo A. Barros-Castro

8.1 Introduction

Design-oriented research usually involves modelling for the purposes of representing, exploring or simulating real-world situations. A widely used approach is agent-based modelling (ABM); agent-based models help us to understand or reveal the effects of collective behaviour by representing the rules governing agent decisions and the influence of these decisions on a virtual (or sometimes actual) real-world environment. A number of different theories guide the design of individual agents, their interactions and the environment on which they act, most of which are focused on representing the currently observed situation or are driven by historical data in order to enable calibration and the interpolation of future scenarios. However, these approaches often lack the in-depth reconstruction of historical trajectories or fail to capture the dynamics that have led to collective patterns, whether successful or not. Furthermore, most of these strategies treat agents as human actors or institutions, leaving out non-human actors or treating them as part of a different category, despite the fact that both human and non-human actors exercise agency, interaction and collective behaviour. Thus, the main research question addressed in this chapter is how to extract key elements from historical trajectories in socio-technical systems via the identification of networks of human and non-human actors and their past dynamics, such that this information can be used for the design of agent-based modelling and simulation.

Put another way, this chapter proposes the use of actor-network theory (ANT) as an innovative pragmatic approach to the design of agent-based models and simulations with the purpose of supporting policy design or enhancing decision-making. Given that ANT stems from a socio-technical tradition, it provides theoretical foundations, insights and guidelines for the exploration and conceptualization of real socio-technical systems, with the emphasis on intervention. The first part of this chapter introduces the main conceptual elements of ANT

Social Systems Engineering: The Design of Complexity, First Edition. Edited by César García-Díaz and Camilo Olaya.
© 2018 John Wiley & Sons Ltd. Published 2018 by John Wiley & Sons Ltd.

and ABM. From there, the chapter proceeds to discuss possible links between ABM and ANT concepts using an exploratory case study in the context of waste of electrical and electronic equipment (WEEE). Based on the application of ANT in the case study, a methodological framework for integrating ANT and ABM is presented and discussed in the last part of the chapter, along with a brief examination of open issues and concluding remarks.

8.2   Agent-Based Modelling

There is no single, all-encompassing definition of an agent. Agency can be framed within the economic theory of agency (Ross, 1973) or the theory of the firm (Jensen and Meckling, 1976), from a human social cognitive theory perspective (Bandura, 1989). Agency can be understood from a computer science perspective, which combines distributed systems and artificial intelligence into the notion of an intelligent software agent, usually acting on behalf of a human user (Sycara *et al.*, 1996). This last perspective opens a wide range of possibilities (i.e., agents can be simple or complex, autonomous or semi-autonomous, homogeneous or heterogeneous). What is more, from the computer science perspective, agents may form small, collaborative teams or a large, self-organizing social system. In simple terms, agents usually represent social actors, such as individual people, collectives, institutions, businesses, countries or any entity with a certain specific goal (Gilbert, 2007; Gilbert and Troitzsch, 2005; Railsback and Grimm, 2011). As such, agents should be unique and autonomous. To be clear, unique implies that an agent may be different from others (as opposed to homogeneous agents, whose behaviour is closer to that of automata) and autonomous implies that agents pursue their own objectives, that is, act independently of each other (Railsback and Grimm, 2011).

A benefit of this computer science understanding of agents lies in the fact that it requires the focus to extend beyond the agent itself; as a result, it includes communication and interaction among agents, as well as interaction with an environment. The goal, then, becomes not just the individual design of agents, but to design them as part of an artificial society (Epstein and Axtell, 1996), or to design an agent-based simulation in general (Sanchez and Lucas, 2002), which means conceptualizing, designing and implementing ABM and simulations. In ABM, agents interact within an environment, and this interaction denotes the ability to send messages to each other and affect each other's actions. Messages may represent a spoken conversation between real actors or information flows via one agent's observation of another. Agent-to-agent interaction of this nature represents the primary difference with respect to other computational models (Gilbert, 2007). Moreover, agent interaction generates emergence or collective patterns that can be visualized in the system's macro level, thereby suggesting that ABM has as primary goal the study of complex, collective emergent structures or patterns that emerge from simple rules defined at the individual level (Macy and Willer, 2002).

ABM allows for the modelling of individual heterogeneity through explicit representations of the rules governing each agent's decisions by situating agents in different places within an environment (i.e., the virtual world in which agents act) (Gilbert, 2007). An ABM, then, is a system composed of individual agents (and the rules governing their simple decisions), the environment (whether geographical/natural or artificial, such as a knowledge space) and the social structure (made up of relationships and rules governing agent interaction) – see Figure 8.1. The effects of collective agent action in relation to the environment and the social

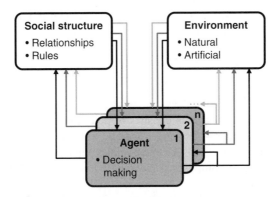

Figure 8.1   Agent-based model representation. *Source:* Adapted from Knoeri *et al.* (2010).

structure generate emergent patterns, although such patterns can, in turn, affect the agents individually and even affect their future decision-making, akin to the co-evolution of agency and structure in Giddens' structuration theory (Fuchs, 2003; Knoeri *et al.*, 2010). Therefore, ABM presents two levels (or more) of interactions: system-wide changes and individual adaptations to different external conditions (Railsback and Grimm, 2011).

One of ABM's most salient features is the ontological correspondence between computational agents in the model and real-world actors (Gilbert, 2007). This correspondence is associated with the level of similarity between the two, and, when this correspondence is high, ABM is more easily implemented, making the outcomes easier to interpret.

8.2.1   ABM Approaches

When setting up the agents and their social structure, we find a number of theories, some of which are rooted in the social sciences (e.g., the theory of interpersonal behaviour) (Triandis, 1979), to understand curtailment behaviours such as switching off appliances or driving behaviours in transport systems. Looking at a different time scale, the theory of planned behaviour (Ajzen, 1991; Feola and Binder, 2010) can be used for one-off behaviours, such as heating system installation or refrigerator replacement.

Some ABMs and simulations, such as SocLab, developed by Sibertin-Blanc *et al.* (2013), have used the sociology of 'organized action' (Crozier and Friedberg, 1983) and are also referred to as strategic analysis. Strategic analysis is a theoretical framework for the analysis of organizational dynamics based on power relations. Within the same field, we also have the institutional analysis and development framework proposed by Ostrom (2005), in which individuals are the key driving components of a social system. In Ostrom's theoretical framework, as well as in the framework proposed by Giddens' structuration theory (Held and Thompson, 2008), the social context not only encompasses the behaviour of individuals, but is also structured by individual interactions. For their part, individual interactions are shaped by institutional settings (Ghorbani *et al.*, 2012).

In addition, there are several theories stemming from the computer science side of ABM. Most notably, Zambonelli *et al.* (2003) have proposed a model for multi-agent systems that is based on computational organization: software agents assume different roles (with their own permissions and responsibilities), interact with each other and are modelled separately from the environment in which they interact.

8.2.2   Agent Interactions

The simplest interaction among agents consists of data transfer from one agent to another (e.g., one agent informs another of its gender and age, enabling the second agent – receiving the information – to make a decision). A more complex interaction would be the exchange of messages, whereby an agent composes a message within the confines of a language and another agent interprets it.

Interactions can be included in the ABM as part of the (individual) agent decision-making process and as rules, norms or shared strategies in the social structure.

A similar phenomenon to the decision-making process is agent decision and action according to a reasonable set of rules aimed at optimizing the agent's own utility (seen in many models). Agents, however, may also act irrationally, or randomly, without optimizing their welfare. Some economists assume that individuals are *hyper-rational*, that is, they assume people always select optimal courses of action. Yet, when real-world actors, such as policymakers, are forced to make decisions that involve environmental or social criteria, their own interests, like economic benefits, may not play as important a role. Herbert Simon expressed this as a distance between managerial practice and models of expected utility (Pomerol and Adam, 2008), a fact that made boundedly rational agents the most applied logic to the design of agents.

Physical objects are usually seen as mediators between agents, establishing a clear divide between non-human elements and human actors. However, in the study of complexity and socio-technical systems, the possible agency of objects themselves – such as artefacts, documents, information systems, laws, technologies, physical tools, among others – should be taken into account. Moreover, interactions between agents are temporally affected by past (real) and future (possible) collective effects of their individual decisions. These phenomena should be more explicitly incorporated into ABM simulations. To this end, ANT (which is presented in the next part of this chapter) may help designers deal with the aforementioned ambiguities and tensions stemming from the existing multiplicity of conceptual framing.

8.3   Actor-Network Theory

ANT is a conceptual framework for exploring socio-technical processes that entail a symmetry between human and non-human actors (referred to as *actants*) in networks (Correa-Moreira, 2011). ANT was developed within the discipline of science and technology studies (STS); in STS, scientists and their colleagues, in addition to the materials and artefacts available to them (documents, laboratory materials, literature and information technologies) are intertwined and constantly reshape each other. Recent examples include the analysis of the role played by the main information system in a national science and technology system (Gonzalez, 2010), the role of network dynamics in the implementation of e-government, and the application of information and communication technologies (ICT) in the public sector of developing countries (Stanforth, 2006).

The concept of an 'actor-network' (A-N) was developed by Michel Callon, Bruno Latour and John Law in the context of STS in the 1980s. A-N, with the two words (and concepts) linked by a hyphen, aims to *bypass* the distinction between agent and structure. That is to say, A-N implies recognition of the fact that actors build networks by combining social and technical elements. Moreover, a network's elements are, simultaneously, constituted and shaped within the network (Stanforth, 2006). ANT, as well as anthropological theories such as Bennett's (1976) human adaptive dynamics cited in Bharwhani (2004), does not isolate individual actions from the relations and connections that give them purpose. Like Giddens' structuration theory, ANT regards actions as framed by institutional and other relations, yet ANT goes further insofar as it disrupts the dichotomy between structure and agency altogether (Steen *et al.*, 2006). Thus, agency is not only assigned to humans, but also to non-humans, which is a *de facto* invitation to conceive of a network as a heterogeneous mix of textual, conceptual, social and technical *actants* (Ritzer, 2004). According to Latour, an agent is '*any thing* that does modify a state of affairs by making a difference' (Latour, 2005, p. 71). Agency is not an essence inherent to humans; rather, it is a capacity realized through the association of actors (human or non-human) and is, consequently, relational, emergent and shifting (Orlikowski, 2007). From an ANT perspective, non-humans are more than passive resources at the disposal of humans: they are active, vibrant agents that also exert power (Dwiartama and Rosin, 2014). Latour goes on to add that 'ANT is not the empty claim that objects do things "instead" of human actors: it simply says that no science of the social can even begin if the question of who and what participates in the action is not first of all thoroughly explored, even though it might mean letting elements in which, for lack of a better term, we would call non-humans. [...] The project of ANT is simply to extend the list and modify the shapes and figures of those assembled as participants and to design a way to make them act as a durable whole' (Latour, 2005, p. 72).

To put ANT into action, the history of the observed phenomenon (with its concomitant micro-phenomena), the main facts (milestones) from which emergent patterns of behaviour have occurred (Méndez-Fajardo and Gonzalez, 2014), should be identified. Using the resulting evolutionary timeline, ANT presents four constitutive elements. First and foremost, the obligatory passage point (OPP) between the problem to be solved and the solution. This can be an A-N (e.g., the *actant* that mobilizes the system through authorizing actions), or a process to be developed by all actors in order to achieve both the system and individual goals. Second, the main A-N in a local network is determined by virtue of its relation to milestones. Third, the main A-N is identified for the global network based on its ability to interfere with the system or even impact the local network if the OPP is weakened. To avoid confusion, readers should note that A-Ns in a local network can interact directly with A-Ns in a global network.

The fourth element, referred to as 'moments of translation', involves describing the dynamics between local and global networks. To wit, moments of translation are how the A-Ns align their interests, how they focus to generate successful action (Callon, 1986; Gonzalez, 2010). These translations constitute the mechanisms through which networks progressively arise around power relations (Stanforth, 2006). Callon (1986) proposed four moments of translation: (1) *problematization*, or how the OPP makes itself indispensable; (2) *interessement*, or how allies are locked into place; (3) *enrolment*, or how roles are defined and coordinated; and (4) *mobilization*, or how the principal A-Ns borrow the force from more passive A-Ns and become their representatives or spokespeople. Moments of translation constitute 'episodes' from which milestones emerge over the course of a part of the history.

Following these episodes, the ways in which A-Ns increase or decrease the gap between individual interests and the OPP's goal can be established. The closing or distancing of this gap affects the aggregation level or attachment in the global network (control over the global) or the level of mobilization in the local network (control over the local).

8.4   Towards an ANT-Based Approach to ABM

Approaches to the design of ABM (Section 8.2.1) allow us to understand networks of actors and their dynamics; however, such approaches do not explicitly account for the role of objects (tools, artefacts, documents, etc.) in networks. For example, the aforementioned theory of planned behaviour could be used to set up consumer interests when deciding whether or not to replace a piece of equipment with a new one. This theory may even apply to the decision regarding disposal of the old equipment. Yet, the influence of the appliance itself is not included in the consumer's decision-making process. One of the main questions remaining unanswered, which is in fact addressed here, is: Perhaps that object has agency?

Likewise, theories of organizational dynamics fail to include the actual role of organizational norms or technological infrastructure (computers and software) in the achievement, or lack thereof, of organizational goals. Agent-based simulation tools, such as SocLab (Sibertin-Blanc et al., 2013), hold that ANT does not place power relations at the heart of their analysis, as, they claim, is the case for Crozier's sociology of organized action (Crozier, 2009). However, in the sociology of translation (Callon, 1986), moments of translation do indeed represent power relations: power relations are shown by the ways in which actor-networks are defined, associated and obligated to remain faithful to their alliances (Stanforth, 2006).

Different studies have integrated ABM into the analysis of social networks (Baber et al., 2013; El-Sayed et al., 2012; Hamill and Gilbert, 2010) or collaborative social networks (Madey et al., 2003). Nevertheless, to date, the studies conducted have not treated non-humans as *actants*, nor have they sufficiently investigated the moments of translations (*problematization, interessement, enrolment, mobilization*) as per the tenets of ANT.

8.4.1   ANT Concepts Related to ABM

ANT proscribes the division between agency and structure, human and non-human or micro-level and macro-level phenomena. ANT is premised on the intertwining of these facets through collective dynamic activity (Ritzer, 2004). In particular, the A-N concept allows model designers to eliminate the human/non-human distinction, in that an actor is itself a network made up of human and material elements. In contrast to other theories supporting the design of ABMs focused on human agency, both human and non-human A-Ns can become agents in ABM. In fact, they need not be distinguished, for the emphasis is placed on networks, not on individuals. As a result, designers can translate directly between human A-N *interessement* and agent decision criteria, although there is a difference when treating non-humans: ANT proposes the agency of artefacts, which can be understood not as 'things making decisions', but rather as 'things generating human decisions' when there is some sort of relationship between them – mainly when non-humans are members of moral and political associations (Sayes, 2014).

In ANT, macro-level phenomena are considered 'more stable' networks (see 'Solid projects' in Figure 8.3 later). In such networks, nodes are semiotically derived, making networks local, variable and contingent (Ritzer, 2004). This can be translated as emergence in ABM. In other words, the mobilization process in ANT can be translated as an emergent pattern, and this translation dynamic in ANT reveals the key elements needed for an A-N to cooperate (*enrolment*) or defect (low degree of attachment in the global network). Cooperation has been linked to the 'shadow of the future' concept (Bó, 2005; Cohen *et al.*, 2001), in which a rational agent expects its present actions to affect the future behaviour of other agents. ANT employs ABM as an explicit framework for the study of interactions between agents, not only in the form of message exchanges, but also as extensions of an agent to their network, thereby redefining the agent in relation to the network. In so doing, ANT provides a theoretical support for the design of ABM in socio-material settings.

8.5   Design Guidelines

To find solutions to real problems, designers use modelling as part of their research methodologies. They seek to represent the real world in order to understand phenomena and contextual factors, such as related issues, actors, actor relations and causes and effects, among others. Models also serve to stimulate discussion and learning processes for stakeholders by explaining phenomena, raising enlightening questions or disciplining the policy dialogue (Epstein, 2008). To understand and describe the context in which a problem manifests, action research and exploratory case study methods are suggested in this chapter as methodological companions to ANT, including, but not limited to, interviews and participatory observation. The following process for applying ANT to the design of an ABM is abstracted from a case study developed below; see Figure 8.2 for more details. The first step entailed the definition of the initial focal problem corresponding to the main problem to be addressed. Building from initial design aspects, the second step requires the identification and review of documents to reconstruct an initial description of the case history, as well as to link actors to each discovered fact, which is the third step.

On the basis of these data (steps 1 and 2), the structure and schedule of interviews must be developed before commencing fieldwork. Interviews should include, at a minimum, the following elements:

(a) Information about the relevant experience of the interviewee, including timeline.
(b) Facts he or she considers to have changed the situation or a specific project's trajectory within the larger case story, including dates (at least years, though hopefully more precise dates), locations, people and organizations that participated, as well as relevant documents, studies, etc.
(c) Ask for opinions about the system's main elements in order to determine the importance of each, according to the interviewee. Include opinions about the system's ideal future, which may help confirm interviewee interests.

To represent the moments of translation (step 4), a graph is invaluable; see, for instance, the graph in Figure 8.3. Changes in the path mean that some milestone has occurred and generated important changes in the composition of the local and global networks.

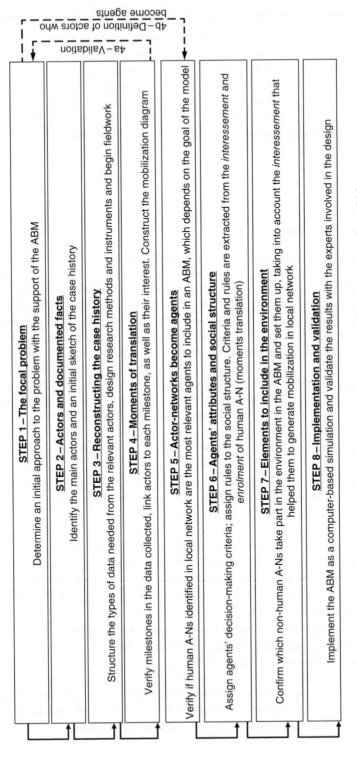

STEP 1 – The focal problem

Determine an initial approach to the problem with the support of the ABM

STEP 2 – Actors and documented facts

Identify the main actors and an initial sketch of the case history

STEP 3 – Reconstructing the case history

Structure the types of data needed from the relevant actors, design research methods and instruments and begin fieldwork

STEP 4 – Moments of translation

Verify milestones in the data collected, link actors to each milestone, as well as their interest. Construct the mobilization diagram

STEP 5 – Actor-networks become agents

Verify if human A-Ns identified in local network are the most relevant agents to include in an ABM, which depends on the goal of the model

STEP 6 – Agents' attributes and social structure

Assign agents' decision-making criteria; assign rules to the social structure. Criteria and rules are extracted from the *interessement* and *enrolment* of human A-N (moments translation)

STEP 7 – Elements to include in the environment

Confirm which non-human A-Ns take part in the environment in the ABM and set them up, taking into account the *interessement* that helped them to generate mobilization in local network

STEP 8 – Implementation and validation

Implement the ABM as a computer-based simulation and validate the results with the experts involved in the design

4a – Validation

4b – Definition of actors who become agents

Figure 8.2 Abstraction of steps for applying ANT to the design of an ABM.

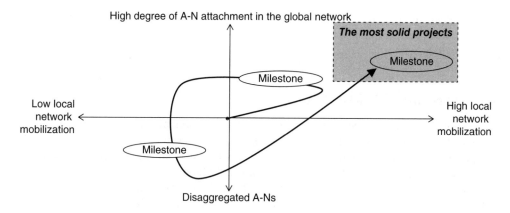

Figure 8.3 Graph of the mobilization of A-Ns in local and global networks. *Source:* Adapted from Méndez-Fajardo and Gonzalez (2014), Stanforth (2006).

As this figure shows, the most solid projects are those in which both higher degree of A-N attachment in a global network and high local network mobilization occur. In such situations, the interests of all actors are strongly aligned, and the result is a successful project (programme, strategy, action). Looking at the same graph, and before deciding which A-Ns become agents (step 5), each milestone must be documented, along with which actors constitute the local network. Then, these A-Ns are selected based on the two most solid moments (i.e., moments of translation in which the A-Ns are both more aggregated and more highly mobilized – more solid moments appear in the upper-right quadrant of the mobilization diagram.

In order to define which A-Ns should be included in ABM, the focal problem should be adjusted via a validation of milestones and identified A-Ns (step 4a), a process carried out with the participation of relevant actors. As a result of this adjustment, we can establish the list of A-Ns to be included (step 4b), allowing for the definition of agent attributes (step 6). To design decision-making aspects, we identify the criteria within the moments of translation (*interessement* and *enrolment*), especially with respect to prioritization. It behoves us to mention that some actors may falsely express their most important need to keep up an image with the rest of the actors.

In step 7 we can see that some non-human A-Ns will form part of the environment (e.g., as physical compounds). The results of steps 6 and 7 are the rules that agents should follow in the model.

To easily validate the ABM as the last part of the methodology proposed (step 8), the model should be implemented as a computer-based simulation. Doing so facilitates evaluation of the rules (through experiments) included in the design by observing the resultant emergent patterns. In this step, it becomes especially relevant to include the participation of experts, in addition to that of the actors involved in the case.

The application of ANT to particular subjects can be useful when looking to discover the potential of ANT for socio-technical studies. In the next part of this chapter, an exploratory case study on WEEE management in Colombia is presented and interpreted using this theory.

Nevertheless, the complete ABM, and its implementation and validation (step 8), are not included here, for our main goal is to debate the possibilities opened up by ANT's conceptualization of design for an ABM.

8.6 The Case of WEEE Management

In developing countries, institutions responsible for territorial planning are faced with rapid urbanization and the concomitant solid waste management problems (Guerrero *et al.*, 2013). As part of the municipal solid waste, WEEE[1] has taken on added importance due to the potential risks to public health and the environment, which can be traced to the toxic components they contain, such as heavy metals (Ongondo and Williams, 2011; Wäger *et al.*, 2011; Widmer *et al.*, 2005). To tackle the negative impact of poor WEEE management, policymakers tend to make decisions from a limited perspective, generally restricted to technical or economic outlooks, while neglecting a systems-based approach that involves additional dimensions (e.g., social, legal, ecological, political and even cultural dimensions) (Marshall and Farahbakhsh, 2013), as well as different management processes, different interests of all actors involved and the cause–effect phenomenon (Méndez-Fajardo and Gonzalez, 2014). In contrast, the principle of extended producer responsibility (EPR) has been widely implemented in developed countries, and EPR requires a systems approach since it demands the participation, cooperation and coordination of all stakeholders (Agamuthu and Victor, 2011; Mayers, 2007). To design more sustainable strategies for EPR in both developed and developing countries, the effects of past decisions should be analysed in an attempt to understand the processes that have generated successful (or at least differential) actions in the past.

WEEE management can be characterized as a typical socio-technical system, insofar as it demands technical artefacts to achieve material goals and is strongly affected by human attitudes and decisions. Therefore, ANT helps us to learn from past WEEE management experiences, identifying the key elements for the design of a more sustainable WEEE management and identifying key elements influencing decision-making, so as to increase sustainability in future actions.

The research herein was carried out in the South American nation of Colombia as an exploratory single-case study to study current phenomena in a real-world context (Maguire *et al.*, 2010; Yin, 2003a,b). The Colombian study is embedded: it includes two units of analysis (Yin, 2003b). These two units are decision-making roles and operational roles within Colombian WEEE management, respectively. The main purpose of the case study was to ascertain whether

---

[1] According to the European WEEE Directive, 'electrical and electronic equipment' or 'EEE' means equipment which is dependent on electric currents or electromagnetic fields in order to work properly and equipment for the generation, transfer and measurement of such currents and fields and designed for use with a voltage rating not exceeding 1000V for alternating current and 1500V for direct current (The European Parliament & The Council on Waste Electrical and Electronic Equipment, 2012). The categories of WEEE can be distinguished as follows: (i) Large household appliances; (ii) Small household appliances; (iii) IT and telecommunications equipment; (iv) Consumer equipment; (v) Lighting equipment; (vi) Electrical and electronic tools (with the exception of large-scale stationary industrial tools); (vii) Toys, leisure and sports equipment; (viii) Medical devices (with the exception of all implanted and infected products); (ix) Monitoring and control instruments; and (x) Automatic dispensers (The European Parliament & The Council on Waste Electrical and Electronic Equipment, 2012).

or not decisions regarding WEEE management in Colombia were made using systemic processes. A corollary goal was to investigate whether or not decision-makers applied a systems-based approach to their own decisions.

8.6.1 Contextualizing the Case Study

As per assessments, in 2013 the generation of WEEE (aggregated EU categories; see footnote 1) in Colombia was around 120,000 tons, which corresponds to 2.5 kg/inhabitant/year. The breakdown of Colombian WEEE was as follows: large household appliances (24%), IT and telecommunications equipment (16%), consumer equipment (38%) and lighting equipment (13%). Although not explicitly included in these categories, Colombia has implemented a post-consumer programme for batteries (9%).

The main related processes inside and outside Colombia are shown in Figure 8.4. In this figure, we see that production involves international producers and importers of electrical and electronic equipment (EEE), as well as local (domestic) assemblers (which account for less than 10% of EEE). When obsolete EEE is discarded, it becomes WEEE. Some EEE pre-treatment materials are recovered locally or disposed of in local (regional) landfills, while post-treatment elements are exported and processed in other countries. In the same graph, primary distribution refers to the sale of new equipment – either imported or locally assembled – that could be developed directly by the companies producing them, or by large and small market chains. In addition, disassembling, also called dismantling (León, 2010), is the manual process by which devices or parts (e.g., CD drives, memory cards, etc.) are separated for subsequent material recovery (such as ferrous metals).

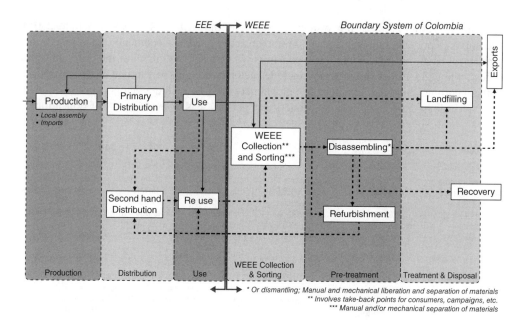

Figure 8.4 WEEE management processes in Colombia.

The processes illustrated in this figure are restricted to the formal system (i.e., processes authorized by the environmental authority). In Colombia, formal systems only exist for waste from computers, batteries, lighting and mobile phones; however, unknown amounts of small (irons, hairdryers, blenders, etc.) and large household appliances (refrigerators, washing machines, etc.) have also been pre-treated in an informal chain. Informality is present in most waste management activities of developing countries (Chi *et al.*, 2001; Guerrero *et al.*, 2013): equipment is dismantled by pounding the objects against the ground in public areas, private households or warehouses (León, 2010; Streicher-Porte *et al.*, 2005; Widmer *et al.*, 2005). These activities increase public health and environmental risks.

8.6.2 ANT Applied to WEEE Management in Colombia

The historical evolution of WEEE management in Colombia is described in Figure 8.5. It is important to highlight the relevance of regulations and working groups or committees in this evolution.

Generic human A-Ns in WEEE management refer to producers, distributors, consumers, recyclers and the government at the national, regional and/or local levels (including national and international technical support, an innovative legal strategy). National governmental organizations include the Ministry of Environment and Sustainable Development (MESD), the Ministry of Health and Social Protection, the Ministry of Information and Communication Technologies and the Ministry of Commerce, Industry and Tourism. Informal recycling is not legally included, despite its relevance as an A-N in the system. In the same vein, associations that bring together producers, distributors or recyclers, such as FENALCO (National Federation of Traders) for dealers or ANDI (Industry Group of Colombia) for producers must also be considered A-Ns.

Table 8.1 shows A-Ns in the system. To reiterate, the term A-N in ANT (Latour, 2005) encompasses not only human but also non-human actors, such as documents and laws. For the present research, local networks involved actors who participated in all of the different 'episodes', while the global network only includes actors who took part in one or two episodes, without current active participation. The global network also includes actors expected to play an active role in the system (e.g., consumers as a generic A-N or the Ministry of Education).

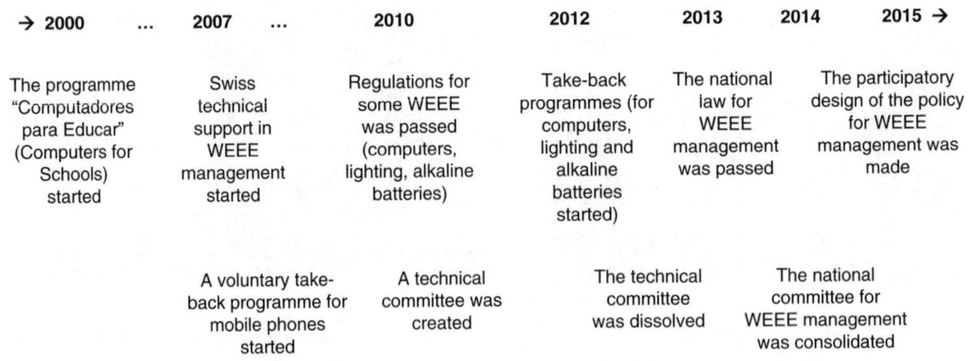

→ 2000	...	2007	...	2010	2012	2013	2014	2015 →
The programme "Computadores para Educar" (Computers for Schools) started		Swiss technical support in WEEE management started		Regulations for some WEEE was passed (computers, lighting, alkaline batteries)	Take-back programmes (for computers, lighting and alkaline batteries started)	The national law for WEEE management was passed	The participatory design of the policy for WEEE management was made	
		A voluntary take-back programme for mobile phones started		A technical committee was created		The technical committee was dissolved	The national committee for WEEE management was consolidated	

Figure 8.5   The general timeline of WEEE management in Colombia.

Table 8.1   Actor-networks in Colombia's WEEE management system*

Ministry of the Environment and Sustainable Development (MESD)	Computers for Schools programme and CENARE (2000–present)
Ministry of Commerce, Industry and Tourism	Mobile phone service operators
Ministry of Education	Post-consumer programmes to collect computers, lighting and batteries (2012)
Ministry of Social Protection	Voluntary agreement regarding the collection and management of mobile phones (2010)
Regulations for the management of computers, lighting and batteries (2010)	Recyclers (formal and informal)
National Law for WEEE management (2013)	National Cleaner Production Center (CNPML)
Colombian Chamber of Informatics and Telecommunications (CCIT)	Academia
Decree on Hazardous Waste Management in Colombia (2005)	Swiss Federal Laboratory for Materials Sciences (EMPA)/State Secretariat for Economic Affairs (SECO)
FENALCO (National Federation of Traders)	Assessments of WEEE management in Colombia (2008–2010)
ANDI (Industry Group of Colombia)	World Resources Forum (WRF)
National WEEE Committee (NWC) (2014)**	Consumers

* Documents, groups, programmes, laws and regulations are assigned the date of their creation.
** Involves government representatives, producers, distributors, recyclers and experts.

Explicit relationships or links between all actors are not displayed due to the complexity of meaningful presentation in the graph. Furthermore, relationships change, have different strengths, include feedback loops and result in high interaction density, such that any presentation of all links would be subject to modification. Nevertheless, in the description of the moments of translation, the main relations between actors are explicitly stated.

8.6.2.1   Moments of Translation

The structured interviews of relevant actors included questions about the most important facts in Colombia's WEEE management history, and which actors were involved at each stage. Based on interviews and a thorough document review (e.g., reports and assessments), the milestones (or mobilization moments) in the evolution of the WEEE management system and the A-N involved for each milestone were identified (Figure 8.6). The system's OPP was determined to be the MESD.

The starting point (Milestone 0 in Figure 8.6) was the creation of the Computers for Schools programme in 2000. In the following years, two parallel courses were charted: on the one hand, industry mobilization; on the other hand, government mobilization. As for the former, industry interest in donating computers was driven by four benefits: (1) the social and community benefits of improving technological access for children and their future development options; (2) the positive environmental impact of preventing toxic compounds

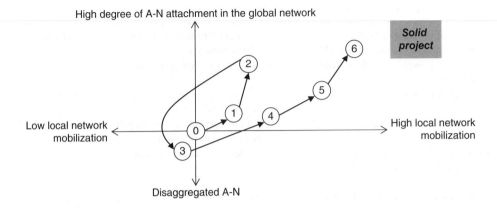

High degree of A-N attachment in the global network

Solid project

Low local network mobilization

High local network mobilization

Disaggregated A-N

Milestone	General Description
0	Computers for Schools programme was created
1	The Switzerland-Colombia cooperation started
2	Computer and mobile-phone working groups started
3	Regulations for computers and lighting were passed
4	The post-consumer programmes *EcoComputo, Lumina* and *Pilas con el Ambiente* were implemented
5	The national law for WEEE management was passed
6	Increased involvement of stakeholders in the participatory design of national policy on WEEE management

Figure 8.6 Mobilization of actors in local and global networks.

from being disposed of in sanitary landfills; (3) the economic incentives stemming from not having to pay for the management of this waste; and (4) the financial savings stemming from new processes adapted to comply with internal environmental management strategies.

With regard to the last point, the government began to design regulations for hazardous waste management in the face of mounting pressure related to international agreements such as the Basel Convention (1992) or the Kyoto Protocol (1997), as well as the pressure generated by domestic laws, such as the National Environmental and Natural Resources Management Law (1993) and the National Hazardous Waste Law passed in 2005. This 2005 law was instrumental in spurring industry action. As a result, an agreement was forged between the Swiss and Colombian governments, ratified in 2007 (Milestone 1 in Figure 8.6); this agreement called for the participation of the most relevant A-Ns. Local network mobilization increased as a function of the *interessement* or attraction of interests which was, for the OPP, the prevention of negative environmental impacts, compliance with hazardous waste laws and international sustainability accords and, for the Swiss Federal Laboratory for Materials Sciences

(EMPA) and the State Secretariat for Economic Affairs (SECO), assisting in WEEE management in developing countries. The Swiss initiative had previously been carried out in China, India and South Africa through the 'Swiss e-Waste' programme; its focus turned to Colombia and Peru from 2007 onwards. During the first two years of this collaboration, not only did local network mobilization of A-Ns increase, but so too did global network attachment. Results included the creation of the computer and mobile-phone working groups (Milestone 2 in Figure 8.6) and assessments of EEE and subsequent WEEE. These technical documents (assessments) were critical for WEEE-related decision-making in Colombia (2008–2013).

In 2009, the local network visited Switzerland as part of a 'study tour' of the Swiss WEEE management locations with an eye towards obtaining primary information to be used in Colombian WEEE management design strategies. Nevertheless, after this trip, the working groups fell into an internal crisis caused by divergent member interests. As a result, global network A-N aggregation and local network mobilization decreased; despite the emergence of this crisis, we witness one especially important positive outcome – the MESD passed regulations (2010) to achieve mandatory collection rates of computers and lighting (Milestone 3 in Figure 8.6).

A leadership shift, which saw producer representation move to ANDI, in combination with a change in the composition of the local network, eventually led to increased mobilization and participation. Three post-consumer programmes for collecting waste from computers (*EcoComputo*), lighting (*Lumina*) and alkaline batteries (*Pilas con el Ambiente*) emerged as a result (Milestone 4 in Figure 8.6). The *interresement* to achieve this milestone was based on two targets: producer legal compliance and avoidance of penalties and economic opportunities for authorized recyclers, which represented financial savings for producers. These developments are different from the experience of the second working group, for the mobile-phone working group engendered a voluntary agreement by some mobile-phone service operators to set up take-back points for consumer equipment (i.e., to collect mobile-phone WEEE from subscribers or users of the telephone service). The equipment collected was then passed on to authorized recyclers.

The National WEEE Management Law was passed in 2013 (Milestone 5 in Figure 8.6), consolidating the process begun in 2010 by the A-N in the local network and spearheaded by the OPP. The passage of this law laid the foundations for the creation of the National WEEE Committee (2014). Implementation of the regulations passed from 2010 on has demonstrated the importance of public laws when it comes to increasing rates of WEEE collection. The dynamics of post-consumer programmes proved to be an instructive learning process to avoid (when possible) failures in the implementation of the system.

However, the system still has not reached stability: currently, the MESD – supported by the CNPML, EMPA/SECO and the Javeriana University (Bogota, Colombia) – is designing the instruments and regulations to implement the law and establishing a control system. The National WEEE Committee has been involved in this design process, above all as pertains to policy design (Milestone 6 in Figure 8.6), in order to constantly encourage interest in and motivate the rest of the actors in the local network (A-N *enrolment*). The use of participatory methodologies has allowed us to identify the causes and effects of the current insufficient WEEE management system in Colombia, as well as identify the relations among causes, define structural causes, design strategies and the policy's action plan. Likewise, it has increased confidence and motivation in A-Ns, in turn strengthening the local network.

To recap, the most relevant milestone (the most solid project in Figure 8.6) has been the passing of laws, given that these laws also include previous episodes aimed at aligning the interests of relevant actors. Indeed, the implementation of regulations has illustrated the importance of mandatory strategies in terms of increasing WEEE collection rates and more responsible WEEE management. In addition to these regulations, technical and methodological support has played a key role in getting the focal actor, MESD, to take crucial actions. That being said, there are other elements observed during fieldwork that should not be ignored when implementing sustainable WEEE management: (1) actors in local networks should designate the same representative for the duration of the process; (2) this representative should take an active interest in the topic, rather than view it as an obligation; (3) for non-human A-Ns, detailed reports should be made of all strategies designed and implemented, including key elements (actors, actions, type of WEEE, failures, successes, possible future problems and consequences, etc.); and (4) it is crucial to have complementary strategies to ensure project sustainability on schedule and the availability of resources.

8.6.3   ANT–ABM Translation Based on the Case Study

In ANT, human and non-human A-Ns are ontologically understood as anything that modifies another A-N through a series of *actions*; they are referred to as *actants* (by Latour, 2005), whereas *actor* has been used exclusively to refer to humans. In the same sense, an *agent* in ABM is anyone who makes decisions and acts accordingly. As this chapter has demonstrated, the human A-Ns in local and global networks may become agents, while non-human A-Ns may form part of the ABM environment, in spite of the fact that their setup should feature mobilized actions in the (ANT) networks.

The case study has clearly shown that regulations (incentives or penalties) and take-back mechanisms (post-consumer programmes in Table 8.1) should be incorporated into the environment. Furthermore, decision-making processes in agents should be related to these two aspects. In this model, regulations, take-back mechanisms and educational programmes are included in the design of *sustainability strategies* (see Figure 8.7).

The relevant actors, although they should become agents, may not be explicitly accounted for in the ANT analysis, depending on the goal. For example, the goal of the exploratory case study in WEEE management was focused on policymakers and strategies implemented over the last few years, essentially ignoring consumer behaviours related to these strategies. Given that WEEE recovery is the primary common interest of WEEE management actors, consumers play an important role, so they should be included in the model, regardless of whether they were explicitly included or not in the local networks. However, this is a question of boundary and scope, which is not addressed directly by the present proposal, beyond the suggestions of ANT analysis in terms of what is meaningful for mobilizing and aligning A-Ns.

The moments of translation become the relationships and rules within the social structure and the criteria and weights for agent decision-making. For instance, in light of ANT analysis results, producers decided to participate (that is, create a post-consumer mechanism), for it ensured legal compliance, which was the criterion assigned the most significant weight (i.e., interest). Hence, the passage and enforcement of the new law highlights the relationship between producers and the government, represented in this case by the environmental authority

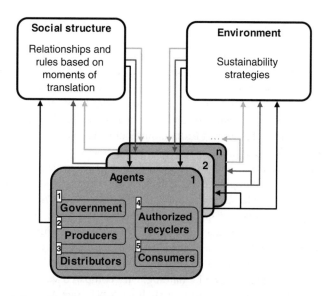

Figure 8.7   ABM elements based on the application of ANT for the case study.

(MESD). In this example, the law's crafting, its publication and the instruments to ensure compliance are part of the *interessement* used by the system's OPP.

The interests of human A-Ns (that is, their criteria) and the level of participation in global and local networks in relation to the milestones were identified and taken from the ANT analysis to be used in the design of two parts of the ABM: on the one hand, the processes of the decision-making tools in agents and on the other hand, the relations and rules in the social structure (Figure 8.7). These data are based on local/global networks and mobilization dynamics (Figure 8.6).

8.6.4   Open Issues and Reflections

While not every ABM is destined to support real-world interventions, the aim of the present research has been to properly apply ANT to the design of an ABM that can be used to enact policy changes. Therefore, the interest is not to endow ABM with theoretical validity or provide a conceptual framework for ABM to make theoretical contributions. Rather, the goal here is utility (i.e., to be pragmatic with a focus on problem-solving). The connection between ANT and pragmatism has already been discussed: according to Rimpiläinen (2013), ANT and pragmatism undermine the distinction between subjects and objects, given that knowledge emerges from transactions (Dewey and Bentley, 1949) or events and practices (Mol, 1999, 2005). On the basis of the recognition that knowledge emerges from practice, this chapter has sought not to build theory, but to apply ANT in ABM design, framing it as a heuristic for intervention.

Multiple studies have evinced the relevance of using decision-support systems with ABM in policy design. The present research contributes further evidence of its relevance by emphasizing the importance of including past milestones documented through action research with the principal actors (ideally policymakers) and by involving the agencies of non-humans. Armed with an understanding of these dynamics arrived at via ABM, designers and policy-makers can form more successful alliances in order to achieve more sustainable solutions.

The ABM designed using ANT has been implemented in a computational tool that supports decision-making processes for WEEE management. In such processes, the agency of regula-tions (such as policies) and technical artefacts (such as take-back mechanisms) merit further discussion. Both the computational tool and participatory validation in a decision-making exercise represent the subject matter of forthcoming publications.

One open issue worth highlighting is a deeper engagement with the philosophy of works such as Latour's actor-network theory (Latour, 2005) in order to delineate more clearly the agency of objects from an ontological point of view. A good example would be a close reading of Barad's agential realism (Barad, 1996, 2003, 2007), which holds that 'the adjectival form of the word "agency" modifies and specifies the form that realism takes here, in defiance of traditional forms of realism that deny any active participation on the part of the knower. Agency is a matter of intra-acting, that is, agency is an enactment, it is not something someone has' (Barad, 1996, p. 183). Similarly, the socio-materiality developed by Orlikowski (2007) or Leonardi (2013), among others, could be ontologically compared with the A-N concept. For example, in the proposed methodology, which links ANT and ABM, we assigned some agencies to non-humans (think of take-back mechanisms); this lines up with the logic of the agential realism as a pragmatic decision. However, further studies are needed to extend this pragmatic decision, to position the 'agency' concept within other ontological discussions.

A last open issue is related to this ontological approach: the importance of reviewing the goal-seeking concept in the management science contribution of Simon (1955, 1997) and purposeful design (Kroes, 2012). Both theories should be applied to the artefact design process (e.g., take-back mechanisms in WEEE management). In this regard, the ontology of the objects, a foundational concept in ANT, has been studied by some contemporary philosophers, but many elements remain linked to environmental systems, considered as socio-technical systems.

8.7 Conclusions

This chapter proposes the use of ANT as an innovative theoretical approach to the design of an ABM for socio-material contexts. In so doing, we employed ANT theoretical foundations (i.e., the representation of historical trajectories in socio-technical systems). This allowed for the identification of networks of human and non-human actors, as well as their past interaction dynamics, which can either mobilize and align or distance these actors. Furthermore, this chapter relies on ANT to provide an initial advance in the achievement of design guidelines that express the way that results of ANT-based studies can be developed further for ABMs.

An exploratory case study in the context of WEEE provides support for applying the abstracted process described here to design. ANT and agent-based modelling may be combined in cases where non-human agency is relevant and merits inclusion in the resulting ABMs. Problematic situations for which this proposal could facilitate a generalization of the method are: industrial processes in which people and machines are continuously interacting; the design and construction of physical infrastructure that may generate social and environmental impacts (e.g., bridges, highways, prisons or sanitary landfills), among others. Therefore, linking ANT and agent-based modelling can strengthen the ability of ABMs and other simulations in terms of informing decision-making processes and policymakers where interactions between human and non-human actors generate emergent collective behaviour. In addition, by incorporating an ANT approach to the reconstruction of historical trajectories, we can further explore and/or explain previously identified collective patterns that, in socio-technical network terms, are aligned or misaligned. Extensions of the present work include the implementation of computational agent-based simulations, as well as empirical validation of the decision-making enhancement that they provide. Future work may also look to deepen the dialogue with *socio-materiality* as an evolving topic in information system design and practice.

References

Agamuthu, P. and Victor, D. (2011) Policy trends of extended producer responsibility in Malaysia RID B-8145-2010. *Waste Management & Research*, **29**(9), 945–953.

Ajzen, I. (1991) The theory of planned behavior. *Organizational Behavior and Human Decision Processes*, **50**(2), 179–211.

Baber, C., Stanton, N.A., Atkinson, J., McMaster, R. and Houghton, R.J. (2013) Using social network analysis and agent-based modelling to explore information flow using common operational pictures for maritime search and rescue operations. *Ergonomics*, **56**(6), 889–905.

Bandura, A. (1989) Human agency in social cognitive theory. *American Psychologist*, **44**(9), 1175–1184.

Barad, K. (1996) Meeting the universe halfway: Realism and social constructivism without contradiction, in L.H. Nelson and J. Nelson (eds), *Feminism, Science, and the Philosophy of Science*, Vol. **256**, Springer-Verlag, Amsterdam.

Barad, K. (2003) Posthumanist performativity: Toward an understanding of how matter comes to matter. *Signs*, **28**(3), 801–831.

Barad, K. (2007) *Meeting the Universe Halfway: Quantum physics and the entanglement of matter and meaning*, Duke University Press Books, Durham, NC.

Bennett, J.W. (1976) *The Ecological Transition*, Transaction Publishers, New York, NY.

Bharwhani, S. (2004) Adaptive knowledge dynamics and emergent artificial societies: Ethnographically based multi-agent simulations of behavioural adaptation in agro-climatic systems, PhD thesis, University of Kent, Canterbury.

Bó, P.D. (2005) Cooperation under the shadow of the future: Experimental evidence from infinitely repeated games. *American Economic Review*, **95**(5), 1591–1604.

Callon, M. (1986) Some elements of a sociology of translation: Domestication of the scallops and the fishermen of St Brieuc Bay. *Sociological Review Monograph, Vol. 32*.

Chi, X., Streicher-Porte, M., Wang, M.Y.L. and Reuter, M.A. (2011) Informal electronic waste recycling: A sector review with special focus on China. *Waste Management*, **31**(4), 731–742.

Cohen, M.D., Riolo, R.L. and Axelrod, R. (2001) The role of social structure in the maintenance of cooperative regimes. *Rationality and Society*, **13**(1), 5–32.

Correa-Moreira, G.M. (2011) El concepto de mediación tecnológica en Bruno Latour. Una aproximación a la Teoría del Actor Red. *Psicología, Conocimiento y Sociedad*, **2**(1), 54–79.

Crozier, M. (2009) *The Bureaucratic Phenomenon*, Transaction Publishers, New Brunswick, NJ.

Crozier, M. and Friedberg, E. (1983) Actors and systems. The politics of collective action. *Érudit Relations Industrielles*, **38**(2), 448–452.

Dewey, J. and Bentley, A.F. (1949) *Knowing and the Known*, Beacon Press, Boston, MA.

Dwiartama, A. and Rosin, C. (2014) Exploring agency beyond humans: The compatibility of actor-network theory (ANT) and resilience thinking. *Ecology and Society*, **19**(3), 28.

El-Sayed, A.M., Scarborough, P., Seemann, L. and Galea, S. (2012) Social network analysis and agent based modeling in social epidemiology. *Epidemiologic Perspectives & Innovations: EP + I*, **9**(1).

Epstein, J.M. (2008) Why to model? *Journal of Artificial Societies and Social Simulation*, **11**(4), 12.

Epstein, J. and Axtell, R. (1996) *Growing Artificial Societies: Social science from the bottom up*, MIT Press, Cambridge, MA.

Feola, G. and Binder, C.R. (2010) Towards an improved understanding of farmers' behaviour: The integrative agent-centred (IAC) framework. *Ecological Economics*, **69**(12), 2323–2333.

Fuchs, C. (2003) Structuration theory and self-organization. *Systemic Practice and Action Research*, **16**(2), 133–167.

Ghorbani, A., Bots, P., Dignum, V. and Dijkema, G. (2012) MAIA: A framework for developing agent-based social simulations. *Journal of Artificial Societies and Social Simulation*, **16**(2), 9.

Gilbert, N. (2007) *Agent-Based Models*, Sage Publications, Thousand Oaks, CA.

Gilbert, N. and Troitzsch, K.G. (2005) *Simulation for the Social Scientist* (2nd edn), Open University Press, Berkshire.

Gonzalez, R.A. (2010) Sistema de información ScienTI Artefacto central del Sistema Nacional de Ciencia y Tecnología: COLCIENCIAS, in *Colciencias cuarenta años entre la legitimidad, la normatividad y la práctica*, Observatorio Nacional de Ciencia y Tecnología, Universidad Nacional de Colombia, Universidad del Rosario, Colombia.

Guerrero, L.A., Maas, G. and Hogland, W. (2013) Solid waste management challenges for cities in developing countries. *Waste Management*, **33**(1), 220–232.

Hamill, L. and Gilbert, N. (2010) Simulating large social networks in agent-based models: A social circle model. *Emergence: Complexity & Organization*, **12**(4), 78–94.

Held, D. and Thompson, J.B. (2008) *Social Theory of Modern Societies: Anthony Giddens and his critics*, Cambridge University Press, Cambridge.

Jensen, M.C. and Meckling, W. (1976) Theory of the firm: Managerial behavior, agency costs and ownership structure. *Journal of Financial Economics*, **3**(4), 305–360.

Knoeri, C., Binder, C.R. and Althaus, H.J. (2010) An agent operationalization approach for context specific agent-based modeling. *Journal of Artificial Societies and Social Simulation*, **14**(2), 4.

Kroes, P. (2012) *Technical Artefacts Creations of Mind and Matter: A philosophy of engineering design*, Springer-Verlag, Dordrecht.

Latour, B. (2005) *Reassembling the Social: An introduction to actor-network theory*, Oxford University Press, Oxford.

León, J. (2010) *Modelling Computer Waste Flows in the Formal and the Informal Sector – A case study in Colombia*, Swiss Federal Institute of Technology (EPFL)/Swiss Federal Laboratories for Materials Testing and Research (EMPA), Lausanne/St. Gallen.

Leonardi, P.M. (2013) Theoretical foundations for the study of sociomateriality. *Information and Organization*, **23**(2), 59–76.

Macy, M. and Willer, R. (2002) From factors to actors: Computational sociology and agent-based modelling. *Annual Review of Sociology*, **28**(1), 143–166.

Madey, G., Gao, Y., Freeh, V., Tynan, R. and Hoffman, C. (2003) Agent-based modeling and simulation of collaborative social networks, in *Association for Information Systems*, AIS Electronic Library (AISeL).

Maguire, S., Ojiako, U. and Said, A. (2010) ERP implementation in Omantel: A case study. *Industrial Management & Data Systems*, **110**(1), 78–92.

Marshall, R.E. and Farahbakhsh, K. (2013) Systems approaches to integrated solid waste management in developing countries. *Waste Management*, **33**(4), 988–1003.

Mayers, C.K. (2007) Strategic, financial, and design implications of extended producer responsibility in Europe: A producer case study. *Journal of Industrial Ecology*, **11**(3), 113–131.

Méndez-Fajardo, S. and González, R.A. (2014) Actor-network theory on waste management: A university case study. *International Journal of Actor-Network Theory and Technological Innovation*, **6**(4), 13–25.

Mol, A. (1999) Ontological politics. A word and some questions. *Sociological Review*, **47**(S1), 74–89.

Mol, A. (2005) The body multiple: Ontology in medical practice. *Acta Sociologica*, **48**(3), 266–268.

Ongondo, F.O. and Williams, I.D. (2011) Mobile phone collection, reuse and recycling in the UK. *Waste Management*, **31**(6), 1307–1315.

Orlikowski, W.J. (2007) Sociomaterial practices: Exploring technology at work. *Organization Studies*, **28**(9), 1435–1448.

Ostrom, E. (2005) *Understanding Institutional Diversity*, Princeton University Press, Princeton, NJ.

Pomerol, J.C. and Adam, F. (2008) Understanding human decision making – a fundamental step towards effective intelligent decision support, in D.G. Phillips-Wren, D.N. Ichalkaranje and P.L. Jain (eds), *Intelligent Decision Making: An AI-based approach*, Springer-Verlag, Berlin, pp. 3–40.

Railsback, S.F. and Grimm, V. (2011) *Agent-Based and Individual-Based Modeling: A practical introduction*, Princeton University Press, Princeton, NJ.

Rimpiläinen, S. (2013) Knowledge in networks: Knowing in transactions?, in *Social and Professional Applications of Actor-Network Theory for Technology Development*, Information Science Reference, Victoria University, Australia, pp. 46–56.

Ritzer, G. (ed.) (2004) *Encyclopedia of Social Theory* (1st edn), Sage Publications, Thousand Oaks, CA.

Ross, S.A. (1973) The economic theory of agency: The principal's problem. *American Economic Review*, **63**(2), 134–139.

Sanchez, S.M. and Lucas, T.W. (2002) Exploring the world of agent-based simulations: Simple models, complex analysis. *Proceedings of the 2002 Winter Simulation Conference*, San Diego, CA.

Sayes, E. (2014) Actor-network theory and methodology: Just what does it mean to say that nonhumans have agency? *Social Studies of Science*, **44**(1), 134–149.

Sibertin-Blanc, C., Roggero, P., Adreit, F., Baldet, B., Chapron, P., El-Gemayel, J. and Sandri, S. (2013) SocLab: A framework for the modeling, simulation and analysis of power in social organizations. *Journal of Artificial Societies and Social Simulation*, **16**(4), 8.

Simon, H.A. (1955) A behavioral model of rational choice. *Quarterly Journal of Economics*, **69**(1), 99–118.

Simon, H.A. (1997) *Administrative Behavior* (4th edn), Free Press, New York, NY.

Stanforth, C. (2006) Using actor-network theory to analyze e-government implementation in developing countries. *Information Technologies & International Development*, **3**(3), 35–60.

Steen, J., Coopmans, C. and Whyte, J. (2006) Structure and agency? Actor-network theory and strategic organization. *Strategic Organization*, **4**(3), 303–312.

Streicher-Porte, M., Widmer, R., Jain, A., Bader, H.-P., Scheidegger, R. and Kytzia, S. (2005) Key drivers of the e-waste recycling system: Assessing and modelling e-waste processing in the informal sector in Delhi. *Environmental Impact Assessment Review*, **25**(5), 472–491.

Sycara, K., Pannu, A., Williamson, M., Zeng, D. and Decker, K. (1996) Distributed intelligent agents. *IEEE Expert-Intelligent Systems and Their Applications*, **11**(6), 36–46.

The European Parliament & The Council on Waste Electrical and Electronic Equipment (2012) Directive 2012/19/EU.

Triandis, H.C. (1979) Values, attitudes, and interpersonal behaviour. *Nebraska Symposium on Motivation*, Vol. **27**, pp. 195–259.

Wäger, P.A., Hischier, R. and Eugster, M. (2011) Environmental impacts of the Swiss collection and recovery systems for waste electrical and electronic equipment (WEEE): A follow-up. *The Science of the Total Environment*, **409**(10), 1746–1756.

Widmer, R., Oswald-Krapf, H., Sinha-Khetriwal, D., Schnellmann, M. and Böni, H. (2005) Global perspectives on e-waste. *Environmental Impact Assessment Review*, **25**(5), 436–458.

Yin, R.K. (2003a) *Applications of Case Study Research* (3rd edn), Sage Publications, Thousand Oaks, CA.

Yin, R.K. (2003b) *Case Study Research: Design and methods* (3rd edn), Sage Publications, Thousand Oaks, CA.

Zambonelli, F., Jennings, N.R. and Wooldridge, M. (2003) Developing multiagent systems: The Gaia methodology. *ACM Transactions on Software Engineering and Methodology (TOSEM)*, **12**(3), 317–370.

9

Engineering the Process of Institutional Innovation in Contested Territory

Russell C. Thomas and John S. Gero

9.1 Introduction

Typically, when we talk about an 'engineering approach' to a given social system, we are referring to the transformation of the current state of that social system to some future state that meets needs, requirements or goals. The 'end product' of social systems engineering is the transformed social system. But what if the social system is an innovation community and the 'end product' is innovation or discovery of new knowledge? In this case, we are no longer applying an established body of knowledge and methods to solve a given problem, but instead the social system is trying to expand the state of knowledge and generate inventions that are novel and distinctly beneficial. Given the fundamental uncertainties of innovation and discovery, we can't reliably predict the end results or even the trajectory of innovation that will unfold.

Further, we consider settings that are *contested territory* (i.e., there are rival world views regarding the nature of the problems and the nature of the innovations needed to address those problems). With two or more rival views ('schools of thought'), there might be diverging innovation trajectories where each school of thought develops and exploits knowledge that fits that school but not rival schools. For an example of this in scientific knowledge, consider the rival theories of infectious disease in the early to mid-1800s: the 'germ theory' versus the long-established 'miasma theory' (i.e., noxious air due to rotting organic matter, etc.). In this case, there was very little synergy or mutual benefit in their knowledge and methods. Microscopes, microbe cultures and statistical epidemiology were all tools and methods that advanced germ theory but had no relevance to people who were interested in advancing miasma theory. The same divergence can happen in institutional innovation.

Social Systems Engineering: The Design of Complexity, First Edition. Edited by César García-Díaz and Camilo Olaya.
© 2018 John Wiley & Sons Ltd. Published 2018 by John Wiley & Sons Ltd.

Is it possible to apply an engineering approach to the *process* of institutional innovation? If so, how might it help to promote innovation or influence the trajectory? Cyber security will be our motivating case, described in some detail in Section 9.2. Cyber security is one of a large class of complex socio-technical systems that are characterized by low-probability/high-cost loss events and interdependent risk. Policymakers have recognized since 2003 that we do not have adequate institutions to manage cyber security and mitigate risk, and therefore they have called for 'leap-ahead' innovation, especially in incentive institutions and quantitative measurement systems. Unfortunately, the pace of institutional innovation has been slow – much slower than the pace of innovation in information technology and by threat agents.

Institutions are norms and 'rules of the game' that support and enable social life (North, 1990; Scott, 2007). They can be explicit or tacit, formal or informal. Chisholm (1995) gives these examples: legislatures, bureaucracies, corporations, marriage, insurance, wage labour, the vacation, academic tenure and elections. An *institutional innovation* is a significant change or improvement in an institution such that it has novel functional or performance characteristics. An example is consumer credit scoring (Marron, 2007; Ryan *et al.*, 2011). The incumbent institution had been in place for hundreds of years: credit managers exercising subjective judgement of each consumer's 'character' and personal history. In its mature form, the institution of consumer credit scoring replaced personal judgement with algorithmic judgement, and replaced individual evaluation rules with population and portfolio evaluation rules. It was seen as legitimate, supported the values of consumerism and the information age, fitted well with other institutions (e.g., credit reporting, financial risk management, etc.) and provided stability that enabled fast growth in both consumer credit and firms that depend on consumer credit.

Compared with technological innovation, much less is known about institutional innovation. Based on historical cases, we can say that institutional innovation can happen intentionally and by design, unintentionally through collective or accidental processes, or by a combination of the two. *Institutional entrepreneurs* are actors aiming to achieve innovation intentionally, but they rarely control or determine the entire process, in contrast to technology inventors or entrepreneurs. To succeed, any institutional innovation must fulfil these desiderata (Battilana *et al.*, 2009; Leca *et al.*, 2008; North, 1990; Scott, 2007; Weik, 2011):

1. *Functional.* Does it work? Does it perform?
2. *Feasibility.* Is there a viable evolutionary path from 'here' to 'there'?
3. *Legitimacy.* Does it flow from legitimate authorities, as seen by social actors?
4. *Cultural fit.* Does it fit and reinforce society's values?
5. *Coherence.* Does it interrelate effectively with other institutions?
6. *Uncertainty reduction.* Does it make social life more predictable or less risky?
7. *Stability.* Ultimately, does it make society more stable and resilient?

Given that institutional entrepreneurs are relatively 'blind' regarding most feasible or desirable paths of innovation, how do they take meaningful action? The thesis of this chapter is that institutional entrepreneurs *engineer the process* rather than the end result. By this we mean that they apply an engineering approach to design, build and use *knowledge artefacts* (e.g., dictionaries, taxonomies, conceptual frameworks, formal procedures, digital information systems, tools, instruments, etc.) as cognitive and social scaffolding to support iterative refinement and the development of partially developed ideas. Our hypothesis is that the rate of innovation progress will depend critically on the quality and nature of the knowledge artefacts at their disposal.

The plan for the chapter is as follows. Section 9.2 describes the institutional innovation problem to be solved and describes the two schools of thought regarding how best to solve the problem. Section 9.3 presents a theoretical model of the social processes of innovation, and explains why rival schools of thought arise in nascent fields where there is no established base of scientific knowledge or methods. Sections 9.4 and 9.5 present a computational model of innovation that will be used to illuminate the dynamics of social innovation aided by knowledge artefacts and to identify conditions under which one or the other rival school of thought is likely to prevail. The chapter closes with a discussion of the main findings and implications for institutional entrepreneurs.

9.2   Can Cyber Security and Risk be Quantified?

Although there is no settled definition for 'cyber security', for our purposes we will define it as the confluence of information security, digital privacy, digital civil rights, digital (trusted) identity, digital (content) rights management, digital intellectual property, and the digital aspects of homeland and national security. Given the pervasive and vital role of information and communication technology (ICT) in modern life, cyber security affects every organization and government, plus a large and growing proportion of individuals worldwide.

Cyber security is a vexing problem. Many problematic aspects of cyber security are sociological, economic, political and cultural. This has been well known for over a decade, leading to many policy reports and research funding solicitations that call for research and innovation in these domains (Department of Homeland Security, 2011; National Science and Technology Council, 2011). Many institutional innovations have been discussed or proposed, including some based on analogies with existing institutions in other domains.[1] Unfortunately, innovation progress to date has not been satisfactory.

9.2.1   Schools of Thought

With some oversimplification, we can identify two broad schools of thought regarding institutional innovation in cyber security: (1) the 'quants', who believe that cyber security and risk can and should be quantified in ways similar to other domains involving socioeconomic technical risk (Geer *et al.*, 2003) and (2) the 'non-quants', who believe that cyber security and risk either cannot be quantified or that there is no net benefit compared with alternative methods of guiding or structuring action, decisions, rules, etc. Examples of non-quantitative methods include checklists, audit questions and procedures, policy and practice guidelines, and situational professional judgement (Langner and Pederson, 2013). There is also a third school of thought we might call 'hybrid', which believes that some degree of quantification can be usefully combined with non-quantitative approaches.

---

[1] For example, 'Cyber CDC': Center for Disease Control; 'Cyber UL': Underwriters Laboratory; and 'Cyber NTSB': National Transportation Safety Board.

The degree of difference between these schools of thought varies by what is being quantified and how that quantification is used in analysis and decision-making. Where they differ least is in operational security (e.g., the uptime of a network firewall, the false positive rate for spam filters, etc.). Where they differ most is on risk quantification; that is, can we measure risk in economic units in a way that will guide investment decisions or serve as a foundation for cyber insurance or other incentive contracts? We will focus our attention on risk quantification because it makes vivid the contest between these schools of thought. The quantification of cyber security and risk is an intellectual and social domain where control and influence are contested by interest groups. Focusing on risk quantification, examples of interest groups associated with the quant school include the Society of Information Risk Analysts and companies who specialize in measuring or modelling risk. Examples of non-quant interest groups include some security consultants (Langner and Pederson, 2013) and most regulators.[2] Examples of 'hybrid' interest groups include most large information security companies, the U.S. National Institute of Standards and Technologies (NIST) in the Department of Commerce and ISACA, a professional organization formerly known as the Information Systems Audit and Control Association. Regarding evidence of innovation success and adoption, the quants have been struggling for more than a decade. Verendel (2009) presents a comprehensive survey of academic research up to that time, and finds that quantified cyber security is still a weakly supported hypothesis. As of 2015, no one can yet say that cyber security risk can be effectively and efficiently quantified. Even so, there has been some progress in some areas and growth in the number of people and organizations actively working on new ways to quantify risk.

The non-quants have frequently pointed to this lack of success as evidence that quantified risk is impossible in principle (i.e., in the same way that perpetual motion machines are impossible) or, at least, too complicated and expensive to invest in. Additional negative arguments come from the financial crisis of 2008, where sophisticated/complicated risk models have been widely blamed as one of the aggravating factors. However, the most frequent and fundamental argument against quantified cyber security risk is based on the complicating factor mentioned above – intelligent, adaptive adversaries. Though taken from a report on risk analysis for physical security of nuclear weapons complexes and not cyber security, this quote nicely summarizes the argument against quantified risk in cyber security too:

> The committee concluded that the solution to balancing cost, security, and operations at facilities in the nuclear weapons complex is *not to assess security risks more quantitatively or more precisely.* This is primarily because *there is no comprehensive analytical basis for defining the attack strategies that a malicious, creative, and deliberate adversary might employ or the probabilities associated with them.* [emphasis added] (National Research Council, 2011, p. 1)

The quants respond to this negative argument in a variety of ways, including a claim that lack of progress or complete success over ten years is not sufficient evidence that it cannot be successful given enough time and effort. To support this claim, references are made to historical cases of the development and adoption of quantitative methods in similar domains.

---

[2] In the United States, one example of a non-quant regulator is the Federal Financial Institutions Examination Council (www.ffiec.gov/cybersecurity.htm).

9.3 Social Processes of Innovation in Pre-paradigmatic Fields

Generalizing from the cyber security case, we now turn our attention to social processes of innovation in nascent fields – those that Thomas Kuhn called 'pre-paradigmatic'. In Kuhn's model of scientific revolutions, an established field of science is characterized by a 'paradigm', which is an 'entire constellation of beliefs, values, techniques, and so on shared by members of a given community' (Kuhn, 1970, p. 175). Established paradigms feature exemplars that serve as ideal models or templates to be emulated by subsequent research. For example, Newton's laws of motion served as exemplars in physics until the early twentieth century. In contrast, 'pre-paradigmatic' fields are those where there are no established, widely accepted paradigms and, therefore, lack of clarity over what constitutes 'good' or 'normal' scientific research. Without the normative influence of paradigms, the discourse and debate can be unproductive. Kuhn describes it this way: 'the pre-paradigm period, in particular, is regularly marked by frequent and deep debates over legitimate methods, problems, and standards of solution, though these serve rather to define schools [of thought] than to produce agreement' (Kuhn, 1970, pp. 47–48). Even though quantified cyber security and risk is not purely about science or scientific research, we can characterize it as being in a 'pre-paradigmatic' state of development.

9.3.1 Epistemic and Ontological Rivalry

Recall that we said that institutional entrepreneurs aim to achieve innovation *on purpose*, not just by chance events or through collective processes of change. Furthermore, they aim to achieve innovation in a particular direction, not just anywhere. To achieve this, they need a way of thinking about problems and solutions that enables progress. They also need to have a model of reality that enables progress. In philosophical language, we can say that institutional entrepreneurs need both an epistemology and an ontology that are *beneficial* and *instrumental* to their teleological (i.e., goal-driven) approach to innovation. In pre-paradigmatic fields such as quantified cyber security and risk, the schools of thought often feature rival or even mutually exclusive epistemologies and ontologies.

In the case of quantified cyber security and risk, the two rivalrous schools of thought – quants vs. non-quants – differ sharply over the ontology of cyber security and risk (i.e., what is real and what is not real). For example, some non-quants argue that quantifying cyber risk is impossible in principle because of the non-reality of hypothetical or counterfactual events: 'How is it possible, they say, to quantify what didn't happen?' (Borg, 2009, p. 107). There is considerable disagreement over the ontological status of 'intangible' losses such as reputation. Also, there is ontological debate carried over from mathematics and statistics concerning the reality or non-reality of subjectivist interpretations of probability. There is even dispute over the methods or possibility of ever resolving these ontological debates. For example, many quants are in favour of computer simulations as tools to explore counterfactual or hypothetical situations, while many non-quants argue against the validity of computer simulations as evidence in ontological arguments, given the intelligent, adaptive, creative and malicious adversaries.

Likewise, there is sharp disagreement between quants and non-quants regarding the best way to think about cyber security and risk. Quants argue that quantification and quantitative analysis can be a powerful tool to make better decisions and achieve better outcomes, much in the same way that statistical process control and total quality management has helped revolutionize product and service quality across many industries from the 1980s to the present. Some non-quants counter with the argument that attempting to quantify abstract and non-real entities such as 'risk' is not only a waste of time and effort, but leads to *worse* outcomes through 'analysis paralysis' or mistaken efforts to 'manage' risk (Langner and Pederson, 2013).

From the perspective of sociology of innovation, we are less concerned with the ultimate truth of any of these positions than we are with their functional and instrumental effects (i.e., *are they effective in helping the actors to achieve progress?*). This leads to the next topic: What *knowledge artefacts* do institutional entrepreneurs develop and use during the innovation process and how do they promote progress?

9.3.2   Knowledge Artefacts

A *knowledge artefact* is something created by actors informed by their knowledge and makes that knowledge useful or productive. A knowledge artefact can be a thing (i.e., a message, book, tool, design, etc.) or a realizable process (i.e., a training process, production process, communication process, information processing process, utilization process, etc.). Thus, it is through knowledge artefacts that people create, transform and use knowledge for practical aims. This is not meant to reify knowledge. Instead, knowledge artefacts can be seen as the tangible, observable instantiations of knowledge, much like a circle drawn on a sheet of paper is an instantiation of the Platonic idea of a 'circle'. Boisot (1999) uses a similar term – 'knowledge assets' – which he defines as 'knowledge that yields an appropriable stream of benefits over time' (p. 155). However, this definition presumes that we can point to, instantiate, define or specify the knowledge in question, which can be problematic. Instead, we prefer to use the term 'knowledge *artefact*', to highlight the point that knowledge artefacts are products of human intention and effort and that they can be observed and instantiated, at least in principle. We retain Boisot's notion of an 'appropriable stream of benefits over time' through the emphasis on instrumentality.

Boisot (1995) developed the information space (I-space) framework for characterizing knowledge and knowledge artefacts along three dimensions: (1) codification; (2) abstraction; and (3) diffusion. The main purpose of the I-space framework is to study the transformation of knowledge through lifecycles of discovery, learning and diffusion. Our focus will be on the first two dimensions. The *codification* dimension evaluates knowledge in terms of its degree of compression or abbreviation within some coding scheme – such as categories, taxonomies, variables, conditions, relations, and so on. For any given knowledge, expressing it in a highly codified way will be very compressed and economical. In contrast, uncodified knowledge may take many more words to express, or may even be only learned through experience or example (i.e., 'tacit knowledge'). The *abstraction* dimension evaluates knowledge in terms of the inferences you can draw from it, and the degree of generality regarding inferences, ranging from concrete (highly specific and contextual) to abstract (highly general and free of context).

9.3.3   Implications of Theory

Before moving on to the next section, we can summarize the implications of these theories in relation to our case and also to the general study of institutional innovation in pre-paradigmatic fields. First, institutional entrepreneurs in rival schools of thought are engaged in a contest between each other and also with Nature regarding who has the best way to think about problems and solutions (epistemology) and whose model of reality is most effective (ontology). While this contest plays out in many ways that are mostly or purely social, there is also a contest in the practical world of realizing inventions and innovation. It is not enough to talk a good game or convince many others. Eventually, some inventions work and others do not. Those that work and can be used and understood by the masses will get widely adopted. But institutional entrepreneurs often start in a fog of uncertainty and ignorance, even if some insights, intuitions, role models or goals guide them. Therefore, they create and use knowledge artefacts that have several uses at once. They solve some immediate problem while providing some foundation or platform for further invention or knowledge creation/transformation. In this way, knowledge artefacts can serve as *cognitive scaffolding* (Lane and Maxfield, 2005) to help the institutional entrepreneurs make progress in the face of ignorance and uncertainty. We can usefully characterize their knowledge artefacts along the dimensions of codification and abstraction. Boisot's social learning cycle (SLC) theory predicts that insights that trigger inno- vation cycles start out as tacit (hard to explain) and concrete/specific. SLC predicts that knowledge artefacts will be developed and used in a specific sequence: *first*, they will be increasingly codified (i.e., through formal definitions, taxonomies and measurement systems) and *then*, they will be increasingly generalized thorough more abstract sign systems, relation systems and inference systems.

In the case of quantified cyber security and risk, we can position specific knowledge arte- facts in the I-space, as shown in Table 9.1.

Here is the rationale, starting with ad hoc security metrics. Many medium-sized and large companies have a dedicated information security department and many of these collect and report 'security metrics' to company executives in regularly scheduled reports. Evaluated as artefacts for knowledge relating to quantified security and risk, most of these reports are low in codification because they do not follow any well-defined taxonomy for what should be measured and reported. Also, they are relatively low in abstraction because the rules or logic as to how the different metrics might be combined or interpreted together are mostly in the form of heuristics. Regarding the NIST Cyber Security Framework (CSF), we can locate it as moderate in codification because it does attempt to define key phenomena and conditions in cyber security and risk, but in itself it does not attempt to quantify security or risk. It is mod- erately low in abstraction, since it mostly points to classes of phenomena and does not embody any specific theory or knowledge as to how security is achieved or risk reduced through the implementation of 'best practices'. Finally, regarding risk analysis software, there are several commercial software products and services that quantify some aspect of cyber security and risk. In terms of the I-space we can locate them as high in codification, certainly much higher than the NIST-CSF, and moderately high in abstraction, since they embody formalized knowledge regarding how quantitative inferences are to be drawn from the evidence (i.e., 'ground-truth data').

Table 9.1 List of example knowledge artefacts and their position in I-space

Artefact	School of thought	Codification	Abstraction
1. Ad hoc security metrics	Quant	Low	Low
2. NIST Cyber Security Framework (CSF)	Non-quant	Moderate	Moderate–low
3. Risk analysis software	Quant	High	Moderate–high

From the point of view of the SLC, we can see that there are (at least) two innovation and learning cycles at work. The non-quant learning cycle is represented by the NIST-CSF, and is explicitly aimed to be a viable stage of refinement, performance improvement, and usability that can support widespread diffusion and adoption. The quant learning cycle is aimed for a much higher goal in terms of codification and abstraction, and thus widespread diffusion is not yet happening, or maybe it is just beginning.

In summary, the contest between rival schools of thought in a pre-paradigmatic field such as cyber security can be viewed as different navigational strategies through I-space. The non-quant school is aiming for a lower region in I-space (i.e., less codification, less abstraction), betting that this will be more feasible and will achieve practical success and wider adoption compared with the higher road of the quants. Conversely, the quants are betting that the high road (i.e., more codification, more abstraction), though more difficult to traverse, will ultimately lead to more compelling results – better security, lower risk and better use of society's resources. Note that the I-space framework and SLC theory do not represent the space of possible inventions, because they do not account for the specific traits, characteristics or dependencies of each invention. Therefore, I-space and SLC do not facilitate analysis of how difficult it may be to go from any point 'A' to any other point 'B' in the space of possible inventions. We address this in a computational model, presented in the next section.

9.4   A Computational Model of Innovation

In this section our goal is to demonstrate how computational modelling can be used to investigate institutional innovation in contested territory, and the effects of knowledge artefacts. In the specific case of quantified cyber security, we don't yet know who will win: the quants or the non-quants. Given the time span of institutional innovation, we may not know for many years. By using computer simulation, we can examine a generalized abstract model of innovation and perhaps learn more about the conditions under which one or the other rival school of thought is likely to prevail.

To model the phenomena of interest, we need a way to model the space of possible inventions. We also need a way to model the relative difficulty of achieving each invention, both with respect to making the final discoveries or solving the final problems, but also with respect to the inventions that came before (i.e., precursors and dependencies). Finally, we need a way to model the relative effects of knowledge artefacts as characterized by I-space.

9.4.1   Base Model: Innovation as Percolation

To meet these requirements in a parsimonious way, we chose to develop a *percolation model of innovation* based on the model presented by Silverberg and Verspagen (2005) (S&V). 'Percolation' is the phenomenon of fluid moving or filtering through porous materials. Percolation modelling originates in the fields of physics, chemistry and materials science, and has been abstracted in mathematics as percolation theory. In the model of Silverberg and Verspagen (2005), the 'porous material' is taken to represent the space of possible inventions, the 'fluid' is taken to represent the advancing front of innovation ('best practice frontier') in that space, and the local dynamics of percolation are taken to represent the local dynamics of innovation. We adopt the S&V term 'technology' to mean any solution, method, process, procedure, tool or machine, and also the term 'R&D' means inventive activity, whether formal or informal. S&V use the term 'firms', but we prefer the more general and abstract term 'agents' to refer to localized bundles of inventive activity, be it a person, a team of people, a firm or some mixture.

For readers not familiar with agent-based modelling (ABM), here is a basic overview. ABM consists of an environment and a set of agents that operate and interact within that environment. An 'agent' is a simple program that has a set of behaviour rules, a memory (i.e., internal state), a position within the environment, and runs once each time step. Generally, all agents run the same program, with the only difference being the agent's internal state, its location and the state of the local environment. Figure 9.1 shows pseudo-code for a generic agent (single step).

In our model, the 'agents' represent R&D effort in a particular technology type (column). There is one agent per technology type. Each possible technology is a cell connected in a discrete two-dimensional lattice[3] (i.e., a grid; see Figure 9.2, in detail in Figure 9.3). Each cell in the lattice has horizontal neighbours that are very similar and interrelated, and vertical neighbours that are slightly more or less sophisticated. Overall, the neighbourhood structure reflects technological interrelatedness. Considering the horizontal dimension, each column represents a 'technology type', all sorted so that the most similar types are next to each other. Considering the vertical dimension, each row represents a degree of sophistication, from the minimal 'baseline' at the bottom, rising monotonically without bound in principle, but limited to a maximum size to fit the constraints of computer processing. The lattice is connected with a periodic boundary in the horizontal dimension, so that it has a cylindrical topology. This allows every technology type (column) to have exactly two neighbours and eliminates horizontal boundary effects in the model.

Formally, we define a lattice A with h columns and a periodic boundary in the horizontal dimension (i.e., cylinder topology), v rows and $h \times v = N$ cells indexed by i and j, $0 < i < h$ and $0 < i < v$. Parameters $h > 0$, $v > 0$ are set by the experimenter to be large enough so that the boundaries of the lattice do not influence the results involving rates of innovation and distribution of sizes of innovation. Each lattice cell $a_{i,j}$ can be in one of four states: $0 =$ impossible (black); $1 =$ possible but not yet discovered (white); $2 =$ discovered but not yet viable (light grey); $3 =$ discovered and viable (mid-grey).

```
input: Internal_state, Location, Environment
output: Internal_state, Location, Environment
begin
   local_state ← Sense_local_environment (Environment, Location)
   foreach Behavior_rules
     if (match(local_state, rule_condition)
        then rule_behavior(Internal_state, Location, Environment)
   end
end
```

Figure 9.1　Generic agent algorithm for a single time step.

---

[3] A 2D lattice is chosen for simplicity, but Silverberg and Verspagen (2005) also say that they believe their main results will hold for more general topologies.

Figure 9.2 The 2D lattice of 'technologies' in our percolation model of innovation after 1200 simulation steps. One region in the lattice around the best-practice frontier (BPF) is magnified. See Figure 9.3 for details. Greyscale code: black=impossible, white=possible but not yet discovered; very dark grey=possible but not reachable; lightest grey=discovered but not yet viable; medium-light grey=discovered and viable; medium-dark grey=discovered, viable and on best-practice frontier (BPF). Medium-dark grey cells are the loci of R&D. The two horizontal lines show the 'average' level of innovation (i.e., BPF) across all 'technologies' (columns). The dark line is the average (i.e., mean) BPF, while the dark-grey line is the mean+standard deviation of the BPF.

Through R&D activity of agents, some cells near the baseline are discovered and therefore move from state 1 to state 2. These newly discovered cells become the 'adjacent possible' (Kauffman, 1996), meaning that they are the next candidates for becoming viable technologies (state 3). Any discovered cell (state 2) becomes viable (state 3) when there is a contiguous Manhattan[4] path from it to the baseline. Sites initialized as impossible (state 0) can never be converted into any other state.

The probability that any cell will be initialized in state=1 (possible) rather than state=0 (impossible) is given by parameter p. If all paths from a given possible cell to the baseline are blocked by impossible cells, then we say that these cells are *not accessible*. There is a critical value of $p \approx 0.6$. Much above 0.6 and nearly every possible cell becomes accessible. Much below 0.6 and nearly every possible cell is not accessible. When $0.6 \leq p \leq 0.65$, each random

---

[4] A 'Manhattan' path is a series of up-or-down and left-or-right steps. This implies a Von Neumann neighbourhood for each cell (i.e., only the cells reachable with Manhattan distance of n steps). An alternative definition for a path is 'chessboard distance', which includes any combination of up-or-down, left-or-right, or diagonal steps. This would imply a Moore neighbourhood for each cell. If we switch to chessboard distance and Moore neighbourhood, this provides more connectivity between cells and increases the number of possible-but-not-yet-discovered technologies (state=1, white) to explore. The main effect is to increase the overall rate of innovation, because 'impossible' regions are more easily traversed. However, the qualitative results are no different from the current model with Manhattan path and Von Neumann neighbourhood.

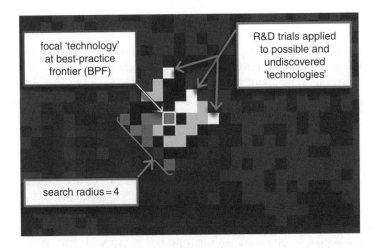

focal 'technology'
at best-practice
frontier (BPF)

R&D trials applied
to possible and
undiscovered
'technologies'

search radius = 4

Figure 9.3 Detailed view of the magnified region in Figure 9.2.

realization produces a different complex pattern of accessibility. For all of our simulation runs, we set $p=0.62$.

Every technology type (column) has an individual technology (cell) that is the most advanced (highest in the column), and this cell has BPF=true. All other cells have BPF=false. All R&D activity takes place in technologies (cells) around the BPF (i.e., BPF=true). The BPF moves vertically in each column as R&D is successful. All R&D takes place within the search radius around the technologies (cells) at the BPF (see Figure 9.5 later). R&D activity consists of each agent expending a budgeted amount of effort to 'discover' cells that were previously 'undiscovered' within their search radius. This is realized through the probability of discovery, which is the search effort e divided by the size of the search area. [In our simulation runs, $e=0.5$ and radius$=4$, which yields a search area of 40 technologies (cells) and a probability of discovery of $p=0.0125$ per R&D attempt.] Thus, in the base model of S&V, an 'R&D attempt' is realized as a random draw from $(0, 1)$, and if this is less than the probability of discovery, then the technology is 'discovered'. Figure 9.4 shows the pseudo-code for this algorithm.

```
1    input: Location, Environment
2    output: Environment
3    begin
4      local_cells ← get_neighboring_cells (Environment, Location)
5      possible_cells ← { }
6      foreach cell in local_cells
7        if (get_state(cell) = 1) then append(possible_cells, cell)
8      end
9      foreach cell in possible_cells
10       draw ← random-uniform(0,1)
11       if (draw ≤ 0.0125) then set_state(cell, 2)
12   end
```

Figure 9.4 Algorithm for R&D activity in a single time step.

After all agents have executed this algorithm, an environment program runs to identify newly discovered technologies (cells with state=2) that are now viable. If any are found, those cells have their state set to 3. The BPF is adjusted by setting *BPF*=true for the highest viable cell in each column. Agents are moved to the BPF cell in their column.

The 'size' of any innovation is the number of rows between the newly discovered technology and the previous BPF technology in that column. In other words, large innovations make a big 'leap' in the vertical dimension, while small innovations might only move up one cell.

9.4.2   Full Model: Innovation with Knowledge Artefacts

We made several extensions to the base model to simulate the effects[5] of knowledge artefacts and learning. The effect of knowledge artefacts is to improve the effectiveness of R&D, but only if the knowledge appropriately matches the domain (i.e., effectiveness is proportional to their *fidelity* relative to Nature and also relative to the social and technical context of innovation).

In our extension, there is only one knowledge artefact available to all agents with parameters set at initialization time corresponding to their characteristics in I-space: codification $c \in (0, ..., 5)$ and abstraction $a \in (0, ..., 5)$. For simplicity, these parameters take integer values. We model the effects of knowledge artefacts through a submodel of the R&D activity: a random draw from multiple balls-in-urns instead of the uniform random number draw in the base model. Each undiscovered technology (cell) starts the simulation with a large but finite number of 'balls' (N=1000), which are possible solutions that are either 'successful' (13 balls) or 'unsuccessful' (987 balls). These balls are distributed in a number of urns determined by the codification parameter c. If c=0, then all the balls are in a single urn, and this is equivalent to a blind 'trial and error' search. If c=5 (maximum value), then the balls are allocated to 32 urns, with all the 'successful' balls in one urn (13 out of 32 balls in that 'lucky urn'). The effect of abstraction is that it increases the probability that the agent will select the 'lucky urn'. If a given agent is aided by a high-fidelity knowledge artefact with high abstraction, then they will most likely select the 'lucky' urn containing all the 'successful' balls. But if either the fidelity parameter $f \in (0, ..., 1)$ is zero or the abstraction a=0, then the agent will be choosing among the 32 urns with uniform probability, which again is equivalent to a blind 'trial and error' search. While both c and a are fixed for the duration of a simulation run, f can change if the learning rate parameter $l \in (-1, ..., 1)$ is not zero. If l>0, then the fidelity f increases during the run as a function of the height of the BPF, and conversely if l<0 then the fidelity f decreases during the run as a function of the height of the BPF. This allows us to simulate scenarios where a knowledge artefact is initially appropriate and effective in guiding innovation but decreases in appropriateness and effectiveness as technologies get more advanced/sophisticated.

9.4.3   Experiment

We design three experimental treatments that, in an abstract way, represent the different schools of thought (quant vs. non-quant) and the differences in the knowledge artefacts they are attempting to create and use. Recall that the non-quants are building and using knowledge

---

[5] For simplicity, we are only simulating the *effects* of knowledge artefacts with different I-space characteristics rather than the specific contents or traits of the knowledge artefacts themselves.

artefacts with less codification and less abstraction, believing that this will be more feasible and will achieve practical success soon and therefore wide adoption soon, too. Conversely, the quants are developing and using knowledge artefacts with higher codification and higher abstraction, which will be more successful to promote innovation, though progress may be more difficult to achieve initially. We do not know whether the quants will learn rapidly or slowly (i.e., adapt and refine their knowledge artefacts), and therefore we define separate experimental treatments for each scenario. We add a 'control' treatment with no knowledge artefact, resulting in four treatments in total.

1. *Trial-and-error* with no knowledge artefact.
2. *Non-quant* with initial knowledge artefact parameters: codification=2, abstraction=2, fidelity=1.0, learning rate=0.0.
3. *Quant – slow learning* with initial knowledge artefact parameters: codification=4, abstraction=4, fidelity=0.2, learning rate=0.2.
4. *Quant – fast learning* with initial knowledge artefact parameters: codification=4, abstraction=4, fidelity=0.0, learning rate=1.0.

Figure 9.5 shows a single run and a series of screenshots at different time steps, along with a time-series chart of the innovation rate (i.e., change in BPF per time step). Figure 9.6 compares two experimental treatments on the same initial lattice configuration.

Each of the four experimental treatments were tested with the same random initial lattices, ten in total, with twenty runs for each lattice condition using different random seeds for each run. Each run ended when the BPF reached the highest row in the lattice.

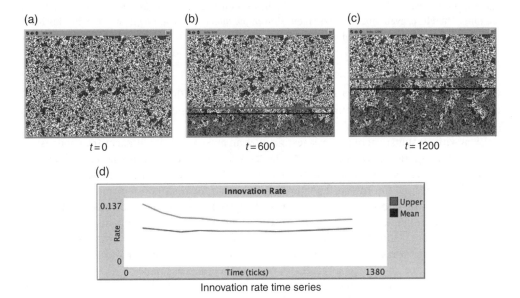

Figure 9.5 A single run at different time steps, given an initial state (a), with a time-series chart (d) showing innovation rate for mean BPF (black horizontal line in b and c) and upper BPF (mean+standard deviation, dark-grey horizontal line in b and c).

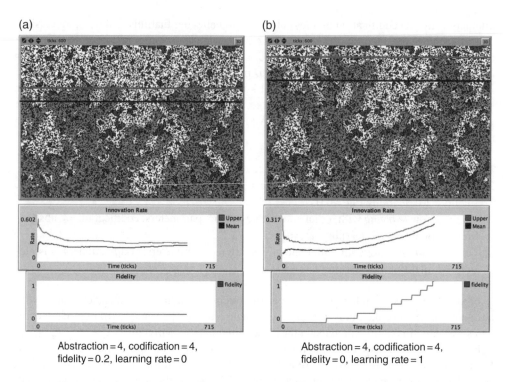

(a)

(b)

Abstraction = 4, codification = 4,
fidelity = 0.2, learning rate = 0

Abstraction = 4, codification = 4,
fidelity = 0, learning rate = 1

Figure 9.6 Screenshots of two treatments tested on the same lattice configuration: (a) quant–slow learning vs. (b) quant–fast learning. In (b), the innovation rate increases as fidelity increases due to learning. Therefore, even though it started out slower, the quant– fast learning treatment wins this innovation race. (The black horizontal lines are mean BPF. The dark-grey horizontal lines are upper BPF = mean + standard deviation.)

Figure 9.7 shows violin plots for the experimental results with two dependent variables: (1) innovation rate at the end of the run and (2) time to complete a run (i.e., the BPF reaches the top of the lattice).

Even without statistical hypothesis testing, we can make several inferences from the results shown in this figure. As we might expect, the 'control' treatment of trial-and-error R&D had the lowest innovation rate at the end of each run and the slowest time to complete a run (i.e., the BPF reaches the top row in the lattice). Notice that the distribution of time to complete for trial-and-error is not symmetrical; instead, it is skewed with some runs taking much longer than average. In comparison, the distributions for the other treatments are much less skewed and more symmetrical. From this we can infer that blind trial-and-error R&D is more prone to getting 'stuck' on difficult landscapes compared with the other treatments, which have the benefit of knowledge artefacts to improve their success rate.

The second result is that the non-quant treatment is noticeably better than trial-and-error, both in innovation rate and time to complete, but not by a large margin. Thus, even though fidelity = 1.0 (i.e., the knowledge artefact was an ideal match to nature), the relatively low codification and abstraction characteristics provide only a modest improvement in innovation rates.

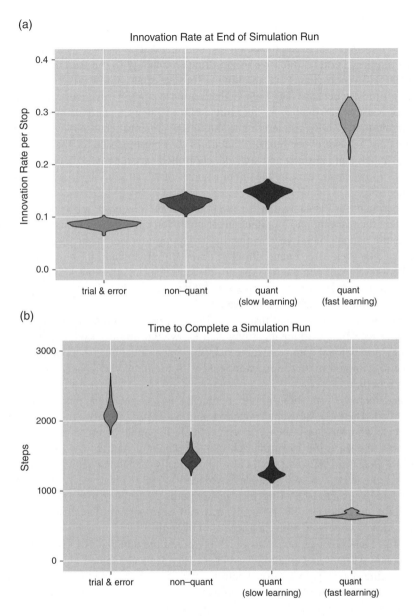

Figure 9.7 Violin plot of experimental results for four treatments, each tested with ten random lattice initial conditions and twenty runs per lattice condition: (a) innovation rate at the end of the simulation run; (b) time to complete a simulation run.

The third result is that the relative success of the quant approach depends critically on the learning rate. With 'slow learning' (initial fidelity=0.2, learning rate=0.2), the quant innovation results are only slightly better than the non-quant results, both in final innovation rate and in time to complete. However, with 'fast learning' (initial fidelity=0, learning rate=1.0), the quant treatment wins the race by a wide margin, both in final innovation rate and in time to complete.

9.5   Discussion

If we frame the contest between the quant and non-quant schools as an innovation race (perhaps on a slippery surface), then we might draw an analogy to the parable of the Tortoise and the Hare. The non-quants are adopting a Tortoise strategy towards knowledge – slow and steady – mostly because they believe that no more aggressive strategy is feasible given the state of Nature. The quants are adopting a Hare strategy towards knowledge – starting very slow and then rapidly accelerating to the finish – mostly because they believe that this will ultimately achieve cyber security outcomes that are much better than less ambitious methods. Though quite abstract and stylized, controlled experiments with our computational model have allowed us to explore the circumstances of which school will win the innovation race, if any.

The experiment results show that the likelihood of the quant school winning is critically dependent on the learning rate (i.e., the rate of improvement in how well its knowledge artefacts fit nature and are therefore effective in facilitating innovation success). If the learning rate is slow, then even if the quant school achieves slightly higher innovation rates, the non-quant school might still win the race due to substantial social advantages. It appears that the viability of the quant school can only be assured if it achieves a high learning rate and becomes demonstrably effective at achieving innovation.

Therefore, it is imperative that institutional entrepreneurs within the quant school adopt practices that accelerate learning regarding their knowledge artefacts. While this advice could apply to any professional community, the risky approach of the quant school means that they have more to gain and more to lose compared with the more conservative approach of the non-quant school.

Of course, there are limitations to our approach. Our computational model is both abstract and simplified. Therefore, it excludes many important factors and dynamics that might ultimately decide who wins, or if there is a winner at all. In a more complete analysis, we would like to assess the scientific merit of each of the schools of thought (i.e., their explanatory coherence) (Thagard, 1992). We would also like to analyse social dynamics such as legitimization (Nicholls, 2010), power struggles (Aronowitz, 1988), rivalry over discourse frames (Hoffman and Ventresca, 1999; Torgersen and Schmidt, 2013; Werner and Cornelissen, 2014) and structuration (Giddens, 1984). Finally, it would be important to analyse the institutional structure of R&D associated with each school of thought.

This holistic analysis would give us a rich picture of the dynamics of institutional innovation in a contested field like quantified cyber security. It would shine light on the challenges and opportunities faced by institutional entrepreneurs who are trying to accelerate innovation in particular directions. As illustrated in this chapter, computational modelling can complement other methods of analysis and make unique contributions to research.

Acknowledgements

This chapter is based upon work supported by the National Science Foundation under Grant No. CMMI-1400466. Any opinions, findings, conclusions or recommendations expressed in this material are those of the authors and do not necessarily reflect the views of the National Science Foundation.

References

Aronowitz, S. (1988) *Science as Power: Discourse and ideology in modern society*, University of Minnesota Press, Minneapolis, MN.

Battilana, J., Leca, B. and Boxenbaum, E. (2009) How actors change institutions: Towards a theory of institutional entrepreneurship. *The Academy of Management Annals*, **3**(1), 65–107.

Boisot, M. (1995) *Information Space: A framework for learning in organizations, institutions and culture*, Routledge, London.

Boisot, M.H. (1999) *Knowledge Assets: Securing competitive advantage in the information economy*, Oxford University Press, Oxford.

Borg, S. (2009) The economics of loss, in C.W. Axelrod, J.L. Bayuk and D. Schutzer (eds), *Enterprise Information Security and Privacy*, Artech House, Norwood, MA, pp. 103–114.

Chisholm, D. (1995) Problem solving and institutional design. *Journal of Public Administration Research and Theory*, **5**(4), 451–492.

Department of Homeland Security (2011) Cyber Security Research and Development Broad Agency Announcement (BAA) 11-02. Solicitation, U.S. Department of Homeland Security.

Geer, D., Soo Hoo, K. and Jaquith, A. (2003) Information security: Why the future belongs to the quants. *IEEE Security Privacy*, **1**(4), 24–32.

Giddens, A. (1984) *The Constitution of Society: Outline of the theory of structuration*, University of California Press, Oakland, CA.

Hoffman, A.J. and Ventresca, M.J. (1999) The institutional framing of policy debates economics versus the environment. *American Behavioral Scientist*, **42**(8), 1368–1392.

Kauffman, S. (1996) *At Home in the Universe: The search for the laws of self-organization and complexity*, Oxford University Press, Oxford.

Kuhn, T.S. (1970) *The Structure of Scientific Revolutions*, University of Chicago Press, Chicago, IL.

Lane, D.A. and Maxfield, R.R. (2005) Ontological uncertainty and innovation. *Journal of Evolutionary Economics*, **15**(1), 3–50.

Langner, R. and Pederson, P. (2013) Bound to fail: Why cyber security risk cannot be 'managed' away. Paper, The Brookings Institution, Washington, D.C.

Leca, B., Battilana, J., Boxenbaum, E. and School, H.B. (2008) Agency and institutions: A review of institutional entrepreneurship. Working Papers Vol. 8(96), Harvard Business School, Cambridge, MA.

Marron, D. (2007) Lending by numbers: Credit scoring and the constitution of risk within American consumer credit. *Economy and Society*, **36**(1), 103–133.

National Research Council (2011) *Understanding and Managing Risk in Security Systems for the DOE Nuclear Weapons Complex*, National Academies Press, Washington, D.C.

National Science and Technology Council (2011) Trustworthy cyberspace: Strategic Plan for the Federal Cybersecurity Research and Development Program. Official policy, United States Government, Executive Office of the President, National Science and Technology Council.

Nicholls, A. (2010) The legitimacy of social entrepreneurship: Reflexive isomorphism in a pre-paradigmatic field. *Entrepreneurship Theory and Practice*, **34**(4), 611–633.

North, D.C. (1990) *Institutions, Institutional Change and Economic Performance*, Cambridge University Press, Cambridge.

Ryan, A., Trumbull, G. and Tufano, P. (2011) A brief postwar history of U.S. consumer finance. *Business History Review*, **85**(3), 461–498.

Scott, W.R. (2007) *Institutions and Organizations: Ideas and interests* (3rd edn), Sage Publications, Thousand Oaks, CA.

Silverberg, G. and Verspagen, B. (2005) A percolation model of innovation in complex technology spaces. *Journal of Economic Dynamics and Control*, **29**(1&2), 225–244.

Thagard, P. (1992) *Conceptual Revolutions*, Princeton University Press, Princeton, NJ.

Torgersen, H. and Schmidt, M. (2013) Frames and comparators: How might a debate on synthetic biology evolve? *Futures*, **48**, 44–54.

Verendel, V. (2009) Quantified security is a weak hypothesis: A critical survey of results and assumptions. *Proceedings of the 2009 New Security Paradigms Workshop, NSPW '09*, ACM, New York, NY, pp. 37–50.

Weik, E. (2011) Institutional entrepreneurship and agency. *Journal for the Theory of Social Behaviour*, **41**(4), 466–481.

Werner, M.D. and Cornelissen, J.P. (2014) Framing the change: Switching and blending frames and their role in instigating institutional change. *Organization Studies*, **35**(10), 1449–1472.

Part III

Cases and Applications

10

Agent-Based Explorations of Environmental Consumption in Segregated Networks

Adam Douglas Henry and Heike I. Brugger

10.1   Introduction

What leads individuals to adopt more environmentally sustainable patterns of consumption? Developing a better understanding of the factors that shape consumption choices that influence the environment – behaviours we refer to broadly as 'environmental consumption' – is of great practical importance. Theories of environmental consumption can inform strategies to help reduce human ecological footprints, or the negative influences that humans have on natural systems (Dietz *et al.*, 2007). Reducing our collective footprint is essential to protect fragile natural systems in the face of rapid (and inevitable) population growth, urbanization and economic development (Clark and Dickson, 2003; Kates *et al.*, 2001; Rosa *et al.*, 2009).

But when it comes to promoting more desirable patterns of environmental consumption, such as increased investment in energy-efficient technologies, we face a fundamental problem: one cannot directory engineer desired human behaviours. At best we can design interventions that provide the right sets of conditions and incentives that are needed to encourage the desired outcomes. So, although it is generally not possible to force humans to behave in more sustainable ways, one can intervene in the complex systems that determine environmental behaviours in an attempt to incentivize or otherwise spur more sustainable consumption patterns. This underscores the need to ground the design of interventions in a robust understanding of the systems that give rise to certain types of behaviour. Indeed, many of the factors driving environmental consumption are not well understood, which makes it difficult for us to know how policy interventions should be designed – that is, how one might appropriately engineer more desirable patterns of environmental consumption.

Social Systems Engineering: The Design of Complexity, First Edition. Edited by César García-Díaz and Camilo Olaya.
© 2018 John Wiley & Sons Ltd. Published 2018 by John Wiley & Sons Ltd.

Agent-based modelling (ABM) has an important role to play in the study of environmental consumption and the development of policies to promote more sustainable consumption behaviours (Rai and Henry, 2016). Agent-based models are a general class of models that represent systems as collections of discrete decision-makers (referred to generically as 'agents') that behave according to a set of programmed rules. These rules are generally relatively simple, however even very simple rules can produce complex outcomes over time and space – producing 'emergent' behaviours that are difficult or impossible to predict ex-ante (Schelling, 2006).

ABM can inform policy in many ways, two of which are particularly relevant in this chapter. First, ABM enables us to seek out micro-level explanations for aggregate patterns of behaviour that serve as proxies for policy problems or programme efficacy, such as the overall number of people who have adopted a given technology in a particular place. ABM is thus a tool for theory building, in that we can explore the degree to which certain theoretical assumptions produce emergent patterns of behaviour that are observed in the real world. Understanding these micro-drivers is important, because they inform policy interventions. For instance, some governmental programmes seek to promote the widespread use of solar photovoltaics (PV) by residential households by providing financial incentives in the form of tax credits in the year that a new solar system is installed. Such programmes assume that the primary barrier to adoption are the capital costs that adopters must pay for a new solar system; while this is likely one important barrier, it is not necessarily the only barrier. Other barriers may include an agent's lack of interest in technologies that are perceived as new and unproven, concerns about the long-term costs and benefits of solar or (as we explore in this chapter) a lack of access to third-party agents that provide information and services surrounding these high-cost technologies. Understanding these alternative barriers to adoption can inform a wider, more robust set of strategies that may achieve the desired outcome.

A second way that ABM can inform policy is to provide a platform to test the efficacy of programmes that are meant to influence outcomes – for instance, the adoption of solar PV over space and time. With a realistic representation of behavioural rules, ABM allows us to explore in a virtual environment the degree to which interventions in agent behaviours create varying outcomes in the system. Even more importantly, ABM allows planners to see the implications of certain policies in terms of evaluative criteria that are difficult to measure in real-world systems, but readily observed within a controlled computational simulation. Following on with the example of solar PV adoption, we might be able to observe how many people in a region have adopted solar, however it is very difficult to know the degree to which adoption is clustered in particular neighbourhoods such that certain segments of a market are saturated and others have few if any adopters. In this way, ABM enables us to do more than just refine theories and make predictions, but also engage in an informed dialogue about policy interventions and the ethics of what these interventions are meant to achieve (Basart and Serra, 2011; Davis, 2012; Robison, 2010).

In the remainder of this chapter, we consider a particular form of environmental consumption: the adoption of high-cost, environmentally friendly technologies. The adoption of such technologies – such as solar PV systems, electric vehicles or rainwater-harvesting systems – is generally a major financial decision for households, creating strong incentives for potential adopters to carefully evaluate the varied costs and benefits of such technologies. Adding to the complexity of these decisions, agents are also influenced by exogenous factors such as government incentive programmes, social interactions with other adopters, as well as the strategic behaviours of third-party firms that market these technologies.

These and other factors behind technology adoption are explored in the following sections. We then present a theoretical model that examines how social network structures interact with firm strategies to produce unique adoption curves in a hypothetical social system. This shows that the strategy used by a firm to target potential adopters might create short-term benefits in that overall adoption rates are very high, but with increasing disparities in levels of adoption between rich and poor communities – creating a problem of environmental injustice (Bickerstaff *et al.*, 2013; Jenkins *et al.*, 2016; Sovacool and Dworkin, 2014). Thus, the model we present here not only illustrates a potential policy problem, but also provides a platform for discussion about what we should (and can feasibly) try to achieve with public policy.

10.1.1 *Micro-drivers of Technology Adoption*

The adoption of 'environmentally friendly' technologies, or technologies that have the potential to dramatically decrease human ecological footprints, is generally a choice that occurs at the level of individuals or households. However, much of the work on the diffusion of environmentally friendly technologies examines aggregate market trends rather than individual behaviours. For example, Lobel and Perakis (2011) propose a proportional hazard model that estimates the effectiveness of subsidy policies for solar PV adoption as a function of cost, aggregate consumer behaviour and social influence effects. This is just one of many examples of research on technology adoption in general, and solar PV adoption in particular, at the aggregate level (Drury *et al.*, 2012; Gallo and De Bonis, 2013; Guidolin and Mortarino, 2010; Kahn and Vaughn, 2009; Kwan, 2012; Rai and Robinson, 2013; Rai and Sigrin, 2013; van Benthem *et al.*, 2008; Vasseur, 2012).

A relatively distinct thread of research on environmental consumption focuses on the decision-making processes of individuals, supported by the survey work of adopters and the general public to understand patterns in why some people adopt a given technology and others do not. Much of this literature has focused specifically on solar energy (Arkesteijn and Oerlemans, 2005; Berger, 2001; Claudy *et al.*, 2013; Faiers and Neame, 2006; Faiers *et al.*, 2007; Jager, 2006; Kaplan, 1999; Labay and Kinnear, 1981; Taylor, 2008; Yuan *et al.*, 2011; Zhai and Williams, 2012). One common, emerging finding of this literature is that *social influence* is an important determinant of environmental consumption, meaning that potential consumers are influenced by the behaviours of those they are socially close to – this 'closeness' might involve geographic closeness (such as spatial proximity) or it might involve closeness in terms of social relations (such as frequency of contact or discussion). There is a growing literature showing the importance of social influence, often referred to as *peer effects*, on environmental consumption. For example, Axsen *et al.* (2009) identify the importance of peer influences in the adoption of hybrid-electric vehicles, based on the stated and revealed preferences of new-car owners. Claudy *et al.* (2011) show that willingness to pay for micro-generation technologies increases when one's social contacts have already adopted.

These are not just one-off empirical findings. The idea that individual behaviour is dependent upon the behaviours of one's social contacts is a central idea in social science theory. Much of the emerging science of networks is built on the idea that our behaviours – who we form ties with, how we behave, how we view the world – are tightly linked to what others do and our embeddedness within social networks (Henry, 2009; Henry and Vollan, 2014; Newman, 2003). One prominent approach to understanding the effects of social ties on environmental consumption is the 'diffusion of innovation', which is a broad set of models and theoretical perspectives on how certain types of behaviour spread through a social system. The diffusion of innovation perspective emphasizes the importance of social networks (particularly communication networks) in the decision to adopt a given technology. By allowing potential adopters to communicate with prior adopters, communication networks help to reduce uncertainty and non-monetary costs associated with technology adoption (Jacobsson and Johnson, 2000; Rogers, 2003).

10.1.2   The Problem of Network Segregation

Existing theories and empirical work on social influence has motivated scholars to give social influence effects a central position in mathematical and computational models of environmental consumption (Bollinger and Gillingham, 2012; Eppstein et al., 2011; McCoy and Lyons, 2014; Zhao et al., 2011). The prevalence of social influence also underscores the importance of underlying social structures in adoption dynamics, and particularly the structure of social networks that link individuals together through relationships such as friendship, kinship, contact at school or the office, or any other relationship through which people obtain information or observe their social surroundings. Social networks are important because they allow for a social multiplier effect, where a single adopter might create additional adopters by virtue of their position within a social space (Banerjee et al., 2013).

While social networks can enhance the diffusion of positive environmental consumption behaviours due to a social multiplier effect, the structure of many real-world networks may also be an impediment to the spread of environmentally friendly technologies. This is because many networks are characterized by *network segregation*, meaning that individuals with similar views on environmental problems, or similar propensities to behave in a certain way, also tend to be socially close to one another (Freeman, 1978; Henry et al., 2011; Schelling, 1971). Network segregation can prevent the positive social multiplier effect from influencing potential adopters with certain characteristics, making it difficult for technology to penetrate certain markets through the social influence effect alone. In other words, network segregation can slow the diffusion of positive behaviours in social systems.

Knowledge of underlying network structures, including network segregation, can provide useful information for firms that seek to market environmentally friendly technologies to private households. The fact that many systems are segregated provides a useful heuristic for technology firms to systematically target individuals with higher propensities to invest – an important consideration given that firms invest a great deal of capital in trying to generate potential leads and close sales. Because of segregation, individuals who tend to be socially close to other technology adopters are likely to share the same underlying characteristics that led to the adoption in others. Thus, firms can more effectively target customers with a higher propensity for adoption by 'following' other adopters – that is, market environmentally friendly technologies to those who already share social connections with prior adopters. This leads technology firms to pursue strategies such as referral programmes or canvassing neighbourhoods that already have high adoption rates.

But while network segregation can be useful for firms to sell more technology systems more quickly, the business of following adopters can also have negative implications for adoption dynamics over time. Consider, for example, Bollinger and Gillingham's (2012) findings that first, peer effects are highly localized geographically and second, the influence of communication with adopters is especially large in low-income households. The conclusion is that marketing efforts in regions where the technology is already visible (i.e., where there are existing solar installations or hybrid-electric vehicles on the roads) is a potentially effective strategy for increasing overall rates of technology adoption. While this might be a promising conclusion from a short-term point of view, it may be that targeting the neighbourhoods with higher adoption rates will hinder the overall adoption process because those neighbourhoods without the technology – which will also tend to be lower-income areas – lack the strong positive effects of social influence.

In this way, following adopters can lead to a fast uptake of adoptions early in the process. Over time, however, this may cause certain market segments to be systematically overlooked by firms and marketers, creating a disparity in the diffusion of environmentally friendly technologies among those individuals with high versus low baseline propensities to adopt (propensity as defined by, for example, affluence). Thus, firms and marketers are incentivized to pursue marketing strategies that speed overall adoption rates but also create inequalities in access to these technologies. These dynamics – and the role of governmental interventions in managing the trade-off between speed and equality of adoption across communities – is the focus of the agent-based model described in the following sections.

10.2 Model Overview

The agent-based model described here explores, in a virtual environment, how firm strategies influence adoption dynamics given variation in characteristics of the system. Social structure – which enables and constrains social influences on adoption decisions – is modelled by assuming that agents are connected within a network, as shown in Figure 10.1.

In this figure, agents are represented as the circular 'nodes' of the network (also referred to as 'vertices'). The shading of agents represents their propensity to adopt – one type of agent will have a higher propensity than the other, representing differences along a range of socio-economic variables that influence the probability of adoption. Lines between agents (in network language these are referred to as 'links' or 'edges') represent the connectedness of agents with one another. Agents are assumed to exercise some degree of social influence through the network links, in the sense that any agent connected to other agents that have adopted is assumed to have a higher probability of adoption than they would otherwise have.

In this particular example, the network is segregated in the sense that agents with similar propensities tend to be linked to one another. In other words, low-propensity agents in this setup tend to be influenced by other low-propensity agents, and high-propensity agents tend to be influenced by other high-propensity agents. These are tendencies only, however, and depend on the level of segregation in the network (this is governed by a model parameter introduced below).

This network perspective may be used to represent a wide variety of social structures and notions of 'closeness'. For example, linkages may represent social connections where people communicate, share information and ideas, or otherwise influence one another's behaviour. Linkages may also represent geographic closeness, where those who are linked reside in the same local neighbourhood. In the real world, we often observe segregation within both types of structure. There are different mechanisms leading to the segregation of a network, such as

Figure 10.1   A sample network connecting agents in a segregated system.

self-selection to a specific subgroup (Henry *et al.*, 2011; Schelling, 1971) or social influence between connected agents (Mäs *et al.*, 2010). However, this model focuses on diffusion processes within fixed network structures, assuming that once the network is established, the evolution process of the network is not critical for the social influences that agents exert on one another.

10.2.1   *Synopsis of Model Parameters*

The main variable of interest here is the firm's strategy and, in particular, whether firms choose to reach out to potential adopters who are linked to other adopters [we refer to this as the 'follow the adopter' (FTA) strategy]. The influence of this strategy, as opposed to the selection of potential adopters uniformly at random (i.e., the 'random' strategy), is examined given three key characteristics of the system that are fixed for a given simulation, but that vary across simulations.

Social influence. Agents who are positioned close to other adopters are assumed to have a higher probability of adoption. This is governed by a social influence model parameter (parameter *SI*) in the interval [0, 1]. A parameter value of zero indicates that agents are not influenced by other adopters in the system. The influence of neighbouring adopters increases as parameter *SI* increases.

Segregation. As noted above, segregation is operationalized as a clustering of agents in terms of their propensity to adopt. In a segregated system, linkages tend to be concentrated among agents with similar propensities (see Figure 10.1). The degree of segregation is governed by the segregation parameter (parameter *S*) in the interval [0, 100]. A value of zero indicates no segregation (i.e., the existence of linkages is independent of agent propensities) while a value of 100 indicates extreme segregation, where very few linkages exist between agents with different propensities. Internally, the segregation parameter S is used to generate hypothetical networks and governs the probability of link formation between pairs of agents as a function of the similarity of each agent–agent pair. In this particular model, ties between agents of the same type are S times more likely to exist than links between agents of different types (for example with a segregation parameter of 75, as in the network displayed in Figure 10.1, links are 75 times more likely to exist between agents of the same colour than between agents of different colours).

Propensity difference. So far we have specified that there are two types of agent in the system, but we have not specified how they differ. The propensity difference model parameter (parameter P) captures the difference between the two types of agent's baseline adoption probability. This parameter ranges in the interval $[0, 1]$, where a value of zero means that agents do not differ in their adoption propensities. These differences increase as P increases.

At model setup, 100 agents are assigned uniformly at random to one of two groups: agents with a low baseline propensity to adopt a given technology and agents with a high baseline propensity to adopt. A network connecting these agents with one another is then generated according to a fixed segregation parameter. Each generated network has density 0.05, meaning that 5% of all possible agent–agent pairs in the system share a link with one another. The density of a network represents the overall connectivity of a social system, and represents the intensity of social interaction (e.g., how often individuals tend to communicate about environmentally friendly technologies, or how often individuals observe one another in their local neighbourhood).

After the initial model setup, the adoption process is allowed to run. The process continues over subsequent time steps t, until all agents in the system have adopted. Adoption dynamics are a result of two stochastic processes: (1) the selection of potential adopters by firms and (2) the decision of an agent to adopt once they are selected. These processes are described in turn.

10.2.2   Agent Selection by Firms

At each time step, firms are assumed to explore the system and select a single non-adopter agent at random. At this point, the selected agent has an opportunity to adopt. In selecting agents as potential adopters, firms are assumed to follow one of two strategies.

FTA strategy. Under the FTA strategy, potential adopters are selected with probability proportional to the number of adopters they are connected to. This is implemented through a lottery system, where an agent who is linked to N adopters has $N+1$ times the probability of being selected as an agent who is linked to no adopters. Thus, the probability that the ith non-adopting agent (agent i) is selected under the FTA strategy at time t, $\Pr_{\mathrm{FTA}}(i, t)$, is given by

$$\Pr_{\mathrm{FTA}}\left(i, t\right) = k\left(N_{i,t} + 1\right), \tag{10.1}$$

where $N_{i,t}$ is the number of adopters that agent i is connected to at time t, and k is a normalizing constant that ensures this is a well-defined probability distribution. This is a stochastic process and agents who are not linked to any adopters may still be chosen, however the probability that they are selected decreases as the number of total adopters in the system increases.

Random strategy. Under a random strategy, potential adopters are chosen from the system uniformly at random – that is, every non-adopter agent has an equal probability of being selected at each time step. Thus, the probability that the ith non-adopting agent (agent i) is selected under a random strategy at time t, $\Pr_R(i, t)$, is given by

$$\Pr_R(i, t) = \frac{1}{Q_t}, \tag{10.2}$$

where Q_t is the number of remaining non-adopters in the system at time t.

10.2.3 Agent Adoption Decisions

Once an agent has been selected by a firm, they make a stochastic decision to adopt or not adopt. We model the probability $A(i, t)$ that agent i adopts after being selected at time t as a logistic function:

$$A(i, t) = \frac{1}{1 + e^{-\left(-2.944 + P*r_i + SI*N_{i,t}\right)}}, \tag{10.3}$$

where r_i is a dummy variable indicating the group (high or low propensity) to which agent i belongs; r_i is coded as one if agent i is in the high-propensity group and zero if agent i is in the low-propensity group.

A few notations about this adoption probability function are needed.

First, the constant coefficient on this logistic function is fixed at −2.944 in order to establish a minimum adoption probability of 5% for any agent in the system once they are selected.[1]

Second, based on this formulation, the propensity difference parameter P becomes a measure of the increase in adoption probability across low- to high-propensity agents. Since P ranges in [0, 1], the maximum difference in adoption probabilities between the two types of agent is roughly threefold. That is, a low-propensity agent with no linkages to adopters will have an adoption probability of 5%, whereas a high-propensity agent with no linkages to adopters will have an adoption probability of about 15% provided that $P=1$.

Third, the formulation also assumes that the marginal linkage to an adopter will influence low-propensity agents more than high-propensity agents, and this difference will also increase as the model parameter P increases. This is simply a feature of the logistic model being used here.

10.3 Results

The above model was coded in R, and adoption dynamics were explored by running a total of 11,060 simulations with firms assumed to follow fixed strategies, either the FTA strategy where potential adopters are chosen with probability proportional to the number of adopters

---

[1] The minimum probability that an agent i will adopt is a situation where i has low propensity (i.e., $r_i=0$) and the agent is not linked to any adopters (i.e., $N_{i,t}=0$). Thus, a constant coefficient of −2.944 fixes this minimum probability of adoption at approximately 0.05, or 5%. We conjecture that the value of this coefficient will not alter adoption trends other than speeding up or slowing down the overall process.

they are connected to (5556 simulations) or a 'random' strategy where potential adopters are chosen uniformly at random (5504 simulations). The influence of other model parameters was explored by assigning, at the setup stage of each simulation, model parameters uniformly at random from their possible ranges noted above. All simulations were run until complete saturation was achieved – that is, until 100% of all agents have adopted. These simulations allow us to characterize the salient ways in which adoption dynamics differ as a result of firm strategy, as well as our assumptions about the structure of the system and agents' behaviour.

10.3.1   Influence of Firm Strategy on Saturation Times

When we evaluate the strategy of following adopters on the basis of saturation times – that is, how long it takes for all agents in the system to adopt – this systematically yields smaller saturation times. This makes intuitive sense, given that the strategy of following adopters will tend to maximize the probability that a single agent will adopt at any given time step. There are two complementary reasons for this: (1) agents chosen under the FTA strategy tend to have more connections to adopters, which increases their adoption probability A and (2) these agents who are connected to other adopters also tend to have higher baseline propensities due to network segregation. The differences in saturation times as a function of firm strategy are depicted in Figure 10.2, which shows the distributions of the average number of time steps it takes for each agent to adopt (a variable we call *average wait time*) over all simulation runs, by FTA and random strategies. The FTA strategy yields significantly smaller average wait times than random strategies, on average ($p<0.001$; t-test with equal variances assumed).

Of course, much of the variation in saturation times seen in Figure 10.2 is also explained by the model parameters that determine the level of segregation in the system (S), the degree to which potential adopters are influenced by their network neighbours (SI) and the difference in adoption propensities between high- and low-propensity agents (P). Table 10.1 summarizes

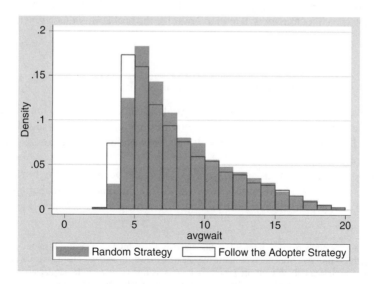

Figure 10.2   Distribution of average wait times by firm strategy.

Table 10.1   Effect of model parameters on saturation times, by firm strategy (DV = average wait time)

Model parameter	FTA strategy	Random strategy
Segregation	−0.001	0.001 *
(S)	(0.001)	(0.001)
Social influence	−10.998 ***	−10.522 ***
(SI)	(0.065)	(0.065)
Propensity difference	−2.679 ***	−2.787 ***
(P)	(0.065)	(0.064)
Constant	14.686 ***	14.734 ***
	(0.059)	(0.059)
N	5556 simulations	5504 simulations
R^2	0.844	0.836

Note: The table reports the results of an ordinary least squares (OLS) regression model with average wait time as the dependent variable.
*** $p < 0.001$; * $p < 0.05$.

the effect of firm strategy, as well as other model parameters, on average wait times. This table presents the results of two linear regression models, conditional on FTA or random strategies, with wait times as the dependent variable and model parameters as independent variables.[2]

In the FTA versus random scenarios, the social influence and propensity difference parameters have a roughly similar effect on average wait times. The estimates for both of these parameters are negative and significant, meaning that as the parameter increases, average wait times tend to decrease. More interesting is the role of network segregation – segregation tends to increase wait times under random strategies, however under the FTA strategy segregation appears to have no effect on average wait time. This supports the notion above that, in a segregated system, the FTA strategy will tend to optimize the probability that a given agent adopts in a particular time step.

10.3.2   *Characterizing Adoption Dynamics*

Examining saturation and average wait times reveals something about the long-term implications of firm strategies; however, this only provides a high-level understanding of model dynamics. Rather than examine the end result only, it is useful to also examine the dynamics of adoption over time.

Figure 10.3 summarizes the adoption dynamics of models run according to random versus follow adopter strategies. The figure shows, over all simulations, the distribution of adoption rates observed at different stages of the process. Since the saturation times (i.e., the times at

---

[2] Here and elsewhere we apply regression models to the analysis of our simulation results. It should be noted that our purpose in doing this is different from the typical use of statistical models to estimate characteristics of the population based on sample data. Instead, these tools are used here to help us find patterns in the large amount of information generated by the computational simulations. Our use of regression models should be viewed as a way to summarize, in an intuitive and simple way, the complex dynamics that emerge from our computational simulations.

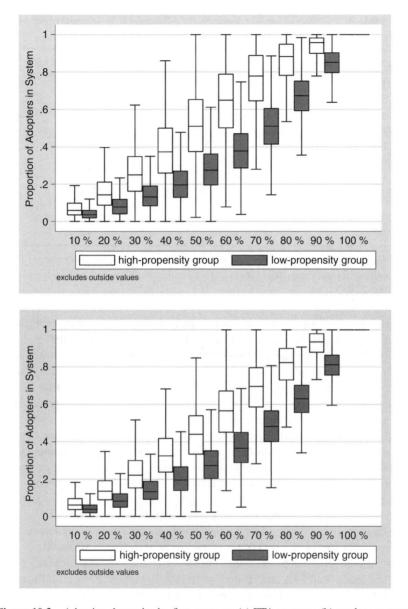

Figure 10.3   Adoption dynamics by firm strategy: (a) FTA strategy; (b) random strategy.

which the adoption rate equals 100%) vary, the time scale of each simulation has been normalized to a percentage of the process measure. On the horizontal axis we have the percentage of the process that has been completed – so, for example, the 20% mark is the time at which a given simulation has reached 20% of the final saturation time. While the actual times will differ across simulations, this approach allows us to observe patterns in the overall shapes of the adoption curves.

Table 10.2 Effect of model parameters on differences in wait times between high- and low-propensity groups (DV = difference in average wait time)

Model parameter	FTA strategy	Random strategy
Segregation	−0.002 ***	0.001 *
(S)	(0.000)	(0.000)
Social influence	−0.638 ***	−0.555 ***
(SI)	(0.029)	(0.021)
Propensity difference	0.765 ***	0.662 ***
(P)	(0.029)	(0.020)
Constant	0.201 ***	0.200 ***
	(0.026)	(0.019)
N	5556 simulations	5504 simulations
R^2	0.180	0.247

Note: The table reports the results of an OLS regression model with equity (difference in average wait time) as the dependent variable. *** $p < 0.001$; * $p < 0.05$.

These graphics allow for a comparison of adoption dynamics within the high- and low-propensity groups. One striking pattern is that, in both FTA and random strategies, the uptake of adoption in the low-propensity group systematically lags behind that of the high-propensity group. This is to be expected. However, random strategies also tend to have a smaller gap between the proportion of low- and high-propensity adopters – meaning that while random strategies tend to yield fewer overall adopters, adopters are more evenly spread out between the high- and low-propensity communities.

Table 10.2 quantifies these differences in the equitable uptake of adoption between high- and low-propensity communities through another set of linear regression analyses. In these models, the 'equity' of adoption in the two communities is operationalized as the difference in average wait times between low and high adopters. Larger values correspond to larger gaps between the average wait time for high-propensity agents and low-propensity agents; therefore, smaller values of this variable suggest more equitable adoption trends in high- and low-propensity communities. Indeed, the adoption trends are significantly more equitable under random versus FTA strategies ($p < 0.001$; t-test with equal variances assumed).

As before, these regression models show consistent effects of social influence and propensity difference under both FTA and random strategies. In particular, social influence tends to increase equity and propensity difference tends to decrease equity, controlling for other factors (keeping in mind that larger values of the dependent variable indicate less equitable adoption trends). Also, as before, the role of segregation differs depending on firm strategy.

10.3.3   Incentivizing Different Strategies

These agent-based models help us to characterize the costs and benefits of two different firm strategies – the FTA strategy tends to yield faster overall adoption rates but less equity in adoption rates between groups, whereas random strategies yield more equitable adoption rates but slower overall saturation.

This suggests a role for governmental intervention to create more desirable outcomes for environmental consumption. Assuming that firms will generally attempt to 'follow adopters', it is possible to strike a desirable balance between the speed and equity of adoption dynamics by incentivizing firms to switch from FTA to random strategies at certain points in the process.

To explore this possibility, an additional set of 969 simulations was run where firms were assumed to start the process following an FTA strategy and then randomly switch to a random strategy before all agents have adopted. The point at which this switch occurs was based on a predetermined, randomly assigned 'trigger point' called the *switch percent*. The switch percent ranges in the interval (0, 1), and defines the proportion of all agents who must adopt before firms switch from an FTA to a random strategy. By allowing the switch percent to vary randomly, we are able to explore the varied effectiveness of incentivizing new strategies at different points in the process.

Table 10.3 summarizes the effect of switching strategies in two regression models – the first with average wait time as the dependent variable and the second with equity as the dependent variable. Independent variables include the random model parameters discussed above, as well as the switch percent variable.

These results show that the longer one waits to incentivize firms to pursue random strategies, the more the overall speed of adoption will increase (i.e., average wait times will decrease) but equity will decrease (i.e., the difference in average wait times between high- and low-propensity groups will increase). This suggests a role for planners to manage the trade-offs between overall adoption speeds and equity by incentivizing random strategies at different points in the process.

The actual point at which this should occur depends on the relative value put on the marginal adopter versus the marginal adopter in an underserved community. These are ultimately political questions; however, these political questions may also be given a quantitative, empirical grounding when coupled with models such as the one presented here.

Table 10.3 Effect of switching from FTA to random strategy at different points during the process on overall saturation times and equity between high- and low-propensity groups

Model parameter	DV = speed (average wait time)	DV = equity (difference in average wait time)
Switch percent	−0.545 **	0.177 **
(FTA to random)	(0.162)	(0.057)
Segregation	0.000	0.001 #
(S)	(0.002)	(0.001)
Social influence	−10.775 ***	−0.590 ***
(SI)	(0.155)	(0.054)
Propensity difference	−2.547 ***	0.734 ***
(P)	(0.158)	(0.055)
Constant	14.872 ***	0.094
	(0.163)	(0.057)
N	969 simulations	969 simulations
R^2	0.841	0.248

Note: The table reports the results of an OLS regression model. *** $p<0.001$; ** $p<0.01$; # $p<0.1$.

10.4   Conclusion

This chapter presents a theoretical, agent-based model of how individuals in a social system make choices about adopting a high-cost, environmentally friendly technology such as solar PV. Underlying this research is the assumption that a switch to such technologies is a preferable outcome for all agents at two scales. At the aggregate level, a transition to more environmentally friendly technologies leads to a more sustainable use of natural resources and will thus improve environmental conditions, for example by lessening our overall reliance on fossil fuels. On the individual level, these technologies can lead to long-term decreases in costs of energy, transportation, water and other basic human needs. While most research on technology adoption focuses on the first outcome of overall adoption rates, the second outcome of the individual access to such technologies – and inequalities in access – requires further study.

Our motivation in studying technology adoption was an apparent paradox: firms are able to use knowledge about segregation in societies and in social networks to more effectively locate individuals with a high propensity to invest in high-cost technologies such as rooftop solar PV or electric vehicles. These strategies serve a social benefit, because they promote more rapid rates of adoption. Over time, however, this strategy can cause certain communities of agents to be systematically overlooked by firms, leading to lower adoption rates within these communities and an unequitable exposure to the benefits of environmentally friendly technologies.

This research has real-world policy implications. We suggested here that firms will have incentives to use a 'follow the adopter' strategy because it maximizes the probability that the cost of reaching out to potential adopters will be rewarded with a successful sale. But if a random strategy is preferable, then it suggests that we will want to incentivize firms to 'seed' underserved communities by reaching out to potential adopters that they otherwise would not market their services to. This work demonstrates that within a segregated network, not all actors will be reached through simple diffusion processes. Thus, we argue that policy incentives could and should be designed in a way that keeps both outcomes in mind: increasing the speed of overall adoption and working towards a more equal distribution of benefits from environmentally friendly technologies. There may also be a role for information campaigns that make certain investments more salient to people in communities with low levels of adoption. And yet, the results of this model also show that strategies have varying effects on adoption trends at different points in the process. Following adopters is not necessarily bad, and is preferable when there are very few adopters in the system. But there is a discernible point in the process where shifts in strategy – through government subsidy, or other policy mechanisms – can allow us to enjoy the best of both worlds by exploiting segregation to identify potential early adopters and overcoming the negative effects of segregation by promoting early adoption in underserved communities.

Acknowledgements

This research was supported by the United States Department of Energy SEEDS Programme (Solar Energy Evolution and Diffusion Studies) and the National Renewable Energy Laboratory, under award #DE-AC36-08GO28308.

References

Arkesteijn, K. and Oerlemans, L. (2005) The early adoption of green power by Dutch households: An empirical exploration of factors influencing the early adoption of green electricity for domestic purposes. *Energy Policy*, **33**(2), 183–196.

Axsen, J., Mountain, D.C. and Jaccard, M. (2009) Combining stated and revealed choice research to simulate the neighbor effect: The case of hybrid-electric vehicles. *Resource and Energy Economics*, **31**(3), 221–238.

Banerjee, A., Chandrasekhar, A.G., Duflo, E. and Jackson, M.O. (2013) The diffusion of microfinance. *Science*, **341**(6144), 1236498-1–1236498-7.

Basart, J.M. and Serra, M. (2011) Engineering ethics beyond engineers' ethics. *Science and Engineering Ethics*, **19**, 179–187.

Berger, W. (2001) Catalysts for the diffusion of photovoltaics: A review of selected programmes. *Progress in Photovoltaics: Research and Applications*, **9**(2), 145–160.

Bickerstaff, K., Walker, G.P. and Bulkeley, H. (2013) *Energy Justice in a Changing Climate: Social equity and low-carbon energy*, Zed Books, London.

Bollinger, B. and Gillingham, K. (2012) Peer effects in the diffusion of solar photovoltaic panels. *Marketing Science*, **31**(6), 900–912.

Clark, W.C. and Dickson, N.M. (2003) Sustainability science: The emerging research program. *PNAS*, **100**(14), 8059–8061.

Claudy, M.C., Michelsen, C. and O'Driscoll, A. (2011) The diffusion of microgeneration technologies – assessing the influence of perceived product characteristics on home owners' willingness to pay. *Energy Policy*, **39**(3), 1459–1469.

Claudy, M.C., Peterson, M. and O'Driscoll, A. (2013) Understanding the attitude–behavior gap for renewable energy systems using behavioral reasoning theory. *Journal of Macromarketing*, **33**, 273–287.

Davis, M. (2012) A plea for judgment. *Science and Engineering Ethics*, **18**, 789–808.

Dietz, T., Rosa, E.A. and York, R. (2007) Driving the human ecological footprint. *Frontiers in Ecology and the Environment*, **5**(1), 13–18.

Drury, E., Miller, M., Macal, C.M., Graziano, D.J., Heimiller, D., Ozik, J. and Perry IV, T.D. (2012) The transformation of southern California's residential photovoltaics market through third-party ownership. *Energy Policy*, **42**, 681–690.

Eppstein, M.J., Grover, D.K., Marshall, J.S. and Rizzo, D.M. (2011) An agent-based model to study market penetration of plug-in hybrid electric vehicles. *Energy Policy*, **39**(6), 3789–3802.

Faiers, A. and Neame, C. (2006) Consumer attitudes towards domestic solar power systems. *Energy Policy*, **34**(14), 1797–1806.

Faiers, A., Neame, C. and Cook, M. (2007) The adoption of domestic solarpower systems: Do consumers assess product attributes in a stepwise process? *Energy Policy*, **35**(6), 3418–3423.

Freeman, L.C. (1978) Segregation in social networks. *Sociological Methods & Research*, **6**, 411–429.

Gallo, C. and De Bonis, M. (2013) A neural network model for forecasting photovoltaic deployment in Italy. *International Journal of Sustainable Energy and Environment*, **1**(1), 1–13.

Guidolin, M. and Mortarino, C. (2010) Cross-country diffusion of photovoltaic systems: Modelling choices and forecasts for national adoption patterns. *Technological Forecasting and Social Change*, **77**(2), 279–296.

Henry, A.D. (2009) The challenge of learning for sustainability: A prolegomenon to theory. *Human Ecology Review*, **16**, 131–140.

Henry, A.D. and Vollan, B. (2014) Networks and the challenge of sustainable development. *Annual Review of Environment and Resources*, **39**, 583–610.

Henry, A.D., Prałat, P. and Zhang, C.Q. (2011) Emergence of segregation in evolving social networks. *Proceedings of the National Academy of Sciences*, **108**, 8605–8610.

Jacobsson, S. and Johnson, A. (2000) The diffusion of renewable energy technology: An analytical framework and key issues for research. *Energy Policy*, **28**(9), 625–640.

Jager, W. (2006) Stimulating the diffusion of photovoltaic systems: A behavioural perspective. *Energy Policy*, **34**(14), 1935–1943.

Jenkins, K., McCauley, D., Heffron, R., Stephan, H. and Rehner, R. (2016) Energy justice: A conceptual review. *Energy Research & Social Science*, **11**, 174–182.

Kahn, M.E. and Vaughn, R.K. (2009) Green market geography: The spatial clustering of hybrid vehicles and LEED registered buildings. *BE Journal of Economic Analysis & Policy*, **9**(2), 1–24.

Kaplan, A.W. (1999) From passive to active about solar electricity: Innovation decision process and photovoltaic interest generation. *Technovation*, **19**(8), 467–481.

Kates, R.W., Clark, W.C., Hall, J.M., Jaeger, C., Lowe, I., McCarthy, J.J. *et al.* (2001) Sustainability science. *Science*, **292**, 641–642.

Kwan, C.L. (2012) Influence of local environmental, social, economic and political variables on the spatial distribution of residential solar PV arrays across the United States. *Energy Policy*, **47**, 332–344.

Labay, D.G. and Kinnear, T.C. (1981) Exploring the consumer decision process in the adoption of solar energy systems. *Journal of Consumer Research*, **8**(3), 271–278.

Lobel, R. and Perakis, G. (2011) Consumer choice model for forecasting demand and designing incentives for solar technology. Available at: ssrn.com/abstract=1748424.

Mäs, M., Flache, A. and Helbing, D. (2010) Individualization as driving force of clustering phenomena in humans. *PLOS Computational Biology*, **6**(10), e1000959 1–8.

McCoy, D. and Lyons, S. (2014) Consumer preferences and the influence of networks in electric vehicle diffusion: An agent-based microsimulation in Ireland. *Energy Research & Social Science*, **3**, 89–101.

Newman, M.E.J. (2003) The structure and function of complex networks. *SIAM Review*, **45**, 167–256.

Rai, V. and Henry, A.D. (2016) Agent-based modelling of consumer energy choices. *Nature Climate Change*, **6**(6), 556–562.

Rai, V. and Robinson, S.A. (2013) Effective information channels for reducing costs of environmentally-friendly technologies: Evidence from residential PV markets. *Environmental Research Letters*, **8**(1), 014044.

Rai, V. and Sigrin, B. (2013) Diffusion of environmentally-friendly energy technologies: Buy versus lease differences in residential PV markets. *Environmental Research Letters*, **8**(1), 014022.

Robison, W. (2010) Design problems and ethics, in I. Poel and D. Goldberg (eds), *Philosophy and Engineering: An emerging agenda*, Springer-Verlag, Dordrecht, pp. 205–214.

Rogers, E.M. (2003) *Diffusion of Innovations* (5th edn), Free Press, New York, NY.

Rosa, E.A., Diekmann, A., Dietz, T. and Jaeger, C. (eds) (2009) *Human Footprints on the Global Environment: Threats to sustainability*, MIT Press, Cambridge, MA.

Schelling, T.C. (1971) Dynamic models of segregation. *Journal of Mathematical Sociology*, **1**, 143–186.

Schelling, T.C. (2006) *Micromotives and Macrobehavior*, W.W. Norton & Co., New York, NY.

Sovacool, B.K. and Dworkin, M.H. (2014) *Global Energy Justice*, Cambridge University Press, Cambridge.

Taylor, M. (2008) Beyond technology-push and demand-pull: Lessons from California's solar policy. *Energy Economics*, **30**(6), 2829–2854.

van Benthem, A., Gillingham, K. and Sweeney, J. (2008) Learning-by-doing and the optimal solar policy in California. *The Energy Journal*, **29**(3), 131–151.

Vasseur, V. (2012) Innovation adopters: A new segmentation model: The case of photovoltaic in the Netherlands. Paper presented at the 2nd World Sustainability Forum.

Yuan, X., Zuo, J. and Ma, C. (2011) Social acceptance of solar energy technologies in China – end users perspective. *Energy Policy*, **39**(3), 1031–1036.

Zhai, P. and Williams, E.D. (2012) Analyzing consumer acceptance of photovoltaics (PV) using fuzzy logic model. *Renewable Energy*, **41**, 350–357.

Zhao, J., Mazhari, E., Celik, N. and Son, Y.J. (2011) Hybrid agent-based simulation for policy evaluation of solar power generation systems. *Simulation Modelling Practice and Theory*, **19**(10), 2189–2205.

11

Modelling in the 'Muddled Middle': A Case Study of Water Service Delivery in Post-Apartheid South Africa

Jai K. Clifford-Holmes, Jill H. Slinger, Chris de Wet and Carolyn G. Palmer

11.1 Introduction

At the centre of the South African water law reform process initiated by the first democratic government in 1994 lay the challenge of managing water differently from the way it was managed under apartheid (Rowlston, 2011). This process culminated in the promulgation of the National Water Act of 1998 and the Water Services Act of 1997, which are regarded internationally as ambitious and forward-thinking instances of legislation that reflect the broad aims of integrated water resource management (IWRM) (Schreiner, 2013). The local government sector was also redesigned in the first decade of democracy, with extensive powers and autonomy granted to the sector under a policy of developmental local government (DLG) (Republic of South Africa, 1998). Both DLG and IWRM aspire towards decentralized decision-making, participatory governance and management, and the integration of multiple issues that have social, environmental and technical dimensions. However, both IWRM and DLG have been criticized for implementation failures in post-apartheid South Africa (Mehta *et al.*, 2014; Siddle and Koelble, 2012), which have led to proposals from national government that water policy needs to be redesigned (Department of Water Affairs, 2013) and that local government powers and functions should be reassessed (Department of Water and Sanitation, 2014).

This chapter focuses on the governance and management of water services (the primary intersection between the legislative frameworks for local government and those for water management and water service delivery) to motivate the use of a modelling approach to

Social Systems Engineering: The Design of Complexity, First Edition. Edited by César García-Díaz and Camilo Olaya.
© 2018 John Wiley & Sons Ltd. Published 2018 by John Wiley & Sons Ltd.

explore the ambiguous 'muddled middle' between policy design, implementation and adaptation. The modelling approach involves an ethnographically embedded form of participatory system dynamics modelling. This chapter applies the modelling approach to a case study, drawing on the authors' extended participation in an action research process involving water services in the Sundays River Valley Municipality (SRVM) in South Africa (SA).

11.2   The Case Study

The SRVM contains a relatively small population of 54,500 people and is located in the impoverished Eastern Cape province (Statistics South Africa, 2014). The SRVM is a primarily rural municipality with a number of small urban settlements interspersed between large commercial farms and nature reserves. The local government authority of the SRVM is responsible for providing water services to all urban water users within its jurisdiction. As of 2010, 47% of the population subsisted on a household income of less than R800 per month (approximately US$80), with unemployment estimated at 44% (Sundays River Valley Municipality, 2010). Almost half of the municipal population is therefore reliant on social grants from national government and on receiving free basic services (including water and sanitation) from local government. Over a third of South African municipalities are of a similar size and socio-economic character to the SRVM. Despite this representative quality, in many respects, the SRVM is also an extreme case: in 2010, national and provincial government departments initiated intervention processes in the SRVM, following an extended period of financial mismanagement and bankruptcy in which the provision of water services became increasingly erratic and unreliable. In spite of extensive government interventions, the area continued to face declining water services with disastrous effect. In September 2014, a series of violent service delivery protests broke out in the main town of Kirkwood, where municipal offices and infrastructure were set alight by protestors and burned to the ground (South African Press Association, 2014). The SRVM case therefore offers both extreme and representative aspects of the water challenges faced by local authorities in post-apartheid SA (Clifford-Holmes, 2015; Clifford-Holmes et al., 2016a), which, following Yin (2009), provided the rationale for employing the 'single-case design' utilized in this study.

The modelling approach described in this chapter evolved out of an action research project funded by the South Africa Netherlands Research Programme for Alternatives in Development (SANPAD), entitled 'From policy to practice: enhancing implementation of water policies for sustainable development' (Palmer et al., 2014). The SANPAD project adopted a transdisciplinary stance and used a range of theories, methods, approaches and practices in novel ways, with the aim of testing their usefulness in breaching barriers impeding the implementation of IWRM in SA. Through the action research process, it became evident that the SRVM was a representative case of a South African region facing what Ohlsson (1999) terms 'second-order scarcity'. Water scarcity is frequently more than a function of demand outstripping supply, resulting in physical 'first-order' scarcity. In contrast to the latter characterization of scarcity, Ohlsson (1999) construes second-order scarcity as a given social entity's lack of adaptive capacity. In applying this concept to water in SA, Tapela (2012) argues that 'social water scarcity' occurs when a confluence of factors – including insufficient finances, human capabilities and political will – results in the provision of water services failing to meet a growing demand. Given the abundance of raw (untreated) water in the SRVM, shortages of potable (i.e., drinking-quality) water are clearly a case of second-order scarcity in general, and what Tapela (2012) refers to as 'social water scarcity' in particular. Given this social water scarcity, modelling approaches that contain multiple social dimensions (as to how the modelling is both undertaken and used) were found to be valuable in the SANPAD action research process in the SRVM. The following section locates the way in which modelling was undertaken and used in this study in relation to the broader literature on modelling in the water sector.

11.3   Contextualizing Modelling in the 'Muddled Middle' in the Water Sector

In reviewing the multiple roles that models perform in the water sector, Hare (2011) uses the distinction of Haag and Kaupenjohann (2001) between modelling for scientific research purposes (primarily for forecasting and prediction) and modelling to support policy- and decision-making, noting that it is the latter category that particularly requires stakeholder participation. Within engineering communities, modelling is often used to support design processes and in doing so, engineering modelling can be distinguished from scientific modelling (Bissell and Dillon, 2012; Elms and Brown, 2012; Epstein, 2008). Modelling is useful to engineers as a tool to support causing 'the best change in an uncertain situation within the available resources' (Koen, 2003, p. 24). System dynamics (SD) is one approach that has emerged as useful for a range of these modelling purposes. SD is defined here as 'a way of modelling people's perceptions of real-world systems based especially on causal relationships and feedback' (Mingers and Rosenhead, 2004, p. 532). Within the water sector, the themes that have traditionally garnered the greatest attention of SD modellers are those of regional planning and river basin management, and flooding and irrigation (as reviewed in Winz et al., 2009). In the last decade, SD has also increasingly been used to investigate the challenges associated with urban water supply [as seen in studies on municipal water conservation policies (Ahmad and Prashar, 2010), urban drinking water supply (Xi and Poh, 2013) and urban wastewater management (Rehan et al., 2014)]. The use of SD modelling is on the increase throughout the eleven countries in Southern Africa, as shown in a systematic review of scientific literature published between 2003 and 2014 (Brent et al., 2016). More specifically, SD has a long and diverse history of use in developmental planning, strategic management, and mediation and brokerage in the South African water sector (Clifford-Holmes et al., 2017).

Practitioners of SD have been criticized for over-simplifying the 'problem definition' stage of the modelling process; for reducing 'problematic situations' (where various problems must be dealt with at the same time) to single 'problems'; and lastly, for offering few methodological details on how policies derived and tested through SD process can be implemented in the real world (Rodríguez-Ulloa et al., 2011). However, a number of approaches to stakeholder-engaged SD modelling have developed in the last decade, partly in response to the above limitations. A well-developed approach is 'group model-building' (Rouwette and Vennix, 2006; Vennix, 1999), which aims to build or come to a group understanding of a complex problem. Other examples of stakeholder-engaged modelling in the water sector that employ SD include:

- participatory model building (Langsdale, 2007; Stave, 2010);
- cooperative modelling (Cockerill et al., 2006; Tidwell et al., 2004);
- mediated modelling (van den Belt, 2004).

Critics argue that whilst the above-mentioned approaches incorporate mental modelling and social learning, SD still remains functionalist in nature: given that SD 'sees system structure as the determining force behind system behaviour and tries to map that structure in terms of the relationships between feedback loops' (Jackson, 2003, p. 81), it fails to account for the 'innate subjectivity of human beings' (Flood and Jackson, 1991, p. 79). Accordingly, the theory, methodology and methods of SD have been judged by some critics as unsuitable to the subject matter of its concern: 'human beings, through their intentions, motivations and actions, shape social systems... [hence] we need to understand the subjective interpretations of the world that individual social actors employ' (Jackson, 2003, p. 80). By trying to study social systems 'objectively', SD 'misses the point [because] social structure emerges through a process of negotiation and renegotiation of meaning' (Jackson, 2003, p. 80).

If the above criticisms hold true, then SD modelling is poorly suited to social systems. The approach taken in this study can be situated in relation to three responses to the latter 'SD is structure-focused' critique. The first response is the one posited by critics from within the broader systems thinking field. Instead of relying on SD to 'do everything', Jackson (2003, p. 80) argues that 'a critical systems thinker is likely to want to combine the strengths of [SD] with what other systems approaches have learned to do better'. One such approach is the soft systems methodology (SSM), which provides qualitative tools and techniques for exploring diverse perspectives on a problematic situation whilst addressing 'the socio-political elements of an intervention' (Lane and Oliva, 1998, p. 214). The SSM approach is based on an interpretative perspective of social settings, in which humans are understood to 'negotiate and re-negotiate with others their perceptions and interpretations of the world outside themselves' (Checkland, 1981, pp. 283–284). Several attempts have been made to integrate SD with SSM. 'Holon dynamics', for example, promotes using SSM to generate multiple perspectives on a problematic situation before studying it further using SD (Lane and Oliva, 1998). Similar approaches have been advocated in SD textbooks (see Maani and Cavana, 2007) and have been formalized into an integrated framework called the 'soft system dynamics methodology' (Rodríguez-Ulloa and Paucar-Caceres, 2005). Rodriquez-Ulloa et al. (2011) is one of the only published applications of the latter integrated approach.

An alternative approach is to use SD in combination with other 'problem structuring methods' (Mingers and Rosenhead, 2004), which is the second response to the 'SD is structure-focused' critique. Problem structuring methods (PSMs) are useful in engaging difficult problems that are characterized by multiple actors and perspectives, with incommensurable or conflicting interests, and key uncertainties. As such, PSMs should 'operate iteratively... [and] enable several alternative perspectives to be brought into conjunction with each other' (Mingers and Rosenhead, 2004, p. 531). Horlick-Jones and Rosenhead (2007) report on research and consulting practice that combined ethnographic tools with various PSMs, referring to this combination as a form of 'methodological hybridisation' (Horlick-Jones and Rosenhead, 2007, p. 588). Ethnographic tools and perspectives call to attention the subjective interpretations of the world that individual social actors employ. This focus on agency can serve as a counterbalance to the focus on structure that is arguably central to SD.

The third response directly rebuts the critique that 'SD is structure-focused'. This rebuttal positions SD as beginning from the perspective of stakeholders' activities, recently exemplified in Olaya's (2015) motivation of 'operational thinking' as being key to the way in which SD modelling is undertaken. Rather than modelling being bound by available data and the requirement to fit a model's simulations to historical behaviour, SD is explicitly stakeholder-centred, drawing data from a broad range of sources (Ford, 2009). Central to this way of modelling is the exploration of questions like 'who are the actors in the dynamics of a complex system and how do their perceptions, pressures and policies interact?' (Richardson, 2011, p. 229). SD practitioners – including Lane (2000), Richardson (2011) and Olaya (2015) – are united in their arguments that SD is a non-deterministic approach that ascribes to human beings the capacity to invent (and reinvent) their own futures, whilst acknowledging that this capacity (i.e., agency) is circumscribed by structural conditions. Seen in this way, the orientation of SD towards the structure–agency debate within the social sciences is consistent with the sociological theory and practice advocated by Layder (1998). Rather than conceiving of a structure/agency dualism, Layder (1998) posits a 'middle way' as a dialectical understanding of the co-existence and interrelationship of structure and agency.

This chapter reports on a modelling approach that drew strands from each of the three responses to the 'SD focuses-on-structure' critique. The approach conceived of stakeholders as influencing 'who gets what water, when and how... between the aspirational policy realm and the practical realities on the ground' (Clifford-Holmes et al., 2016a, p. 1002). This space between the clean lines of policy design (how we want things to work) and the messy operational reality of implementation was conceived as 'the muddled middle' (Clifford-Holmes et al., 2016a, p. 1002). 'Modelling in the muddled middle' referred to the conscious choice to engage with stakeholders in the messy reality they face and to employ modelling tools and research methods in seeking to understand (and where possible, address) their issues. Following Vriens and Achterbergh (2006), SD models were conceived as being made of social systems; built in social systems; and built for social systems, whilst recognizing that modelling can perform many functions and be undertaken for multiple purposes (Epstein, 2008). Rather than conceiving of, and undertaking, modelling as a standalone analytical activity, the approach described in this chapter combined SD with institutional analysis and ethnographic tools in a form of methodological hybridization, as outlined in the following section.

11.4   Methods

The case and context of this study were summarized in Section 11.2, which introduced the broader research project within which this study was nested (described further in Clifford-Holmes, 2015; Molony, 2015; Muller, 2013; Palmer et al., 2014). In order to address the aims of this study, a multi-method research approach was employed, which drew on institutional, ethnographic and systems analyses within an evolving, transdisciplinary methodology (Wickson et al., 2006). As part of the single-case study research design, qualitative and quantitative data were collected via participant observation, direct observation, semi-structured interviews, and from documentary and archival sources. Ethnographic analysis was performed following extended fieldwork in the SRVM, which was undertaken between 2011 and 2012, with follow-up visits for workshops and meetings between 2012 and 2014. This fieldwork and the associated analysis allowed for a detailed and multi-layered understanding of water service challenges in the SRVM to develop (Clifford-Holmes, 2015), with a focus on the practical activities of water management (Agnew, 2011). The challenges of municipal water supply in the SRVM were found to involve multiple interactions of material and informational flows across technical and social systems at different scales. As Forrester (1968, 1970) noted, the manner in which people interact with technical and natural systems is contained in their practices.

Table 11.1 Summary of the four models, displayed in terms of the number (no.) of stock variables (with the relative percentage of the stock variables in relation to total variables in brackets); the number of feedback loops; and references for model documentation

	Model name	Model acronym	Timeframe	No. of stock variables	No. of feedback loops	Model documentation
1	The cooperatives model	*CoOP*	April–Sept. 2012	20 (16.5%)	20	Appendix D of Clifford-Holmes (2015)
2	The Greater Kirkwood water supply model	*GKWS*	Nov. 2012– July 2013	13 (10.3%)	71	D'Hont (2013), D'Hont *et al.* (2013) and Appendix E of Clifford-Holmes (2015)
3	The Kirkwood water demand model	*K-DEM*	Aug. 2013– March 2014	6 (7%)	13	Clifford-Holmes *et al.* (2014) and Appendix F of Clifford-Holmes (2015)
4	The modes of failure model	*MoF*	Jan. 2014– March 2015	9 (9.3%)	143	Clifford-Holmes *et al.* (2015)

System dynamics was selected as an ideal method for exploring these practices and the problems to which they give rise, particularly at the strategic level (Sterman, 2000), for the multiple reasons explored in Section 11.3. Rather than building one large model, a portfolio of small models was developed, using the standard modelling process outlined in SD textbooks of problem scoping; formulating a dynamic hypothesis using qualitative causal loop diagrams and specifying a stock-flow structure; then estimating parameters and initial conditions before building and testing a simulation model (Ford, 2009; Maani and Cavana, 2007; Sterman, 2000). The four SD models developed in this study are summarized in Table 11.1. The first model was primarily undertaken as an initial scoping exercise with little stakeholder engagement; as such, this first model is not discussed here (for details, see Clifford-Holmes, 2015). The next three models were developed through a process of system exploration, where different stakeholder groups were engaged in settings where they were at ease, rather than consensus building in multi-stakeholder workshops (as in group model building). Hence, the SD models were developed as system understanding grew over the course of the research period between April 2012 and March 2015 (see Table 11.1), with the models being co-validated with stakeholder groups rather than being co-constructed. The following section describes the dialectical process by which three out of the four models were constructed and used as part of a broader action research process.

11.5 Results

The results of the 'modelling in the muddled middle' approach are described here in terms of which stakeholders were engaged, with what model, in which settings, and in what part of the modelling process. The models themselves are described and presented in increasing levels of

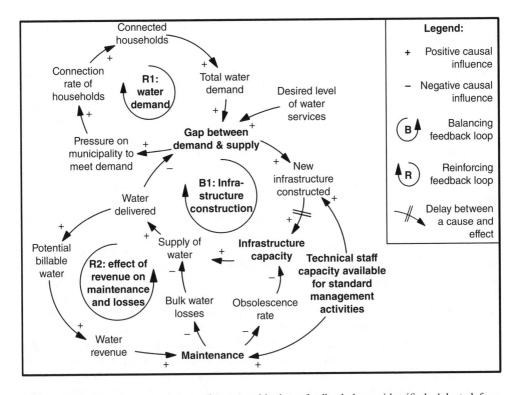

Figure 11.1 Overview causal loop diagram, with three feedback loops identified. Adapted from Clifford-Holmes *et al.* (2015, p. 6).

detail in Clifford-Holmes (2015), as referenced in Table 11.1, and are introduced here with reference to the causal loop diagram (CLD) in Figure 11.1.

The driver of the system presented in Figure 11.1 is the 'gap between demand and supply'. This gap *increases* with the 'total water demand' and *decreases* with 'water delivered'. The standard municipal response to this gap is to adjust the infrastructure capacity, through refurbishing current infrastructure and constructing new infrastructure. With 'new infrastructure constructed', the total 'infrastructure capacity' increases after a delay that accounts for the construction lead time (represented diagrammatically in Figure 11.1 by the two lines on the arrow between these variables). With the capacity increasing, the 'supply of water' and the 'water delivered' can increase. This is the first balancing loop in the CLD (**B1: Infrastructure construction**). The Greater Kirkwood water supply model focused on exploring the operational particularities of infrastructure capacity in the modelled system.

The primary driver of water demand in the region is from households that are connected to the municipal reticulation system for drinking water and sanitation services. Historical data suggests that between 2001 (when the municipality was formed) and 2005, 22% of households in Greater Kirkwood had waterborne sanitation (Kwezi V3 Engineers, 2005). By 2011, the level of service fraction had risen to 77% (Amatola Water, 2014, p. 3), indicating that the gap between demand and supply resulted in 'pressure on [the] municipality to meet demand' by increasing the 'connection rate of households'. The greater the total number of 'connected households', the greater the total water demand and the greater the 'gap between demand and supply', which forms a reinforcing feedback loop (**R1: water demand**). The Kirkwood water demand model explored this aspect.

The rate at which infrastructure decreases in value and function (therefore requiring refurbishing or replacing) is referred to as the 'obsolescence rate', which is influenced by municipal officials undertaking day-to-day maintenance as part of the operational regime of water service delivery. How much maintenance work can be accomplished is influenced both by the attention that the municipal staff can give to maintenance and the revenue dedicated to maintenance, which is subject to the revenue derived by providing water services. The more water is delivered to users, the more the 'potential billable water'. By increasing water revenue, the municipality is able to perform more maintenance, and therefore reduce bulk water losses and the obsolescence rate, which enables more potable water to be delivered and in turn increases the 'potential billable water' (**R2: effect of revenue on maintenance and losses** – explored in the modes of failure model).

At a workshop facilitated by SANPAD researchers in 2012, it emerged that different actors had different understandings of the physical limitations of the Greater Kirkwood water supply scheme. The two central actors were the municipality (the SRVM) and the Lower Sundays River Water User Association (L-WUA), which was the bulk water supply agent to the municipality (see Clifford-Holmes *et al.*, 2016a). The differing views of these actors presented the opportunity to develop a model on the Kirkwood water supply scheme, where the process of model building (and the model itself) could be used for analytical purposes and as part of the ongoing process of facilitation and action research. One of the advantages of SD modelling in this particular situation was that detailed and accurate data on the water supply system (e.g., the evolution in water demand over time and variations in the water treatment pumping capacity) were not prerequisite to modelling. Instead, stakeholders contributed a deep knowledge of system design and experienced behaviour over time, which were captured within the model as operations (including decision rules and the use of information). This showed the importance of operational thinking in the way in which SD modelling was deployed in the case study (Olaya, 2015).

The Greater Kirkwood water supply (GKWS) model explored how water outages have a differential impact (in that all areas of the region do not run dry at the same time). Figure 11.2 demonstrates the relative time periods that the different residential zones are without water, over a one-year simulation period. The reasons for this are detailed in Clifford-Holmes (2015) and D'Hont (2013). What is important to note here is that zone 1 is still primarily inhabited by white residents, who are comparatively wealthier than the poorer (predominantly black) communities residing in zones 2 to 5. The safety mechanisms in the technical design of the pumps, together with the nature of the gravity-feed system to Kirkwood town, collectively result in zone 1 being the last area to be cut off, and the first area to receive water. Municipal residents living in zones 2 to 5 understandably complained that this aspect of system behaviour is inherently unjust and discriminatory, both along class and racial lines (which is especially problematic given the historical development of the town and the politics of the country under apartheid, resulting in the preferential supply of potable water to zone 1 by design). The GKWS model illustrated the inequitable supply system by simulating the 'time without water' across the different zones. In Figure 11.2, for example, zones 4 and 5 (line 3) are shown to lack sufficient access to potable water for more than 80% of the year. The GKWS model usefully demonstrates that the underlying cause of the experience of differentiated service delivery lies

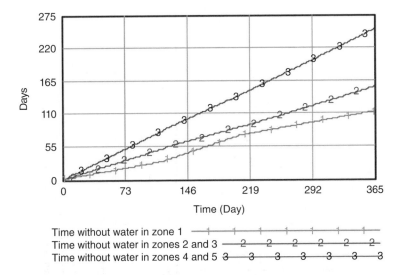

Figure 11.2 Simulation output from the GKWS model illustrating the time without potable water, differentiated across supply zones 1, 2 and 3, 4 and 5 [see D'Hont (2013) and Clifford-Holmes (2015) for details].

more in the physical characteristics of the infrastructural design than in the operational choices made by the municipal employees managing the infrastructure. Apart from the municipal engineering staff, this represented a new insight for all stakeholders and was demonstrative of the objectives of the modelling approach (which focused on systemic exploration rather than blaming individuals or institutions).

Officials from the municipality and the L-WUA involved in the GKWS modelling process acknowledged that infrastructure capacity issues were exacerbated by increasing water demand in the area. However, these demand dynamics were less understood and even more controversial than the supply-side dynamics. Hence, exploring the determinants of the rising demand for potable water was central to the third model, which is detailed in Clifford-Holmes *et al.* (2014). This third model, called the Kirkwood water demand model (K-DEM), modelled the transition from households receiving basic levels of water services to those same households becoming connected to the municipal networks for water supply and waterborne sanitation. The demographic impact of this transition is visible in Figure 11.3(a), which shows that the population living in unconnected households began declining from 2006, as the population living in connected households increased sharply (resulting in a net increase in the total water demand). But whilst this dynamic was modelled in K-DEM at the structural level, the socio-political dynamics affecting this structure were not. One of the most important of these dynamics is driven by the way in which municipal councillors (who are politically elected officials) respond to their constituencies' demands. A senior technical official in the SRVM (interviewed in Clifford-Holmes, 2015, p. 269) described this as follows:

> Our department has been pressured by councillors to get houses up – even without the necessary infrastructural development to service these houses. Then, when we [the technical division] cannot serve the houses, we are blamed for it.

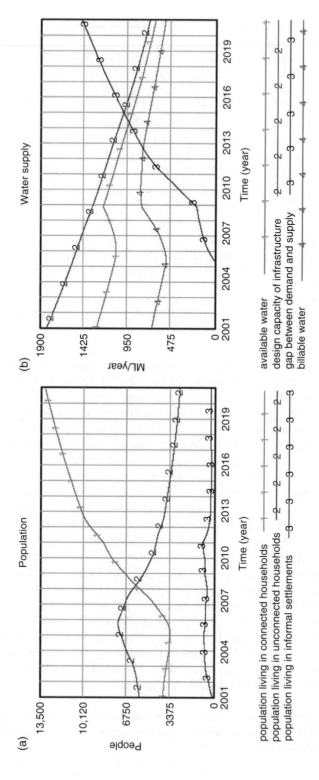

Figure 11.3 The left-hand graph (a) displays the simulated population dynamics in the Greater Kirkwood region. The right-hand graph (b) displays the simulated water supply and the quantity of 'billable water' in relation to the gap between demand and supply [see Clifford-Holmes *et al.* (2015) for full model].

A secondary consequence of this driving demand for potable water is that water conservation and water demand management policies are politically unpopular and therefore operationally difficult to implement (despite being technically feasible). The K-DEM model was used as part of the action research process that engaged national and regional water authorities, in addition to municipal councillors and technical officials, around these issues. The factors pertaining to household water demand were discussed and model outputs from the K-DEM model were compared with engineering assessments and projections over the course of a planning workshop held in the SRVM in early 2014 (described in Clifford-Holmes, 2015). This planning workshop, along with the associated reports and funding proposals, provided the opportunity to validate the dynamics explored in the K-DEM model. However, the process of interacting with the stakeholder groups highlighted what the K-DEM and GKWS models both failed to capture, namely why the SRVM could neither provide adequate quantities of potable water nor effectively manage water demand in the Greater Kirkwood region, and what the different stakeholders could do about these management challenges. The fourth modelling exercise was undertaken to explore the latter issues and did so by synthetically drawing on the earlier modelling work in a process of methodological hybridization (the results of which are described below).

Six interlinked 'modes of failure' were identified through the fourth modelling exercise, which are summarized in Clifford-Holmes *et al.* (2015, p. 1) as:

> The underinvestment in, and over-extension of, water supply infrastructure; the lack of pro-active infrastructure planning combined with the lack of systematic maintenance; the enforced 'fire-fighting' reaction of municipal staff to service delivery crises; and inadequate financial means, infrastructure capacity, and technical staffing capacity.

The CLD in Figure 11.4 provides a 'dynamic hypothesis' of how these interlinked modes were explored in the modes of failure (MoF) model. When the SRVM is unable to increase capacity through constructing new infrastructure, then an alternative way in which it can reduce the demand–supply gap is through over-extending the current infrastructure above its design capacity. The 'use of infrastructure above design capacity' allows the municipality to increase the quantity of potable water produced, and therefore increase the supply of water, which in turn decreases the gap between demand and supply (**B2: effect of infra. [infrastructure] overuse on supply**). The short-term dividend of this policy is visible in the simulations of the MoF model. In Figure 11.3(b), the gap between demand and supply begins to grow in mid-2005 (line 3), as the design capacity of municipal water supply infrastructure continues to decrease (line 2). In order to address this gap, the municipality responds by starting to over-extend the infrastructure, which increases the 'available water' (line 1) between 2006 and late 2009. Given that more houses were connected to the municipal networks for water and sanitation in this period, the proportion of 'available water' from which the municipality could earn revenue (i.e., 'billable water' in line 4) also increased. However, 'water revenue' is determined by the proportion of billable water for which the municipality actually receives payment ('% cost recovery' in Figure 11.4). Similarly, the 'revenue dedicated to maintenance' is subject to the proportion of the 'water revenue' that is reserved for this purpose ('% revenue ring-fenced'). When little to no water revenue is ring-fenced, the municipality can perform no maintenance, which results in water losses and the obsolescence rate increasing, which reduces the water that can be delivered and, in turn, decreases the quantity of billable water.

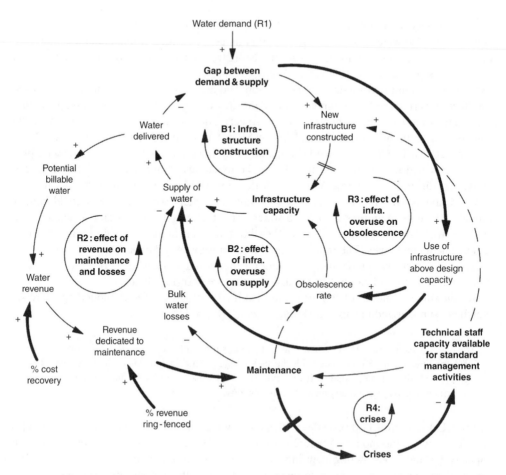

Figure 11.4 Expanded CLD showing an additional three feedback loops (B2, R3 and R4) that build on Figure 11.1. The emboldened arrows show the causal relations between the additional variables, with the two arrows dashed for the sake of clarity where the emboldened arrows overlap with other arrows. *Source:* Adapted from Clifford-Holmes *et al.* (2015, p. 6).

Maintenance is required particularly when the municipality over-extends the water supply infrastructure (given that the longer this infrastructure operates above its design capacity, the quicker it obsolesces, requiring refurbishment and replacement earlier than planned, which has a reinforcing feedback effect (**R3: effect of infra. [infrastructure] overuse on obsolescence**). As noted in Figure 11.1, maintenance is also contingent on technical staff capacity. Over the course of the intervention processes in the SRVM between 2011 and 2012, a municipal technical official argued that 'currently the status of this municipality is that we are running after the emergencies and not planning' (interviewed in Clifford-Holmes, 2015, p. 186). This 'fire-fighting' response of officials reduces their capacity to address standard technical activities, which reduces the amount of maintenance that can be performed. Over time, the accumulated lack of maintenance creates the conditions for new infrastructural crises to occur, which serves to further reduce the municipal staff capacity for standard management activities.

A reduction in these activities influences maintenance and the quantity of 'new infrastructure constructed' (by affecting strategic planning, grant sourcing and other such activities that municipal officials perform in the process of constructing new infrastructure). This feedback loop (**R4: crises** in Figure 11.4) became the primary endogenous driver of municipal crises explored in the MoF model.

The effects of these reinforcing crises were demonstrated in the dramatic events of September 2014, when municipal offices and infrastructure in the Kirkwood region were burned to the ground by protestors. The media reported several reasons for the protests, including 'water cuts that had lasted for about three weeks' (South African Press Association, 2014, p. 1). The timing of these protests presented both challenges and opportunities to the SANPAD action research project. Work had begun on the MoF model in January 2014 (see Table 11.1), with validation workshops and meetings planned for the same period in which the protests occurred. The timing placed constraints on the ways in which the MoF model could be used in multi-stakeholder settings in the SRVM (in the traditional manner of 'group model building'; Vennix, 1999); in contrast, the protests provided the opportunity to use the MoF modelling initiative to clarify the socio-technical problem and to use SD as a tool to effectively communicate the causes of local water services system failure to different audiences (which is closer to the approach advocated in 'mediated modelling'; van den Belt, 2004). Model diagrams and model-related issues were used to pose questions and engage different stakeholders in conversation following the September 2014 protests. The six modes of failure derived through this modelling approach resonated particularly with the participants of the SANPAD project from the municipal technical division. This was demonstrated when a modeller co-presented an adapted version of Figure 11.4 with a senior municipal official in November 2014. The setting was a national-level 'water dialogue' initiated by the South African Water Research Commission and attended by senior representatives of several national government departments (described in Clifford-Holmes, 2015). Following this presentation, the municipal official co-authored a paper on the MoF model with SANPAD researchers (Clifford-Holmes *et al.*, 2015), using this opportunity to review the quantitative simulation model underpinning Figure 11.4.

In summary, all three models explored the effects of practices that met short-term goals in ways that had intermediate- and long-term (negative) implications. For example, the rapid connection of households to the municipal reticulation network (as explored in the K-DEM model) and the over-extending of water supply infrastructure above its design capacity (as explored in the MoF model) are practices that respond to immediate demands in ways that require greater levels of activity in the future. This short-term response is a well-known systems trap, which is referred to within the SD field as a 'better-before-worse' response (see Lyneis and Sterman, 2016). In the SRVM case study, the short-term practices that were operationally caught within the SD models were contextualized with further institutional analysis. For example, technical officials are pressured by elected (i.e., political) officials to connect households as quickly as possible (as noted with K-DEM). A South African infrastructural analyst has accounted for this pressure in terms of new housing developments attracting 'vast amounts of attention and support at the highest level… plans are put in place, resources are made available and importantly the completed projects present "ribbon cutting opportunities"' (Infrastructure Dialogues, 2013, p. 10). Many components of the bulk water supply systems that are required to support these new housing developments are underground and 'out of sight' (Rehan *et al.*, 2014). As such, they offer fewer 'ribbon cutting opportunities' for political elites and therefore typically receive less attention *unless* crises occur. By the time such crises do occur, the damaging conditions that have created many of the crises are already institution-alized (Repenning and Sterman, 2001). An emerging theme of the modelling process reported on here was 'the characterisation of government actors as "fire-fighters" who reactively respond to crises and who are not in control of their own time' (Clifford-Holmes, 2015, p. 384). That 'stamping out spot fires does not stop a major bushfire' (Haslett, 2007, p. 1) is hardly surprising. Indeed, Simon (1996, p. 161) characterized it as a commonplace organiza-tional phenomenon that

> ... attending to the needs of the moment – putting out fires – takes precedence over attending to the needs for new capital investment or new knowledge. The more crowded the total agenda and the more frequently emergencies arise, the more likely it is that the middle-range and long-range decisions will be neglected.

An institutional response to the fire-fighting tendency is to create planning and management groups that are immune to the day-to-day pressures, which provide regulatory oversight and the space to consider longer-term issues along with the capacity to plan and act on these con-siderations. The lack of such institutional separation between operations, regulations and planning was evident in two key places in the SRVM: in the water services managed by the SRVM itself and in the water supply arrangement between the SRVM and its bulk water supplier (Clifford-Holmes *et al.*, 2016a).

The following section discusses the defining characteristics of the 'modelling in the muddled middle' approach, with reference to the results described in this section and with reference to the emerging theory and practice of social systems engineering.

11.6 Discussion

The value of modelling within transdisciplinary research is broadly acknowledged, for example, as a means of understanding and responding to complex real-world problems, syn-thesizing knowledge and providing potential decision support (Badham, 2010). Furthermore, as Voinov and Bousquet (2010) demonstrate in their authoritative review of modelling with stakeholders, a plethora of approaches exist for interacting with stakeholders at different points of the modelling process, and employing different techniques and tools. SD modelling was found to be particularly relevant in the SRVM case for three reasons. First, SD character-izes systems by time and time evolution, which allows for the artificial representation of a given system using differential equations (Luenberger, 1979). This temporal rendering of problem dynamics was useful given the frequent focus on short-term actions carrying secondary consequences in the long term (as noted in Section 11.5). Second, SD modelling – as conceptually underpinned and methodologically employed in this study – contains many social dimensions (see Section 11.3), which were especially pertinent given the second-order water scarcity in the SRVM (as introduced in Section 11.2). Third, SD was selected because it is one of the few modelling approaches that allows for exploring actual practices and experi-ences of stakeholders in applied settings (Maani and Cavana, 2007). Focusing on 'actual practices' supported a grounded analysis of water management challenges in the SRVM, which featured 'actual people who could do or who are doing things that affect other people'

(Agnew, 2011, p. 469). As Olaya (2015, p. 211) notes, SD modelling is a non-empirical approach in that it does not require strict numerical validation to be useful; furthermore, SD modelling encourages researchers 'to go and ask the [relevant] human beings… how they do what they do' (Olaya, 2015, p. 211).

The four SD models, which were developed through the approach described in this chapter, were used to explore scarcity in different ways. In the process of developing the GKWS model, for example, stakeholders were asked for the physical measurements of a supplementary storage canal in the water supply scheme. Stakeholders responded both with different measurements and with differing opinions on the perceived relevance (or irrelevance) of this canal as an additional raw water storage facility. Asking for the measurements of the canal (as parameter data for the GKWS model) created an opportunity to discuss policy options with different stakeholder groups, without personalizing or blaming the operational issues on any one actor. The GKWS model was therefore built from a functionalist perspective that a model could represent 'the real world' (in this case, a specific water supply scheme). The MoF model, in contrast, was consciously employed in discussion with stakeholders as a 'socially-constructed artefact' that was used to facilitate what Howick and Eden (2010) call 'strategic conversations'. In these two respects, the modelling approach described here is similar to that of group model-building (GMB), which conceives of models of social systems as carrying dual identities stemming from functionalist and constructivist perspectives (see Andersen *et al.*, 2007). Where the modelling approach differed from GMB was in the way in which the four SD models were *not* constructed in groups. Instead of bringing a heterogeneous group together and striving to build consensus amongst its members, modellers went from group to group, with different stakeholders interacting at different points in the process. The SD models were then used to support the action research process as 'boundary objects' (Star and Griesemer, 1989), rather than the models themselves being the objects of focus. As described in the above sections, four small SD models were developed through this action research process, rather than one large model being developed at the end of the process. The SRVM case provided further evidence in support of Ford's (2009, p. 305) assertion that developing a 'portfolio of [small] models' frequently offers the opportunity to learn more 'than a single, all-encompassing model', reflecting the fact that participatory modelling is more often about 'the process rather than the product' (Voinov and Bousquet, 2010, p. 1272).

The literature review in Section 11.3 noted that ethnographic tools and perspectives focus on the subjective interpretations of the world that individual social actors employ. The hybrid approach of SD modelling and ethnographic analysis, as employed in this study, used the ethnographic (agency-based) perspectives as a counterbalance to the systems (structure-based) perspectives that SD is frequently accused of privileging. The hybrid approach sought to explore how agents interpreted the problems they faced and found ways of working in the 'muddled middle' between formal institutional structures and informal, practice-based realms. The interpretive perspective reflects a narrative approach to social systems, which conceives of complexity as an attribute of those agents who interact with social systems, rather than solely being an attribute of systems themselves (see Tsoukas and Hatch, 2001). The methodological hybridization of systems, ethnographic and institutional analyses evolved iteratively through the case study. The key benefits of the hybrid approach are consistent with other analyses of the 'added value' of combined approaches. As Horlick-Jones and Rosenhead (2007, p. 599) note with reference to problem-structuring methods, hybrid approaches of ethnographic and systems perspectives provide 'insights into the practical worlds inhabited by the participants and the nature of the problems they [face]'; furthermore, these approaches enable researchers to develop perspectives on problematic situations that transcend 'the subjective understanding of any one actor' (Horlick-Jones and Rosenhead, 2007, p. 595).

Whilst this case study was undertaken, the South African national government acknowledged that the failure of local government to provide adequate water services, as demonstrated dramatically in the SRVM, resides partially in the institutional design. In 2013, revised policy positions were put forward by government in order to resolve these failures (Department of Waters Affairs, 2013). For systems thinkers, institutional failures are opportunities to 'revise, improve, rescind, or better explain the rules' (Meadows, 2011, p. 137), yet doing so within regulated sectors (such as water management) requires broader contextual analyses of the institutions and organizations (as noted in Section 11.3). The ways in which the financial components of municipal water services were modelled in the case study is illustrative. In the MoF model, the 'revenue dedicated to maintenance' was modelled as being subject to the proportion of the 'water revenue' that is reserved for this purpose ('% revenue ring-fenced'). Whilst the MoF model was being validated in interaction with municipal stakeholders in late 2014, the South African Minister for Water and Sanitation announced that legislation would be introduced requiring municipalities to reserve a minimum of 7% of their total budgets for the maintenance and management of infrastructure (infrastructurene.ws, 2014). The broader institutional analysis provided the means to understand the socio-political contexts of these legislative changes (see Clifford-Holmes et al., 2016a). These understandings were then used for communicating the simple scenarios explored within the MoF model, such as the impact of adjusting the parameter of the '% revenue ring-fenced' on the variable 'supply of water' (see Figure 11.4). The choice to not directly model the changing institutional context was motivated by the principle of keeping models 'requisitely simple' (i.e., by only including operational detail that enables a given model to be useful; Stirzaker et al., 2010).

Within the hybrid approach described in this chapter, modelling is positioned as an activity undertaken in the 'muddled middle', which is conceived as the space between the clean lines of policy design and the messy operational reality of implementation. In the case study, modellers exercised judgement about what to model, where and with which stakeholders, in addition to being flexible with how stakeholders were involved in the model-building and validation processes. This approach has subsequently been employed elsewhere in South Africa as part of another project, which is larger both in scope and in geographic scale (see Clifford-Holmes et al., 2016b). The commonalities between the SRVM case and this larger case include transdisciplinary action research, methodological hybridization and a deep interest in social systems design and transformation (Clifford-Holmes et al., 2016b). The modelling approach is held to be particularly relevant for exploring controversial and complex case studies that offer representative and extreme examples of systemic dysfunction, where policy-level analytical objectives co-exist with action research imperatives of employing tools and methods to understand (and where possible, address) stakeholders' issues of concern.

Acknowledgements

We gratefully acknowledge the support of the following funding agencies in this research: the South Africa Netherlands Research Programme for Alternatives in Development; the Mandela Rhodes Foundation; the SKILL programme of the South Africa Vrije Universiteit Amsterdam

Strategic Alliance; and the South African National Research Foundation and the South African Water Research Commission. We also acknowledge the assistance of the following organizations and institutions: the SRVM; the L-WUA; Amatola Water; members of the Eastern Cape Rapid Response Unit of the Department of Water Affairs [now the Department of Water and Sanitation]; the Unilever Centre for Environmental Water Quality at the Institute for Water Research, Rhodes University; and the Policy Analysis section and Multi-actor Systems Research Programme of the Faculty of Technology, Policy and Management, Delft University of Technology. Lastly, we thank Dr Harry Biggs and both of the book editors for their insightful feedback that strengthened this chapter tremendously.

References

Agnew, J. (2011) Waterpower: Politics and the geography of water provision. *Annals of the Association of American Geographers*, **101**, 463–476.

Ahmad, S. and Prashar, D. (2010) Evaluating municipal water conservation policies using a dynamic simulation model. *Water Resources Management*, **24**, 3371–3395.

Amatola Water (2014) *Kirkwood Water Treatment Works: Regional bulk infrastructure grant scoping report rev 01 (11 February 2014)*. Report prepared by Amatola Water for the Department of Water Affairs, South Africa.

Andersen, D.F., Vennix, J.A.M., Richardson, G.P. and Rouwette, E.A.J.A. (2007) Group model building: Problem structuring, policy simulation and decision support. *Journal of the Operational Research Society*, **58**, 691–694.

Badham, J. (2010) A compendium of modelling techniques. *Integration Insights*, **12**, 1–24.

Bissell, C. and Dillon, C. (2012) Preface, in C. Bissell and C. Dillon (eds), *Ways of Thinking, Ways of Seeing. Mathematical and other modelling in engineering and technology*, Springer-Verlag, Berlin, pp. v–vii.

Brent, A.C., Musango, J.K., Smit, S., Pillay, N.S., Botha, A., Roper, S. *et al.* (2016) Utilisation of system dynamics approach in Southern Africa: A systematic review. *Systems Research and Behavioral Science* (in press).

Checkland, P.B. (1981) *Systems Thinking, Systems Practice*, John Wiley and Sons, Chichester.

Clifford-Holmes, J.K. (2015) Fire and water: A transdisciplinary investigation of water governance in the Lower Sundays River, South Africa. PhD thesis, Rhodes University, South Africa.

Clifford-Holmes, J.K., Slinger, J.H., Musango, J.K., Brent, A.C. and Palmer, C.G. (2014) Using system dynamics to explore the water supply and demand dilemmas of a small South African municipality, in *Conference Proceedings of the 32nd International Conference of the System Dynamics Society, 21–24 July 2014*, Delft, the Netherlands. Available at: www.systemdynamics.org/conferences/2014/proceed/index.html.15.

Clifford-Holmes, J.K., Slinger, J.H., Mbulawa, P. and Palmer, C.G. (2015) Modes of failure of South African local government in the water services sector, in *Conference Proceedings of the 33rd International Conference of the System Dynamics Society, 19–23 July 2015*, Cambridge, MA.

Clifford-Holmes, J.K., Palmer, C.G., de Wet, C.J. and Slinger, J.H. (2016a) Operational manifestations of institutional dysfunction in post-apartheid South Africa. *Water Policy*, **18**, 998–1014.

Clifford-Holmes, J.K., Pollard, S., Biggs, H., Chihambakwe, K., Jonker, W., York, T. *et al.* (2016b) Resilient by design: A modelling approach to support scenario and policy analysis in the Olifants River Basin, South Africa, in *Conference Proceedings of the 34th International Conference of the System Dynamics Society, 17-21 July 2016*, Delft, the Netherlands. Available at: www.systemdynamics.org/conferences/2016/proceed/index.html.

Clifford-Holmes, J.K., Slinger, J.H. and Palmer, C.G. (2017) Using system dynamics modeling in South African water management and planning, in T. Simelane and A.C. Brent (eds), *System Dynamics Models for Africa's Developmental Planning*, Africa Institute of South Africa (AISA), Pretoria.

Cockerill, K., Passell, H. and Tidwell, V. (2006) Cooperative modeling: Building bridges between science and the public. *Journal of the American Water Resources Association*, **42**, 457–471.

D'Hont, F. (2013) A system dynamics model for deepening the understanding of Greater Kirkwood's water supply system, South Africa. Bachelor project phase 2, BSc Technische Bestuurskunde, Faculty of Technology, Policy Analysis and Management, Delft University of Technology, the Netherlands.

D'Hont, F., Clifford-Holmes, J.K. and Slinger, J. (2013) Addressing stakeholder conflicts in rural South Africa using a water supply model, in *Conference Proceedings of the 31st International Conference of the System Dynamics Society*, Cambridge, MA. Available at: www.systemdynamics.org/conferences/2013/proceed/index.html.

Department of Water Affairs (2013) *National Water Policy Review: Updated policy positions to overcome the water challenges of our developmental state to provide for improved access to water, equity and sustainability,* Government Gazette No. 36798, Cape Town.

Department of Water and Sanitation (2014) Media release (2 August 2014). Available at: www.dwaf.gov.za/ Communications/PressReleases/2014/Water and Sanitation Summit Media Release.pdf (retrieved 10 August 2014).

Elms, D.G. and Brown, C.B. (2012) Professional decisions (I: The central role of models and II: Responsibilities). *Civil Engineering and Environmental Systems,* **29**, 165–190.

Epstein, J. (2008) Why model? *Journal of Artificial Societies and Social Simulation,* **11**, 1–5.

Flood, R. and Jackson, M. (1991) *Creative Problem Solving: Total systems intervention,* John Wiley and Sons, London.

Ford, A. (2009) *Modeling the Environment* (2nd edn), Island Press, Washington, D.C.

Forrester, J.W. (1968) *Principles of Systems,* Wright-Allen Press, Cambridge, MA.

Forrester, J.W. (1970) Counterintuitive behaviour of social systems, in *Testimony for the Subcommittee on Urban Growth of the Committee on Banking and Currency, U.S. House of Representatives, 7th October 1970.*

Haag, D. and Kaupenjohann, M. (2001) Parameters, prediction, post-normal science and the precautionary principle – a roadmap for modelling for decision-making. *Ecological Modelling,* **144**, 45–60.

Hare, M. (2011) Forms of participatory modelling and its potential for widespread adoption in the water sector. *Environmental Policy and Governance,* **21**, 386–402.

Haslett, T. (2007) Reflections on SD practice, in *Proceedings of the 13th ANZSYS Conference,* Auckland, New Zealand.

Horlick-Jones, T. and Rosenhead, J. (2007) The uses of observation: Combining problem structuring methods and ethnography. *Journal of the Operational Research Society,* **58**, 588–601.

Howick, S. and Eden, C. (2010) Supporting strategic conversations: The significance of a quantitative model building process. *Journal of the Operational Research Society,* **62**, 868–878.

Infrastructure Dialogues (2013) Maintenance of Public Infrastructure: How to master the art of taming contradictory forces. Public event hosted by the Development Bank of Southern Africa, 31 October 2013. Available at: www. infrastructuredialogue.co.za/wp-content/uploads/2013/12/Report-on-Maintenance-of-public-assets-31-Oct-2013. pdf (retrieved 10 November 2014).

infrastructurene.ws (2014) 7% of budgets to go to water infrastructure maintenance. Available at: www.infrastructurene. ws/2014/08/04/7-of-budgets-to-go-to-water-infrastructure-maintenance (retrieved 6 August 2014).

Jackson, M.C. (2003) *Systems Thinking: Creative holism for managers,* John Wiley and Sons, Chichester.

Koen, B.V. (2003) *Discussion of the Method: Conducting the engineer's approach to problem solving,* Oxford University Press, New York, NY.

Kwezi V3 Engineers (2005) *Sundays River Valley Municipality, Kirkwood. Upgrading of water treatment works: additional raw water storage pond.* Technical report, Project No. 211410QOA.

Lane, D.C. (2000) Should system dynamics be described as 'hard' or 'deterministic' systems approach? *Systems Research and Behavioral Science,* **17**, 3–22.

Lane, D.C. and Oliva, R. (1998) The greater whole: Towards a synthesis of system dynamics and soft systems methodology. *European Journal of Operational Research,* **107**, 214–235.

Langsdale, S.M. (2007) Participatory model building for exploring water management and climate change futures in the Okanagan Basin, British Columbia, Canada. PhD thesis, The University of British Columbia, Canada.

Layder, D. (1998) *Sociological Practice: Linking theory and social research,* Sage Publications, London.

Luenberger, D.G. (1979) *Introduction to Dynamic Systems,* John Wiley and Sons, New York, NY.

Lyneis, J. and Sterman, J. (2016) How to save a leaky ship: Capability traps and the failure of win–win investments in sustainability and social responsibility. *Academy of Management Discoveries,* **2**, 7–32.

Maani, K.E. and Cavana, R.Y. (2007) *Systems Thinking, System Dynamics: Managing change and complexity* (2nd edn), Pearson Education, Auckland, New Zealand.

Meadows, D.H. (2011) *Thinking in Systems: A primer,* Earthscan, London.

Mehta, L., Alba, R., Bolding, A., Denby, K., Derman, B., Hove, T. *et al.* (2014) The politics of IWRM in Southern Africa. *International Journal of Water Resources Development,* **30**, 528–542.

Mingers, J. and Rosenhead, J. (2004) Problem structuring methods in action. *European Journal of Operational Research,* **152**, 530–554.

Molony, L. (2015) Water security amongst impoverished households in the Sundays River Valley Municipality: Community experiences and perspectives. Masters thesis, Rhodes University, South Africa.

Muller, M.J. (2013) Linking institutional and ecological provisions for wastewater treatment discharge in a rural municipality, Eastern Cape, South Africa. Masters thesis, Rhodes University, South Africa.

Ohlsson, L. (1999) *Environment, Scarcity and Conflict: A study of Malthusian concerns*, University of Goteborg, Sweden.

Olaya, C. (2015) Cows, agency, and the significance of operational thinking. *System Dynamics Review*, **31**, 183–219.

Palmer, C.G., de Wet, C., Slinger, J., Linnane, S., Burman, C., Clifford-Holmes, J. *et al.* (2014) *Developing a Guide to Research Practice that Engages with Difficult Water Resource Problems*. WRC Report T18138 K8-992-1, Gezina, South Africa.

Rehan, R., Knight, M.A., Unger, A.J.A. and Haas, C.T. (2014) Financially sustainable management strategies for urban wastewater collection infrastructure – development of a system dynamics model. *Tunnelling and Underground Space Technology*, **39**, 116–129.

Repenning, N.P. and Sterman, J.D. (2001) Nobody ever gets credit for fixing problems that never happened: Creating and sustaining process improvement. *California Management Review*, **43**, 64–88.

Republic of South Africa (1998) *The White Paper on Local Government*, CTP Book Printers, Pretoria, South Africa.

Richardson, G.P. (2011) Reflections on the foundations of system dynamics. *System Dynamics Review*, **27**, 219–243.

Rodríguez-Ulloa, R. and Paucar-Caceres, A. (2005) Soft system dynamics methodology (SSDM): Combining soft systems methodology (SSM) and system dynamics (SD). *Systemic Practice and Action Research*, **18**, 303–334.

Rodríguez-Ulloa, R.A., Montbrun, A. and Martínez-Vicente, S. (2011) Soft system dynamics methodology in action: A study of the problem of citizen insecurity in an Argentinean province. *Systemic Practice and Action Research*, **24**, 275–323.

Rouwette, E.A.J.A. and Vennix, J.A.M. (2006) System dynamics and organizational interventions. *Systems Research and Behavioral Science*, **23**, 451–466.

Rowlston, B. (2011) Water law in South Africa: From 1652 to 1998 and beyond, in J. King and H. Pienaar (eds), *Sustainable Use of South Africa's Inland Waters: A situation assessment of resource directed measures 12 years after the 1998 National Water Act*, Water Research Commission, Gezina, South Africa, pp. 19–47.

Schreiner, B. (2013) Why has the South African National Water Act been so difficult to implement? *Water Alternatives*, **6**, 239–245.

Siddle, A. and Koelble, T.A. (2012) *The Failure of Decentralisation in South African Local Government: Complexity and unanticipated consequences*, UCT Press, Cape Town.

Simon, H.A. (1996) *The Sciences of the Artificial* (3rd edn), MIT Press, Boston, MA.

South African Press Association (2014) Chaos in Kirkwood as residents continue to protest. Available at: www.news24.com/SouthAfrica/News/Chaos-in-Kirkwood-as-residents-continue-to-protest-20140923 (retrieved 23 October 2014).

Star, S.L. and Griesemer, J.R. (1989) Institutional ecology, 'translations' and boundary objects: Amateurs and professionals in Berkeley's Museum of Vertebrate Zoology, 1907–39, *Social Studies of Science*, **19**, 387–420.

Statistics South Africa (2014) Sundays River Valley Municipality. Available at: beta2.statssa.gov.za/?page_id=993andid=sundays-river-valley-municipality (retrieved 15 October 2014).

Stave, K.A. (2010) Participatory system dynamics modeling for sustainable environmental management: Observations from four cases. *Sustainability*, **2**, 2762–2784.

Sterman, J.D. (2000) *Business Dynamics: Systems thinking and modeling for a complex world*, Irwin McGraw-Hill, Boston, MA.

Stirzaker, R., Biggs, H., Roux, D. and Cilliers, P. (2010) Requisite simplicities to help negotiate complex problems. *AMBIO*, **39**, 600–607.

Sundays River Valley Municipality (2010) *Water Services Development Plan: January 2010 review*. Internal report prepared by Engineering Advice and Services, Port Elizabeth for the SRVM, Kirkwood, South Africa.

Tapela, B.N. (2012) *Social Water Scarcity and Water Use*. WRC Report No. 1940/1/11, Gezina, South Africa.

Tidwell, V.C., Passell, H.D., Conrad, S.H. and Thomas, R.P. (2004) System dynamics modeling for community-based water planning: Application to the Middle Rio Grande. *Aquatic Sciences*, **66**, 357–372.

Tsoukas, H. and Hatch, M.J. (2001) Complex thinking, complex practice: The case for a narrative approach to organizational complexity. *Human Relations*, **54**, 979–1013.

van den Belt, M. (2004) *Mediated Modeling: A system dynamics approach to environmental consensus building*, Island Press, Washington, D.C.

Vennix, J.A.M. (1999) Group model-building: Tackling messy problems. *System Dynamics Review*, **15**, 379–401.

Voinov, A. and Bousquet, F. (2010) Modelling with stakeholders. *Environmental Modelling and Software*, **25**, 1268–1281.

Vriens, D. and Achterbergh, J. (2006) The social dimension of system dynamics-based modelling. *Systems Research and Behavioral Science*, **23**, 553–563.

Wickson, F., Carew, A. and Russell, A. (2006) Transdisciplinary research: Characteristics, quandaries and quality. *Futures*, **38**, 1046–1059.

Winz, I., Brierley, G. and Trowsdale, S. (2009) The use of system dynamics simulation in water resources management. *Water Resources Management*, **23**, 1301–1323.

Xi, X. and Poh, K.L. (2013) Using system dynamics for sustainable water resources management in Singapore. *Procedia Computer Science*, **16**, 157–166.

Yin, R.K. (2009) *Case Study Research: Design and methods* (4th edn), Sage Publications, Los Angeles, CA.

12

Holistic System Design:
The Oncology Carinthia Study

Markus Schwaninger and Johann Klocker

12.1   The Challenge: Holistic System Design

Over several decades, health-care systems all over the world have been grappling with a formidable challenge. The issue is providing an integral kind of care, with the patient at the centre, rather than technology or doctors. Traditionally, over the last four or five decades, hospitals have increasingly suffered from an orientation that hinges on over-specialization and splintered forms of organization. This orientation threatens the quality of medical care, because patients tend to be treated in a fragmentary way. The perspective on sick people is as if they were conglomerates of organs that can be treated in isolation. The focus is on symptoms. Health-care systems often lack the ability to deal with syndromes that can have multiple causes with complex interrelationships. Normally, the main concern is applying high technology and advanced medication, instead of warranting patients' quality of life. While the pharmaceutical industry has been thriving on expansion and soaring profits, critics of modern health systems diagnose a different trend. They have gone so far as to indict these systems of making patients sick instead of curing them, with iatrogenic effects[1] as a rule rather than an exception (Brownlee, 2007; Illich, 1976). Critiques do not only come from 'outside' (i.e., from sociology, economics, etc.), even representatives of the medical professions deplore overtreatment and a mechanization of medicine going hand in hand with a dehumanizing trend (Allan and Hall, 1988; Hontschik, 2010; Loewit, 2010).

---

[1] Meaning 'disease caused by the doctor', in this case named *pars pro toto* for the health-care system.

Social Systems Engineering: The Design of Complexity, First Edition. Edited by César García-Díaz and Camilo Olaya.
© 2018 John Wiley & Sons Ltd. Published 2018 by John Wiley & Sons Ltd.

These deficits have provoked calls for a holistic kind of treatment. The goal sounds ideal, but it cannot be achieved by giving pride of place to technical or financial means. By now, many leaders in hospitals have understood the need for a holistic approach to care, but very few have been successful in bringing it about. They are in a complexity trap. The stalemate of growing specialization in bureaucratic silos appears to prevent hospitals from coping effectively with increasingly perplexing disease patterns.

Our research question is then: How must health-care systems be designed to provide holistic, patient-centred care, at excellent quality and bearable cost? The main topic of this chapter is organization. This is a transdisciplinary, socio-technical undertaking with far-reaching implications. Organizational innovation is needed, and it can yield better and more abundant fruits than mere technological creations can.

The purpose of this chapter is to explore an exemplar of a long-term process aimed at achieving a holistic system design. The case is from health care. We trace back the process by which a comprehensive oncological care system was built in the Austrian province of Carinthia, with Klagenfurt as its capital. The study covers a period of roughly 30 years, until 2015. It is more than a showpiece, because not only its successes but also the, albeit sparse, downsides along the way are analysed.

The chapter has started with the issue of a holistic system design emerging from the traditional state of established health-care systems. We continue in Section 12.2 with an account of the methodology that guides the enquiry. There then follows the main part of our contribution: we first introduce the case study with a picture of the initial situation in the mid-1980s. Thereupon, we provide an account of the evolution of the oncological care system as it has been designed and built over 30 years. Along the way, several systemic concepts and methods are gradually introduced, as they have been used in the course of that process. In Section 12.3 we then provide an overview of the fruits reaped, and reflect on the strengths as well as the limitations of the health-care system under study. In Section 12.4 we synthesize the analytical components generated. The main question will be what can be learnt from the case. A concise set of insights, teachings and implications concludes the chapter.

12.2   Methodology

We chose the method of a one-case setting. This method is indicated here because the enquiry is longitudinal and revelatory (Yin, 2014). Furthermore, the system under study confronts ubiquitous and overwhelming complexity. Therefore, we have to rely on multiple sources of data to illuminate the approach pursued for dealing with the challenging issues under study. Our account of the development of the oncological care system (OCS) reveals several surprising outcomes achieved by a consequent use of systemic concepts and methods in system design.

To guide the presentation, we rely on a process diagram (Figure 12.1) that stems from a systems approach denominated 'Integrative Systems Methodology' (ISM) (Schwaninger, 1997, 2004).[2]

---

[2] For applications, see also Weber and Schwaninger (2002) and Schwaninger (2013).

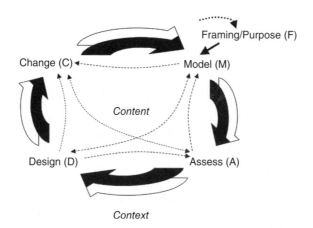

Change (C)

Content

Design (D)

Framing/Purpose (F)

Model (M)

Assess (A)

Context

Figure 12.1   Integrative Systems Methodology for dealing with complex issues – a process diagram.[3]

ISM is a heuristic[4] by which problem-solvers can enhance their repertoire of behaviour (cybernetically: 'variety'; Ashby, 1956) in dealing with complex issues or problems. As shown in Figure 12.1, ISM stresses three dimensions. The first two dimensions are reflected in the two loops on which it is based, namely a content loop and a context loop (hence the double arrows). With 'content' we are referring to the activities or operations of the system under study, and with 'context' to the structural and cultural frame into which it is embedded. Content is about what the system 'does', context about how it organizes itself (or how it is organized). Both of these dimensions require different conceptual tools for dealing with complexity. In the following we will use two systemic methods, qualitative system dynamics modelling at the content level and cybernetic modelling for context. The third dimension is process, in the sense of the sequence of the operations, expressed by the arrows. The two loops in Figure 12.1 are separated only for the purpose of analysis. In fact, they are intertwined and in practice show overlaps. They revolve iteratively, alongside a set of operations. The number of these operations could vary as a function of the notation. Here, a set of four operations is used – modelling, assessing, designing and changing – which can be sufficiently well distinguished and specified.

Given the circular structure of the process, one could start anywhere with its description. Also, in actual practice, the starting point could be anywhere.

Sometimes, actors are suddenly confronted with an assessment or a model from which they have to proceed. We shall take 'framing' (including such aspects as purpose and goals or aims) as the point to start with, framing being a kind of anchor for sense-making. It addresses

---

[3] This scheme was inspired by earlier works by the cybernetician Raúl Espejo (1993), namely his Cybernetic Methodology, and by the postulate to study content, context and process of change, as formulated by the organization scientist Andrew Pettigrew (1985, p. 50).

[4] 'Heuristic' can best be translated as 'the art of finding'. Stafford Beer defines 'heuristic', a contraction of 'heuristic method', as 'a set of instructions for searching out an unknown goal by exploration, which continuously or repeatedly evaluates progress according to some known criterion' (Beer, 1981, p. 402). Engineering is defined by its method: heuristics (Koen, 2003). Hence, ISM is close to engineering methodology.

fundamental issues: Which are the purpose and aim of the process? What is the system-in-focus? Which are the relevant perspectives? These are questions that should be dealt with early on. Modelling then includes tasks such as specifying the goals of stakeholders and the factors critical for attaining those goals, foregrounding issues and elaborating models. Assessing comprises tasks such as apprehending the dynamics of the system and simulating and exploring scenarios, as well as interpreting and evaluating simulation outcomes.

Designing includes tasks such as ascertaining control levers and designing strategies, organizational contexts and action programmes. Under the term 'change', all the tasks that encompass the realization of strategies and action programmes are included. All these operations are about enhancing systemic evolution.

Please note that 'design' in a wider sense encompasses earlier tasks such as modelling and assessment. Anyhow, the diagram shows that the process is not the sequential procedure which the main arrows might suggest. As the additional dashed lines show, the process is characterized by multiple communication processes involving feedback, validation and control. 'Feedback' here occurs when results and insights are fed back to earlier phases (e.g., from design to model). 'Validation' is a process by which the quality of models and strategies is constantly improved. Finally, 'control' is the process by which results are compared to goals, with the ensuing steering and regulation, when necessary. The schema is a simplification, which focuses on the general process characteristics, without any claim to final completeness.

Adhering to the aim of maximizing the space for the substantive issues of the case study, we refrain from bringing in the more detailed scheme of ISM, which has been published elsewhere (Schwaninger, 1997, 2004, 2013). The main point about the diagram in Figure 12.1 in the context of socio-technical system design is the need for proceeding simultaneously at the content and context levels: (a) resolving the issue as regards content and (b) embedding the process in a supportive organizational context. In the following, the abbreviations used for the phases in the diagram – F, M, A, D, C – will be used to denote the sections of the case study.

12.3   Introduction to the Case Study: Oncology Carinthia

12.3.1   Setting the Stage

To start with, a clarification of the role of the authors is needed; for their names, we will use the abbreviations MS and JK. In 1984, JK was put in charge of building an oncological care unit situated in the Department of Internal Medicine at the central hospital (*Landeskrankenhaus*, in short 'LKH') in Klagenfurt, the capital of Carinthia, one of the nine federal states of Austria. Over the ensuing 30+ years JK pioneered and directed the development of an oncological care system covering the whole state, and involving 10 hospitals as well as multiple local physicians. He has managed that system over all these years, and is the main source of information for this case study. Early on, in 1985, he called in MS, who is an organization scientist, to help him conceptually, as a consultant and coach, a role that he continues to hold. According to JK, much of the successful evolution of that health-care system is due to that cooperation. Normally, MS did not appear in front of the staff of the unit, except at certain internal conferences and workshops, for talks and discussions related to the organization and leadership of the system under development.

Oncology, the domain of medicine that deals with tumours, is an interdisciplinary field, by definition. It involves virtually all medical disciplines – internal medicine, radiology, surgery, gynaecology, orthopaedics, neurology, urology, pneumology, haematology, etc. First of all, cancer can appear in any organ of the human body. Second, its therapy often requires a combination of measures, such as medical tumour treatment (chemotherapy, hormone therapy, immunotherapy and antibody therapy), radiotherapy, surgery and psycho-social care. Third, the approaches to therapy are manifold. That has been primarily a consequence of the complexity of the cancer problem, but also of the relative youth of oncology as a field, back in 1985. For most therapies at that time, the physician could not rely on a trusted basis for decision-making, while the progress in pharmacology and medicinal technology kept shaping new recommendations and facilities.

In practice, the need for interdisciplinary therapies was met only rudimentarily or not at all. This was not only the case in the hospitals of small and midsized municipalities. Even in the largest facility of the state, the LKH Klagenfurt, an interdisciplinary mode of operation transcending the borders of different wards was not much used. It occurred in a rather aleatory and sporadic fashion, because it required overcoming bureaucratic barriers. In addition, cooperation could be achieved only if champions succeeded in winning over members of different departments, who were already loaded with tasks.

In that situation the chosen form of therapy was often less a function of the patient's syndrome than of the methods which the respective therapist had mastered or was especially interested in. To formulate it in a pointed way: if the patient landed in the hands of a surgeon, he or she had a surgery. In case they were under treatment with an internist, they had to undergo chemotherapy; under the auspices of the radiotherapist, they would undergo radiotherapy.

In contrast, the desirable approach for an oncological care system would necessarily use all available therapeutic modalities and infrastructure, in a sequential or combined mode, and customized to each specific case.

We will now describe the organizational development process, following the logic of Figure 12.1, with an emphasis on content in Sections 12.3.2 to 12.3.4, on context in Sections 12.3.5 to 12.3.8 and on the overall process in Section 12.3.9. The final Section 12.3.10 is about the attained results.

12.3.2   Framing: Purpose and Overall Goals (F)

JK and the directors of the state health authority, with the directors of the LKH Klagenfurt, shared a common vision. They defined the purpose of Oncology Carinthia as a health system that should provide the highest possible level of oncological care, covering the whole country. They then agreed upon three general goals:

(a) Guarantee of excellent oncological care in the context of the central hospital, using all the resources available within that powerful institution.
(b) Provision of fast and high-level care for oncological patients all over the federal state (i.e., also outside the capital – in small towns and in the countryside, as far as possible 'on the spot').
(c) An increasingly preventive orientation of oncological medicine in Carinthia.

The central oncological care unit was formed.

12.3.3   Mapping the System at the Outset (M)

We now sketch the actual and potential components of the system and the initial steps in the making of the oncological care system. Carinthia covers an area of 9500 km^2 and had about 500,000 inhabitants at the time. The socio-geographic structure was 'healthy', with no excessive urban concentrations. Besides the capital Klagenfurt (86,000 inhabitants), several district centres exhibited their own lively economic, social and cultural activities. Altogether, Carinthia had thirteen hospitals potentially apt to be included in a network of oncological care. Four of them were public (LKH) and the rest private or of a religious order, as in the case of St. Veit (*Krankenhäuser*, in short 'KH' or *Sanatorien*, in short 'SAN'). The number of independent physicians included about twenty internists, of which only one specialized in oncology. Many of them were also candidates for joining the network, to provide supportive care.

Statistically, around 3000 new cancer incidences per year could be expected, 40% of them within the zone of Klagenfurt. The rest were distributed among three regions:

- Villach, Spittal, Laas (30%)
- Wolfsberg (20%)
- Friesach, St. Veit (10%).

These zones are denoted approximately in the map of Figure 12.2. This is already a historical document from the 1980s, drawn for the 'Oncology Case' which we wrote for the University of St. Gallen, Switzerland. There we gave a seminar around that case, at the master level,

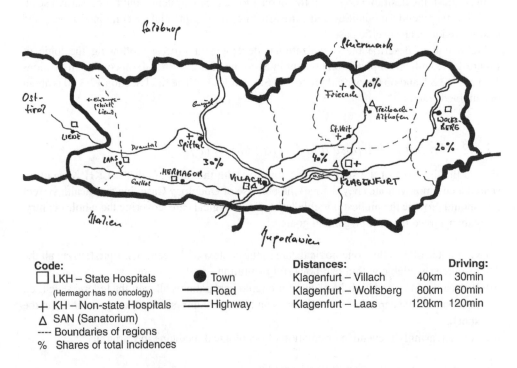

Code:

☐ LKH – State Hospitals
   (Hermagor has no oncology)
+ KH – Non-state Hospitals
△ SAN (Sanatorium)
---- Boundaries of regions
% Shares of total incidences

● Town
— Road
═ Highway

Distances:		Driving:
Klagenfurt – Villach	40km	30min
Klagenfurt – Wolfsberg	80km	60min
Klagenfurt – Laas	120km	120min

Figure 12.2   Map of Carinthia with hospitals.

periodically for about 20 years. These seminars always conveyed useful impulses for the development of Oncology Carinthia.

Assuming a 50% five-year survival rate, the yearly stock of incidences to be cared for was around 11,000,[5] either by acute treatment or by follow-up therapy.

In a conversation between JK and the director of the LKH Klagenfurt towards the end of 1984, some vital issues for the joint endeavour were raised (extract from the conversation, based on notes of JK):

JK: 'There is no infrastructure for a serious oncological treatment…'
Director: 'What do you mean by *serious oncological treatment*?'

JK outlines his concept of an interdisciplinary care, which also reaches the population in the countryside. He pleads for a network organization, encompassing all relevant resources across the state, to ensure excellent care services all over Carinthia. The director finds these plans somewhat 'high-flying', but he is also attracted to the idea. Finally, JK also remembers the initial session with the state authority:

Director: 'For the time being, I cannot promise you large budgets for this project. At the moment, I only see a possibility to hire modest additional staff starting two years from now. But you are free to use the staff of your department (internist ward) and gain the support from other units… As you know, our purchases of new lab and therapy equipment is up to the general technological progress, in both the Klagenfurt LKH and the peripheral hospitals.'
JK: 'You are addressing the formal part of the problem… What I am even more concerned about are the informal aspects… Both nursing staff and doctors know little about oncology. Therefore they do not see the positive side. They also lack know-how, particularly outside Klagenfurt. To be honest, even we oncologists do not know much about the topic.'
Director: 'What do you mean?'
JK: 'It is unclear which therapies are being used and which ones are successful. Strictly speaking, we would have to agree periodically upon precise therapy strategies for the main indications, and apply these consequently. We would also have to examine the results continually, and adapt the strategies in the light of new knowledge.'

Finally, the two agreed on a first realistic objective: the build-up of an oncological care unit centred in Klagenfurt. This would comprise an allotment of ten beds for intensive therapeutic situations and an ambulatory where patients could be treated and therapy concluded, also receiving aftercare. They also agreed that the goal would not be a central oncology clinic:

- The creation of a ghetto of cancer sufferers had to be avoided. Patients should, to the greatest possible extent, be treated in the department that was their 'home'.
- The journey to a distant clinic is time-consuming and wearing. The proximity of family is a major supportive factor in the healing process. For both reasons, sufferers should be treated as close to their homes as possible.

---

[5] The new incidences (3000 p.a.) are a flow, while the incidences to be cared for (11,000) are a stock.

Subsequently, the LKH management endowed JK with the mission to build a consultation service involving other hospitals.

JK had collected practical and scientific experience on the matter of 'cancer' over years, including work at a Swiss oncological centre with a worldwide reputation for both its management and doctors. He knew that in addition to the professional, medicinal aspects, he would be confronted with demanding organizational and leadership issues in order to achieve his goals. He was fully aware that the medical infrastructure lagged behind what he envisioned. Concerning structural aspects of the emerging system, JK was clear about several points:

(a) He calculated that a health-care centre with oncological focus needed to serve a catchment area of between 200,000 and 500,000 to 550,000 inhabitants. Carinthia was in that range.
(b) He had a clear concept of a structure for interdisciplinary work. Two experienced and highly competent oncologists would be needed, one with an internist and one with a radiology background. Such a team could cover the ambulant part as well: treatment, aftercare and consulting, including the maintenance and orchestration of relationships with the specialized departments (surgery, urology, otolaryngology, etc.). But that is only part of the story. For cases of cancer in wards for gynaecology, otolaryngology, paediatrics or pneumology, interdisciplinarity always means that the pertinent specialists call on the oncologists. The treatment should be planned under the lead of the oncologists, while the therapy and aftercare should be carried out by the specialists. In the aftercare phase, local independent physicians can also become active.
(c) Not all activities in the chain *detection→diagnosis→therapy→aftercare* need to be performed in the ambulatories and wards of hospitals. Simpler forms of therapy and certain measures of follow-up care can be delegated to independent general practitioners and specialists.
(d) Prospects should be treated only as in-patients if absolutely necessary (e.g., in case of major surgeries) but then be relegated to the ambulant mode.
(e) He estimated that about two-thirds of the cases would fall under the domain of internists, and the rest to other departments.

12.3.4   A First Model (M) and Assessment (A)

Early on, JK and MS drew a first model, using the methodology of qualitative system dynamics, to provide an overview of the most important factors making up the system-in-focus and their dynamic interrelationships. Their aim was first to understand and assess 'how the system ticks'. Second, they wanted to discover the priorities and levers for the design of the system.

They attempted to elicit the relevant perspectives on the system-in-focus. The schema in Table 12.1 distinguishes the main stakeholder groups ('interest groups') and their goals with respect to the system-in-focus. Then the key factors (i.e., aspects critical for the attainment of these goals) are ascertained.

The goals and key factors stand for components that may constitute a model of the system under study, such as the one in Figure 12.3: that is, a causal loop diagram (CLD)[6] (i.e., a qualitative representation giving a first idea of the dynamics of that system).

---

[6] CLDs are devices stemming from System Dynamics, a methodology for modelling and simulation, going back to Professor Jay Forrester (see Forrester, 1961; Senge, 1990; Sterman, 2000).

Table 12.1 Stakeholders, goals and critical factors

Interest groups	Goals	Key factors
Patients and their families	Be healthy Suffer little	Prevention Quality of life Quality of care
Champions of oncological care	Be excellent professionals Have an interesting job Realize their ideas Lead an effective team	Motivation Research and knowledge management Strong infrastructure Cooperation Effective coordination
Local hospitals and doctors	Qualify in oncology Become members of care network	Training Cooperation
Professional staff	Have an interesting job Have a bearable job Become more qualified	Training Psycho-hygiene Cooperativeness of other units involved
State authorities and central hospital administration	Effectiveness of care system Efficiency of care system	Low incidence of cancer Cancer prevention Success of care High productivity Coordination
Public in general	Stay healthy	Social and ecological consciousness Quality of environment Healthy behaviour

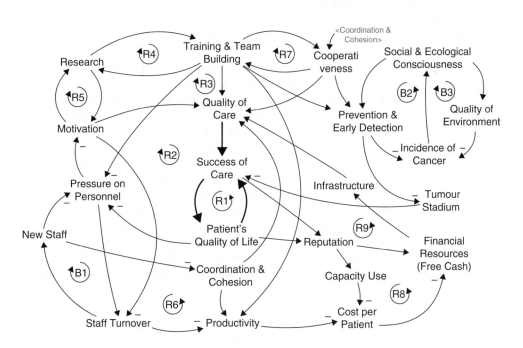

Figure 12.3 Causal loop diagram showing the dynamics of the system-in-focus.

In the diagram, arrows denote causal relationships and signs show the directions of those relationships. All arrows that carry a negative sign denote that the two connected variables point in opposite directions [e.g., more personnel turnover leads to less productivity (and less personnel turnover leads to more productivity)]. All arrows not provided with signs implicitly show connections of variables pointing in the same direction (e.g., more quality of care entails higher success of care).[7] The diagram shows nine reinforcing ('R') and three balancing ('B') loops. The polarity of a loop is the result of a multiplication of all signs in that loop. Loops with even numbers of negative signs are always reinforcing (e.g., loop R1 [zero minuses]: '+' * '+' = '+'; loop R8 [two minuses]: '+' * '−' * '−' * '+'…). Loops with uneven numbers of negative signs are always balancing (e.g., loop B1: '+' * '−' * '+' = '−'). Reinforcing loops promote either increase or decrease, both of which, if not attenuated at some point, will entail destabilization eventually. Balancing loops lead to attenuation, and potentially to equilibrium. The benefit of distinguishing these two kinds of loop is that those dynamics can be identified which make the system develop on the one hand, and which lead to a balance of the system on the other hand. For technical details, see Sterman (2000, chapter 5).

The number of reinforcing loops is higher than that of the balancing loops, because Oncology Carinthia is still in an early phase, heading for development. A brief summary will uncover the meaning of these loops.

- R1: the *core loop*, a 'motor' where a patient's quality of life is a function of quality of care, entailing successful care. The quality of life for patients dynamizes successful care, because patients can contribute more to the healing process and to a positively spirited milieu.
- R2: the *motivation loop*, where motivation drives the quality of care and is driven by successful care and the patient's quality of life.
- R3: the *qualification loop*, where training enhances quality of care, which motivates people. Motivated doctors engage in research, which improves training activities.
- R4: the *knowledge loop*, where research and training dynamize each other.
- R5: the *research loop*, where professionals' motivation thrives on their research engagement, and motivation triggers commitment to that additional work.
- R6: *the productivity loop*, where productivity is strongly affected by quality and success of care, which reduce pressure on personnel and staff turnover. Productivity strengthens the financial position and thereupon the infrastructure, which is a prerequisite for service quality.
- R7: the *cooperation loop*, cooperativeness, which is itself driven by coordination and team cohesion, triggers competence enhancement, expressed in training and team-building.
- R8: the *financial loop*, where the reputation of the oncological care system attracts new patients, leading to higher capacity use. As the cost per patient is diminished, the financial and infrastructural position improves. So do the quality and success of care, which again strengthen the reputation.
- R9: the *external funds loop*, similar to R8, but showing a direct link from reputation to financial resources, denoting an improved position for gaining funds from external sponsors.

---

[7] To make generally correct the statement 'X and Y move in the same [opposite] direction', a more precise formulation is necessary: 'If X increases, Y increases above [below] what it would have been' (Richardson, 1997).

- B1: the *personnel dynamics loop*, where staff turnover implies staff leaving and new staff coming in. The higher that index is, the less pressure on personnel (*new staf→pressure on personnel has a negative link*); thus, less pressure on personnel decreases staff turnover. This is a balancing loop that regulates (balances, controls) turnover through new staff, since new staff decrease pressure.
- B2: the *prevention loop*, where incidence of cancer is regulated (balanced, controlled), to some extent, through consciousness, prevention and early detection, which lead to lower incidences of cancer.
- B3: the *ecological loop*, where incidence of cancer is regulated (balanced, controlled), to some extent, through consciousness and quality of environment, which lead to lower incidences of cancer.

Loops B2 and B3 hypothesize very long-term dynamics, congruent with the vision that prevention should have greater prominence in Carinthia than in the past.

The logical structure and the impact of the CLD in Figure 12.3 highlight the crucial role of quality of care, patients' quality of life and the priority of human resources over financial and infrastructural resources.

This CLD is focused on aspects of content (inner loop of Figure 12.1). It helps in understanding how the system 'ticks', and was also used at various stages when scenarios of the development of the system were discussed. The CLD signalled those 'places' where interventions would be indicated. In Sections 12.3.6 and 12.3.7 we describe how the respective measures were taken.

12.3.5   The Challenge Ahead

The idea of the champions was, first of all, to set the norm to enable an excellent level of care (including quality, reliability and high speed). Strategically, that care had to be delivered locally (i.e., be as decentralized as possible), and central only where absolutely necessary, with intelligent use of all available resources. There were problems ahead:

- resistance of medical departments that should join the effort;
- weak know-how and lack of interest among peripheral hospitals;
- deficits of knowledge among independent physicians;
- fear among doctors and nursing staff over increasing demands and uncertainties;
- low motivation among staff;
- no formal authority among oncologists about parties that should be included;
- scarce budgetary means;
- limited personnel capacity in the central oncology unit;
- low interest, among authorities, in preventive care.

In the face of these issues, the challenges presented themselves as follows:

- winning the cooperation of medical departments at the Klagenfurt and peripheral hospitals;
- multiplying know-how and enhancing knowledge-building in the peripheral hospitals;

- involving and linking multiple resources;
- creating robust and nimble structures to enhance the viability of the oncological care system;
- information management – making data and information available for the control of therapies and the creation of new therapy options;
- balancing decentralized and centralized care;
- balancing the efficiency of care operations and the effectiveness of care strategies.

In sum, the venture ahead was very demanding. A high diversity of tasks had to be achieved, distributed human and technical capacities had to be networked skilfully, flexibility was to be built into the system and the restriction of high scarcity of resources had to be taken into account.

12.3.6   A First Take on Design (D): Ascertaining Levers

The CLD in Figure 12.3 reveals tangible dynamic features. From there the next question follows naturally: Which are the levers to improve the system in line with the purpose and goals as defined at the outset? At that point the goals were already much more concrete and integrated in the model of the whole system, because the variables were derived from the key factors representing all stakeholders.

Model analysis directed our attention to three main levers: (a) psycho-hygiene for the staff; (b) structures, information systems and knowledge management; and (c) leadership (bold parameters in Figure 12.4). The kinds of interventions chosen thereupon were not a result of the CLD per se, but they emerged in the champions' ongoing discussion of challenges and pertinent responses. The CLD made the 'mechanisms' driving the system under study transparent. Hence, it was a vehicle for keeping that discourse going. In the ensuing efforts the identified levers were put into practice in sophisticated ways, as will be shown.

These levers have the character of strategic parameters with great potential:

(a) *Psycho-hygiene.* The staff in oncological care are subject to a stress load that tends to be greater than in other professions. Therefore, introducing psycho-hygienic measures was crucial, to sustain and foster the psychic health of people, adopting both preventive and restitutive measures.

(b) *Structure, information system and knowledge management.* Structure is a powerful device that was considered crucial for strengthening quality of care, coordination and team cohesion. In addition, information systems and knowledge management were prominent in strengthening research.

(c) *Leadership.* Ultimately, everything in an organization is subject to the influence of leadership and hinges on its quality. Motivation as well as coordination and cohesion were identified as two main aspects to be strengthened by that driver. Coordination and cohesion then impinge strongly on cooperation. Equally crucial was a major effort to win the cooperation of all necessary parties.

Moving these levers served to create a context (outer loop in Figure 12.1) that would govern what we call 'content' – the operating activities of the OCS. The context parameters proved highly effectual in changing the dynamics of the content variables (Figure 12.3).

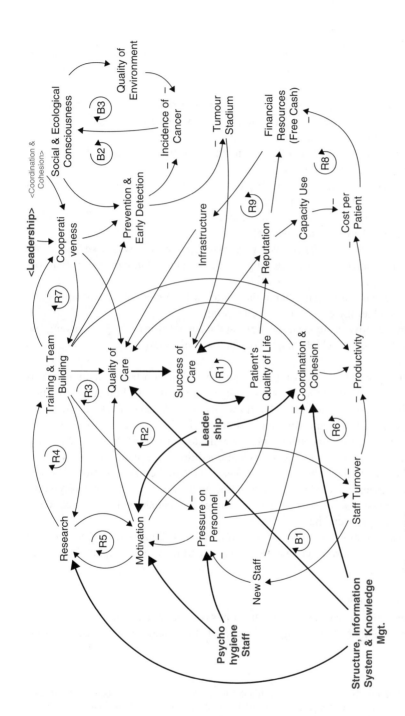

Figure 12.4 Three main levers for the development of the oncological care system.[8]

---

[8] Variables in brackets are 'shadow variables', used to avoid entanglements in the diagram. For example, *<Leadership >* is identical with *Leadership*.

12.3.7   From Design (D) to Change (C)

The levers ascertained in the last section were used early on to develop the system, and they go on being important today. First of all, *leadership* was strong from the start, when the champions took over the mandate to build a new OCS. To make their moves successful, the champions adhered to certain strategic leadership principles (Malik, 2008):

- *Show competence, be modest, reliable and persistent.* Competence in oncological matters, demonstrated humbly, was the way to be accepted for the champions, who gradually gained an image of reliable, enduring partners.
- *Be cooperative and compliant.* Instead of imposing their position, they opened a dialogue, leading to a game in which both sides could win.
- *Learn ambiguity tolerance.* This was necessary to become able to live with the trade-offs that always emerge in situations with multiple players.
- *Explain one's own behaviour.* Some of the champions' moves were bold. They had to explain themselves in order to win acceptance.
- *Training by practice and by attractive events.* Knowledge-building was a key issue. As people embodied that knowledge, both learning-on-the-job and cultural events, which fostered the interchange of people, were imperative.
- *Convince by evidence and results.* The champions started with modest but promising results, avoiding big words and letting events speak for themselves.
- *Constructive tenor.* Negative critique was avoided and the whole discourse led to a build-up of positive energy in the team, which gained cohesion and momentum.

Second, the leadership was not only competent and focused on substantive matters; it was also perpetually concerned with human beings – patients and staff. This humanistic orientation was very effective in salutogenic terms,[9] but it was also very demanding for the staff. The permanent confrontation with difficult situations – strenuous health conditions, tortuous treatments, troublesome fates – tended to exhaust nurses and doctors. The leaders of the OCS were perfectly aware that given these socio-psychological conditions, a special treatment of the staff was required: they called it *psycho-hygiene*, and gave it the highest priority (together with quality of care). Measures taken included scheduled, structured and at the same time easy-going team sessions, team supervision, offers of psycho-oncological seminars with self-experience, yoga, drumming workshops, round dancing, etc.

Third, organizational structure was crucial for the people working in the OCS. The first structural provisions aimed to win the cooperation of the departments and peripheral hospitals. Our approach encompassed:

- offering knowledge and support;
- regular meetings (tumour boards) and ad hoc meetings (tumour councils);
- offering permanent readiness of oncologists for consultation about difficult cases (hotline for inquiries);

---

[9] The term 'salutogenesis' derives from the Latin '*salus*', health and the Greek '*genesis*', generation. The salutogenetic framework (Antonovsky, 1987) conceptualizes an approach to health care that concentrates on the factors supporting health and well-being rather than the factors causing disease.

- active duty 24-7 for emergencies;
- initially focusing on easy, success-prone treatments;
- regular educational events, which included presentations on therapeutic successes.

Two concepts of this design need to be specified. Besides the typical regular department meetings, two forms of teamwork were systematically cultivated, namely tumour boards and tumour councils. A tumour board is an interdisciplinary body at the hospital level, which brings together doctors of the site dealing with tumour cases, with the support of an oncology team member. It takes place in a fixed rhythm and deals with the set of current tumour cases of the hospital: diagnosis and therapy are discussed, and a decision on how to proceed is taken (by consensus). The second kind, tumour councils, are meetings where doctors in charge of one or more cases, supported by oncologists, go to patients and confer on the spot. Normally, such councils are invoked by the patron of a case in one of the specialized departments (e.g., surgery or gynaecology), in those cases which cannot be taken care of by the regular tumour board (e.g., given their urgency). These two concepts are crucial for the evolution of the OCS, and will be taken up later. For more about structural change, see the next section.

12.3.8   Progress in Organizational Design (D)

As this is a chapter in a book about social systems engineering, we are emphasizing the structural aspects of organizing (context loop in Figure 12.1). We are focusing on design, transformation, problem-solving and controlled experimentation. Indeed, the first thing we designed in Oncology Carinthia was organization structure, in order to put the normative principles and strategic orientation into practice.

The structural diagnosis and design of the OCS was of primary importance in the evolution of that system. Structure is not merely the expression of a state, for it changes behaviour. And change was needed if the oncological system was going to take shape. Structure and other levers were the components we could manipulate directly. Many of the factors that constitute the competencies of the organization could be influenced only indirectly. This becomes visible when following the arrows in Figure 12.4. For example, leadership cannot influence quality of care directly, but indirectly (e.g., by strengthening the motivation of the staff).

Soon after embarking on the new venture we used a powerful device of organizational diagnosis and design, Stafford Beer's Viable System Model (VSM) (Beer, 1981, 1984, 1985). That model is extraordinary in that it claims to define not only the necessary but also the sufficient structural preconditions for the viability of any organization. The model has been tested in multiple case studies (for an overview, see Schwaninger, 2009; Schwaninger and Scheef, 2016) and in two surveys (Crisan Tran, 2006; Schwaninger and Scheef, 2016). After Beer's original works, other authors have made methodological contributions to facilitate the application of the VSM (e.g., Espejo and Reyes, 2011; Espejo et al., 1996; Hoverstadt, 2008; Pérez Ríos, 2012).

To facilitate the understanding of our following account of the organization design for Oncology Carinthia, we start with a resumé of the theory of the VSM and its implications for organizational diagnosis and design.

1. ***Components of the model.*** *An organization*[10] *is viable if, and only if, it has a set of components – which are management functions (also called 'systems') – defined as follows (see Figure 12.5):*
   - *Component 1.* The largely autonomous, basic operative units which adapt to change and optimize the ongoing business. Basic units (denoted as circles), with their respective management (square boxes), are called 'primary units'. Examples: a company's business units, a division, a hospital.
   - *Component 2.* This is the *coordination* function, which reduces oscillations and enhances self-regulation. For example, the reporting systems, operative planning, internal service units, standards of behaviour, knowledge bases and a good deal of communication.
   - *Component 3.* The operative management of the organization as a whole. In a company we would have the *executive corporate management* here. It provides overall direction, allocation of resources and striving for an overall performance optimum, which often differs from the optima of the subsystems (primary units).
   - *Relationship components 1–3* (vertical channel). Negotiation of goals and resources – accountability, budgetary control/management by exception, intervention (only if the cohesion of the whole organization is threatened).
   - *Relationship components 1–2–3–3*.* Attenuation of complexity, filtration of messages coming from basic units to inform system 3 and relieve channel 1–3. Enhances organizational cohesion.
   - *Component 3*.* The *auditing channel*, where the information flowing through channels 1–3 and 1–2–3 is complemented and validated via direct access to the basic units. For example, monitoring and management by walking around, informal communication as in social and cultural activities.
   - *Component 4.* Also called the *intelligence* function, it stands for the *long-term orientation* to the future and the overall environment, including exploration, modelling and diagnosis of the organization in its environment. Here we have organizational development, strategy (in interaction with component 3), research and knowledge creation. Component 4 can trigger emergence (of new system properties) via self-reference; that is, the reflection of the system itself and, if indicated, its reframing and redesign.
   - *Relationship components 3–4.* Interaction of short- and long-term as well as internal and external perspectives, processes of strategizing.
   - *Component 5.* The identity as manifest in the supreme norms and values that govern the system – the *ethos* of the organization or *normative management*, also called *policy* function (Espejo and Reyes, 2011). Striking the balance between present and future, keeping the internal and external perspectives in proportion, within a long-term or even timeless horizon. Component 5 is often (partially) codified in corporate charters, credos, value statements, etc.
   - *Relationship components (3–4)–5.* Moderation of the interactions of systems 3 and 4, solution or dissolution of conflicts between the distinct logics of these systems.

---

[10] This theory also holds for other social systems, e.g., families, teams or societies.

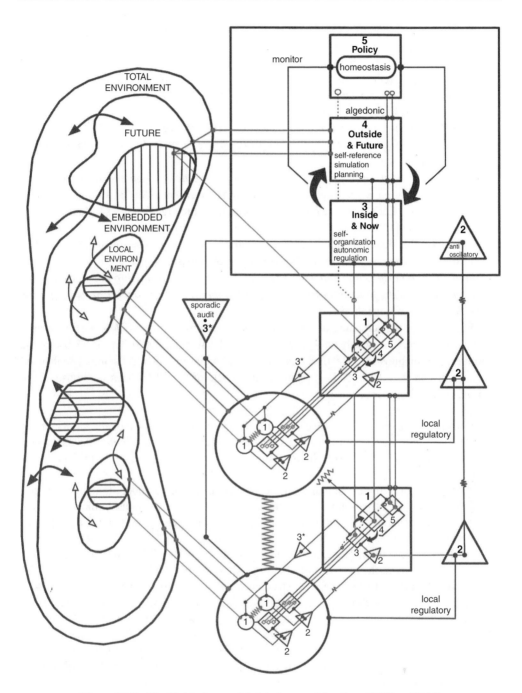

Figure 12.5 The Viable System Model. *Source:* After Beer (1985, p. 136).[11]

---

[11] Drawn by Ivan Ulyanov.

See also the Appendix to this chapter. Components 1–2–3–3* represent the operative system and 2–3–4–5 the meta-system of the organization. In addition, certain alert devices can always be identified in viable systems. Beer (1985, p. 133) calls them 'algedonic signals' (from the Greek *'algos'*, pain and *'hedos'*, pleasure). These warning systems send signals of imminent danger directly up to component 5, to trigger a crisis management. This component will not be analysed further here.

2. ***Principle of recursion.*** *The viability, cohesion and self-organization of a social body depend upon these functions (components 1 to 5) being recursively present at all levels of its organization.*

   A recursive structure comprises autonomous units within autonomous units. Moreover, a viable organization is made up of viable units and is itself embedded in more comprehensive viable units. Each unit, inasmuch as it is producing the organization's task, rather than servicing or supporting this producing, replicates – in structural terms – the totality in which it is embedded. It has all the functions outlined under (1), to be able to manage, from start to finish, the processes for the purpose of which it exists.

3. ***Diagnosis and design.*** *Any deficiencies in this system, such as missing functions, insufficient capacity of the functions or faulty interaction between them, impair or endanger the viability of the organization.*

   Accordingly, the screening of any social system in terms of the VSM almost invariably brings to the surface valuable diagnostic points (Schwaninger, 2006). This is the use of the model in the diagnostic mode. In contrast, a system's viability can be substantially enhanced if it is designed or improved according to the tenets of the VSM. This is the use of the model in the design mode.

Following the principle of recursive structure, an organization can be modelled as a cascade of viable systems embedded in more encompassing viable systems. Figure 12.5 visualizes such a structure with two levels of recursion. In the following we refer to four levels (Figure 12.6).

The organization of the Carinthian health-care system was the product of pragmatic structuring that found its expression in the organization charts as can be found in most organizations today. At that time, the principle of viability was not a category used in the management discourse of that system.

We reflected on the status quo and concluded that the use of the VSM could add great value. In particular, we found the category of viability important. In addition, we were attracted by the fact that the model focuses on deep structures, as opposed to superficial structures. It is suspicious of defining sections based on perfunctory criteria and arbitrary hierarchical positions, as encountered in conventional organization practice. Organizational work with the VSM considers basic functions, relationships and information flows. It brings the environment and the customer into the organization chart. And it builds structures for effective governance. Finally, and most important, it provides structures for the absorption of complexity for the whole organization, along the lines where that complexity emerges.

Figure 12.6 visualizes the structure of the OCS as implemented within three years. The power of recursive organization design is visible from the diagram: the organization unfolds its capacity to absorb environmental complexity along the fronts where that complexity unfolds. Thus, the organization is in a position to respond effectively.

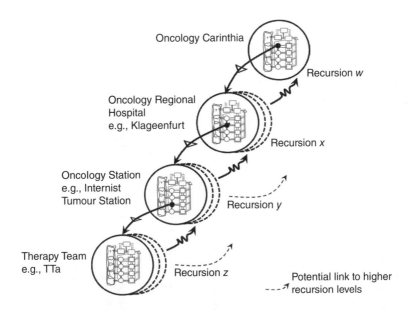

Figure 12.6   Oncology Carinthia as a recursively structured system.

We decided early on to use the VSM mainly in the design mode, and have done so ever since. We also made use of it in the diagnostic mode – when you are trying to design a better organization, you need to be aware of the flaws in the extant one.

Please note that at the second level (recursion *x* – oncology regions), LKH Klagenfurt figures as a regional hospital, just as St. Veit, Villach, etc. would. This is distinct from the central oncological unit, which manages Oncology Carinthia as a whole and which is seated in that hospital as well.

A generic representation of the VSM can be found in the formulas in the Appendix.

In Table 12.2, a schema is presented which reflects a paragon for the structure of Oncology Carinthia, as designed and implemented in the 1990s to put the strategy in place.

A system-in-focus is a unit of oncological care at any level of recursion. In other words, the overall health system is outside the influence of JK, and will only marginally affect his area of responsibility. At the first level (recursion *w*), the system-in-focus is the whole OCS of Carinthia. At the next recursion level (recursion *x*), it is the oncology in a region (e.g., Klagenfurt, Villach or Wolfsberg), where it is centred around the central district hospital (LKH) and other local clinics. Finally, at recursion level *y*, it is an oncology station such as a tumour ambulatory or a subsystem of the internist's ward.

The table discloses a number of unconventional features of the system, as implemented in the OCS. Basic units are the individual hospitals, but individual local oncologists can also have such a function (e.g., when they assume full responsibility for the therapy or aftercare of a patient). In the case of recursion *y*, the structure is based on teams.[12] The basic unit here is not a doctor or a patient, but a connection of four components, namely patient and family,

---

[12] Or systems of interaction, if we take – with Luhmann (1995) – communications as the primary components of systems.

Table 12.2 Recursive distribution of tasks in the oncological care system[13]

	Recursion w: Oncology Carinthia	Recursion x: Oncology regions (e.g., Oncology Klagenfurt)	Recursion y: Oncology stations (e.g., tumour ambulance)
Basic units	Oncology Klagenfurt, Oncology Villach, Oncology Wolfsberg, oncology in six further hospitals	Oncological ambulance, oncology ward at internal station, n local practitioners with oncological competence	N teams (interaction systems, including patient and family, physician and nursing), n local practitioners with oncological competence
Component 1: Local management	Heads of basic units: oncologists	Heads of station: oncologists or internists	Patient & family
Component 2: Coordination	Standard therapies, tumour database, training, oncology circle, doctors' letters, electronic messaging and conferences	Tumour database, tumour boards and councils, coordination sessions (radiotherapy/haematology/pathology), standard therapies, training, doctors' letters	Tumour councils, standard therapy plans, daily station meeting, nursing guidelines, coffee break
Component 3: Executive management	Lead team OCS (2 oncologists, 1 internist-oncologist, 1 radiologist-oncologist)	Management team/local leaders (physicians and nursing staff; 9 persons)	Station leaders (physicians and nursing staff in charge)
Channel between components 1 to 3	Allocation of time and OCS staff, management by objectives, definition/negotiation, standard therapies	Allocation of time and personnel, design/negotiation of therapy plans	Assignments/requests, participation
Component 3*: Audit channel	Visits to local oncology units, phone calls, messaging, special studies, inquiries, tests, informal communication, cultural events	Medical visits, phone calls, messaging, informal communication	Medical visits, continual contact/conversations with patients, informal communication
Component 4: Organizational development/strategic management	OCS leaders team, therapy and prevention strategies, ongoing research, congresses and symposia, networks, strategy workshops	Development plan, future-oriented education, management team, leaders OCS	Station development plan, future-oriented education, leaders of station, leaders OCS, head internal station
Component 5: System ethos/normative management	Ethos OCS – values, principles, vision and mission, management framework OCS, leaders OCS	Values, principles, vision and mission, ethos OCS, management framework OCS, local management team, leaders OCS	Values, principles, vision and mission, ethos OCS, leaders of station, leaders OCS

[13] For relationships not detailed in this table (e.g., 1–2, 3–4, etc.), see the generic descriptions above.

doctor and nurse. Each one is an integral part of the therapy team, but none is only a member of that unit. Doctors and nurses are also members of other similar teams, just as the patient and the family are at the same time members of other social systems.

To highlight some of the features outlined in Table 12.2, we comment mainly on the innovative aspects and revert to all four recursion levels.

Component 1. A remarkable feature of the local management (component 1) in recursion *y* is a reversal of the conventional arrangement. The management function (i.e., the primary regulatory responsibility) is with patient and family. This corresponds to the emancipatory idea of valorizing the role of the patient, who becomes the main agent pursuing his or her health. Making this philosophy real requires – despite this declaration about structure – that the medical and nursing staff take a different view than is common in most health systems. The patient is not a passive object to be manipulated according to expert considerations, but a force aligned with the joint quest for a successful treatment. This novel view did not emerge by itself. The champions played a crucial part in conveying the inherent values, via discourse and acting-by-example.

Component 2. A crucial role in coordination (component 2) is with the tumour boards and councils (recursion *x*). These are virtual units,[14] in which the medical cases are discussed with themes ranging from diagnosis to therapy. The cases treated by a tumour board vary in number and size; usually the therapies are defined by these boards. The tumour council is a kind of individualized tumour board (see below). Other group initiatives are the oncology circle at recursion *w* and the coordination meetings of radiologists, haematologists and pathologists at recursion *x*. This principle greatly increases both the efficiency and effectiveness of oncological care. Finally, standard therapies, training, messaging and the doctors' letters edited by the Klagenfurt oncologists fulfil an important coordinative function.

Component 3. The executive management always involves oncologists and nursing staff. Only at the last recursion (recursion *z*), which is not elaborated in detail in Table 12.1, are the managers patients and family.

The connection between local and executive management (components 1 and 3), at the different recursions, makes use of the precious but very limited capacity of that vertical channel. Here is where the negotiation and control of goals, as well as the allocation of resources, in addition to important feedback mechanisms and participation, take place.

Hospitals that do not have their own specialized oncologists benefit from a new service installed as a mobile unit. This is a resource of recursion *w* deployed for the hospitals at recursion *x*. The oncologists from Klagenfurt visit the peripheral hospitals – physically or virtually[15] – in a constant weekly (LKH Villach, LKH Wolfsberg, KH St. Veit) or two-weekly rhythm (hospitals Spittal and Friesach), all others occurring as needed. They participate in the respective local tumour boards and also now and then, if indicated, in the local tumour ambulatories. In this way they make their expertise available, therewith contributing to the quality of the decentralized operations. The idea here is that the doctor comes to the patient,[16] rather

---

[14] In many of the processes at the OCS, virtual forms of organization are adopted at the team level. Teams in the OCS are flexible in that they are formed as changing casts drawn from resource pools that exist in different locations. The resources are there, but the teams materialize in response to changing needs. They work across space, time and organizational boundaries (Lipnack and Stamps, 2000), reverting to personal contact, other communication media (mainly electronic connections) and information systems (e.g., tumour database).

[15] Until 2015, the journeys of the oncologists were almost entirely replaced by a teleconferencing system.

[16] This idea is also constitutive for the profession of barefoot doctors in India.

than the patient 'feeding the system'. In addition, the mobile doctors are a vehicle for knowledge transfer, and indeed, the oncological know-how at the periphery has made great progress over the years. Meanwhile, the KH in St. Veit has hired its own oncologists.

Component 3*. The audit channel (component 3*) comprehends direct forms of access to the basic units (e.g., at recursion w, the visits to local oncology units). At recursions x and y, the medical visits are crucial, because they give the professionals a first-hand impression of the local care situation and the individual state of the patient. Also, informal communication and cohesion-building socio-cultural activities play a crucial role here at all levels.

Among the cohesion-building measures are the coffee breaks in the wards, the 'onco-lunches' that gather OCS people of Klagenfurt and beyond, and the yearly oncology symposia which bring together oncologists from all over the state plus colleagues from the neighbouring state of Styria. These events fulfil both coordination and auditing functions. In certain cases they might also contribute to the intelligence function.

Component 4. This *intelligence* function fulfils tasks ranging into the long term and the wider environment. These tasks are, in the first place, the concern of the OCS leaders team, whose members are involved in the strategic development at all three levels of recursion: at recursion w as the pioneers and masterminds, at recursions x and y in support roles. In the latter it is the managers/leaders of these recursions who are the designers of the long-term future of their units, making up development plans and providing their staff with education for the future. The development of therapy strategies, going hand in hand with research activities and international activities in knowledge networks, is mainly in the hands of the OCS leaders at recursion w. Knowledge is built up in the process, mainly at recursions w and y.

Component 5. Finally, the system's identity, manifest in the ethos of the system, with normative management has become a systemic braid that connects members and organizational cultures of all recursion planes. Shared values, principles, vision and mission are the same for all three recursion levels, but they need different people to enforce and exemplify them, namely the leaders at each level. In this structure, as shown in Table 12.2, one and the same unit often fulfils different roles with respect to the management components. For example, on recursion w the management team is active in both functions, executive management (component 3) and strategic management (component 4).

The structure outlined here is a network, and so it need not be emphasized that the activities therein involve various forms of networking and communication, from formal to informal and from personal to electronic. That network is crucial for the alignment of the views of multiple purposeful actors with different goals and interests. And it enables building a shared corpus of knowledge over time. Much of that knowledge is tacit know-how (Nonaka and Takeuchi, 1995), embodied in the people and teams of the organization.

In addition to the 'master structure' just described, we will now delve, in more detail, into one crucial organizational feature mentioned – the teams at various levels, which are of three kinds.

First, the therapy team is the nucleus of the structure. These self-regulating teams are formed around each patient, as the primary units at recursion z. The care here is accomplished by the patient, his or her family, a medical doctor and a nurse. Besides its therapeutic function, the team also engages in prevention, as far as possible. While patient and family are members of that team only, doctor and nurse are normally also part of other teams, around other patients. They are always virtually present in each of these teams, but physically present only at certain times.

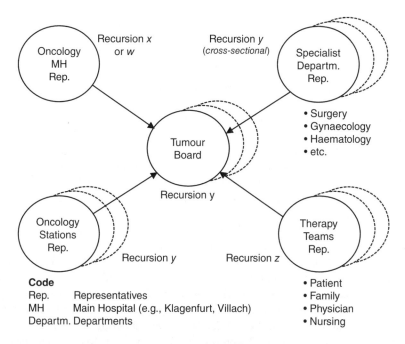

Figure 12.7   The constitution of tumour boards.

Second, the tumour board at recursion x is a platform that manages the continuous flow of cases to be dealt with in a given hospital.[17] It also plays a role in the building of local knowledge. A tumour board is formed by members of different organizational units of different recursions (x and y, or even x, y and z). See Figure 12.7. A tumour board meets regularly to investigate the current cases under treatment. The local oncologist, if extant – and if not, a mobile oncologist from the hub – as well as members of specialist departments (surgery, gynaecology, haematology, etc.), constitute that body. Whenever indicated, representatives of therapy teams join the board. The interaction in tumour boards is supported by the latest communication technologies, for example if (additional) oncologists from the hub need to be involved selectively. The leadership of the board is in the hands of an oncologist.

The arrows in Figure 12.7 denote the provenience and inclusion, in the board, of members of different organizational groups; the multiple communications in the group are not specifically represented. The diagram visualizes how members of several recursions, normally two (x and y), possibly three (x, y, z) constitute such a body. The composition of the board varies as a function of the cases to be treated. The specialized departments are of the support function type, and they are represented as the cases demand.

In the central hospital at Klagenfurt (LKH) – recursion x – more than one tumour board has been formed. Cancer patients have their treatment across different specialized wards, gynaecology and pneumology being two prominent examples. One of the oncologists moves from the LKH oncology unit to these departments, where he or she participates in visits to cancer

---

[17] Tumour boards can also be formed at recursion level x.

patients, in this way forming a local tumour board, together with the local doctors in charge. This approach was later copied in the largest of the peripheral clinics, as it developed internal capacity for oncology.

Finally, the tumour council is an entity that forms itself spontaneously, if a case needs a level of attention that goes beyond the possibilities of the tumour board. The composition of these bodies varies, according to three criteria: (a) an oncologist is always present; (b) the responsible doctor ('patron') and normally the nurse in charge of the case are present; (c) other specialists are on the team as needed. A tumour council can be summoned whenever a doctor or nurse from a specialist department needs assistance in dealing with a difficult case. In that instance, an oncologist visits the respective ward, where he gathers with the local medical and nursing staff and most important, the patient, in an on-site inspection ('ad-hoc meeting'). Hence, the tumour council bundles distributed resources flexibly and effectively, and is a major factor for the quality of care.

These more or less virtual teams are of a non-hierarchical ('heterarchical') type (McCulloch, 1988; Schwaninger, 2009). They have proven to be efficient and powerful: they enhance the quality of care, an optimal use of available resources and growth of the body of knowledge. Also, an increase in the cohesion of those dealing with cancer and the cooperativeness across disciplines and departments has been clearly observable. Cultural events are regularly planned to support this process. For example, once a year, a gathering of medical and nursing staff from the oncology units in Carinthia and Styria, the neighbouring state, takes place in Bad Kleinkirchheim, a beautiful resort. The purpose of these events is to exchange ideas and get information about new developments in the field. Similar events take place, on a smaller scale, for the Carinthian staff, normally combined with a concert.

The mobile doctors as well as the tumour boards and councils are instances of decentralization. Nevertheless, certain technical resources could not be decentralized *ad libitum*. For example, back in the 1990s a laser canon cost 700,000 euros (i.e., close to US$1,000,000). At the beginning, only one of these machines was available, in Klagenfurt. A few years later a second one was installed in Villach. The other hospitals continue without such infrastructure, and patients must travel to get radiologic treatment.

The structures analysed here represent the current state-of-the-art. They were crucial for reaching the goals set at the beginning, and which we have gradually approached. For the future there is still space for improvement. For example, the preventive strand of our activities is still weak, and should be strengthened. We are up-to-date on early detection, but hardly present in public or as supporters of other information agents to promote healthy behaviour.

12.3.9   The Evolution of Oncology Carinthia (C)

Taking a broad view on the evolution of Oncology Carinthia, we can discern remarkable changes which have shaped that system. Gradually, doctors confronted with tumour instances came to understand the advantage of close cooperation with the oncological care unit established at LKH Klagenfurt, consisting of an internist-oncologist and a radiologist-oncologist. Even those who had rejected the new approach initially moved on to cooperative and even supportive behaviour.

The concept of virtual teams was introduced. Doctors increasingly took part in tumour boards, invoked councils and conferred with the oncology hub. The 24-7 on-call service was a major factor in building trust and gave security when a doctor took on a treatment. The concept of mobile units added a new dimension in providing high-quality care covering the whole state. The introduction of these units and the virtual teams approach also added enormous flexibility to the provision of service capacity.

In this way a growing share of the departments, in the LKHs as well as at the KHs, was integrated into the OCS. Increasingly, tumour boards and councils were established. The oncology champions were surprisingly active in research, participating in congresses and professional networks. Following the principle 'as central as necessary and as decentralized as possible', intensive-care patients were medicated at LKHs, under rigorous supervision of the oncological experts, while easy-to-manage cases could be treated at the clinic closest to their homes, frequently in ambulant mode: a patient-friendly but also very economical way to provide oncological care. Altogether, an excellent level of care relative to the scientific state-of-the art was realized at Oncology Carinthia: the OCS had become an effective health-care system.

12.3.10   Results

The case studied here is a showpiece that demonstrates two things. At the content level, it shows the huge potential of service industries for increasing quality and productivity, even in cases of severe resource restrictions. At the context level, it makes the strength of holistic system design palpable. We were successful in conceiving Oncology Carinthia as a viable and adaptive whole, by supporting the evolution of the OCS conceptually and methodologically with systemic methods.

The first result is the organization design that has been accomplished. It is conceived around the patient as the focal point, with all features of a network organization. Patients and their families, often factored out from organizational plans, are the prime agents of the system in Oncology Carinthia. Both are crucial in the process of recovery. The central hospital and nine more clinics, as well as registered doctors, are part of the care network, with a pivotal oncology unit as the main knowledge hub and coordinating agent. Among the innovative features of the structure are cross-sectional teams, transdisciplinary collaboration, the concept of mobile units that bring doctors to the patient rather than the other way around, and networks both inside and across the hospitals. The care process covers all phases from prevention to medical treatment to follow-up care and psycho-social accompaniment. It follows essentially a saluto-genetic orientation. One of the strengths of the arrangement is that both the design and management of the process are governed conceptually, with a heavy dose of theory. The implementation is an infinite learning process. In sum, an intricate, systemic path of dealing with the enormous complexity at hand has been discovered.

A second result is the stunning performance of the OCS. Despite an extreme scarcity of financial resources, both the quality and success of oncological care have been increased. The system under study has become a showcase of holistic medical treatment that has evoked sustained interest in professional circles all over Europe and beyond.

Third, Oncology Carinthia stands as an exemplar for the successful management of expertise. The influence of organizing and managing in general on the evolution of the system has become tangible. Yet it has not provoked the likely conflict between medical and managerial logics, which often deteriorates the qualification of professionals (Boos and Mitterer, 2014).

Instead it brought to fruition a constructive force for the system's viability. The reason is twofold. On the one hand, management in this case has never become a pathologically auto-poietic system (Beer, 1979) – in contrast, it has been instrumental in pursuing the purpose of the whole system-in-focus: a state-wide, excellent level of care, enabled by transdisciplinary collaboration. On the other hand, management, including leadership, was radically decentralized and, furthermore, integrated or 'dissolved' into heterarchical, mobile and virtual structures. This has entailed a powerful rise in the repertoires of behaviour ('variety') at all levels of the organization, enabling multiple agents to cope with complexity forcefully. Oncology Carinthia has become more agile in both time and space.

The concerns and needs raised initially (Sections 12.3.1 to 12.3.3) were met fully by the organization developed within roughly ten years (1985 to 1995): by then the new system was running at 'full steam'. In fact, the outcomes exceeded all expectations. A major factor was the substantial freedom granted to the champions by the state health authority of Carinthia.

Some of the results were unexpected. For example, the evidence that (a) organizational structures can be nimble and robust at the same time and (b) a more complex structure is not necessarily more expensive; it can even be more economical.

The results referred to here are not an endpoint. On the contrary, the OCS team has seen itself confronted with new challenges along the way. Over time, competencies evolved and structures had to develop as well. Transformations of structure took place for meeting new needs. For example, the establishment of two transdisciplinary units, a central ambulance and an oncological ward for special cases, as well as the foundation of an intensive care unit for oncology.

We were not dealing with a machine, but with a social system. Hence, the design approach was both formal and informal. The results were emergent. Culture and structure were always 'in progress', adapting and evolving. We have claimed that the OCS Carinthia has become a highly effective system of health care. Can this claim be upheld in view of empirical data?

We consulted the 'Tumorregister Kärnten' (Tumour Database Carinthia) in Klagenfurt, to examine if there was any evidence of medicinal effects of the OCS, over the period covered by our study. We received long-term data series on the evolution of five-year survival rates, in Carinthia, for the five main entities of cancer indications (Figure 12.8). Five-year survival rates are the most important indicator of effectiveness in oncological care (Ziegler et al., 2007).

The axes in the five graphs show the period of survival after the treatment (from 1 to 5 years) and the percentage of patients surviving (1.00 being 100%). Each graph shows two curves, one for the period 2005–2004 and the other for 2005–2013 (in the case of stomach cancer, for 2001–2006 and 2007–2013).[18] The graphs tell us two things: (a) for prostate, lung and stomach cancer (Figure 12.8, left), there is a highly significant improvement in survival rates from the first to the second period. The respective p-values of the log-rank test are $p < 0.0001$ for the first indication and $p < 0.01$ for the second and third indications.[19] (b) For colorectal cancer and breast cancer (Figure 12.8, right), there is a trend indicating improvement, even though the level of significance is less impressive. Here, the p-values are at $p < 0.1$.

[18] Initial values n (number of patients): prostate, 3828 (2005), 3819 (2005); lung, 2734 (1995), 2908 (2005); stomach, 783 (2001), 848 (2007); colorectal, 3259 (1995), 2953 (2005); breast, 3579 (1995), 3760 (2005).
[19] According to the null hypothesis, there is no difference between the two survival curves. Given the results of the log-rank test, the null hypothesis is refuted.

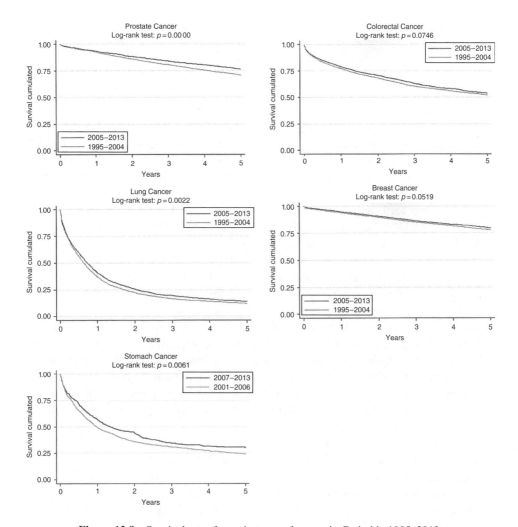

Figure 12.8   Survival rates for main types of cancer in Carinthia 1995–2013.

This analysis testifies success and that success could only be achieved through high quality of care. The numbers underpin the claim made above: the OCS became a highly effective health-care organization.

12.4   Insights, Teachings and Implications

What can be learnt from this case? Many insights and lessons have emerged throughout the sections of this chapter. As indicated at the outset, the purpose of this contribution is to explore an exemplar of a long-term process aimed at achieving a holistic system design for a patient-centred system of health care. In closing we shall try to condense our findings in a brief resumé, which cannot lay claim to being a full-fledged theory, being instead a set of crucial aspects conducive to ongoing learning. As these aspects have been tried and tested extensively over a long period, we dare to switch, at times, from the descriptive mode to 'normative' propositions (i.e., suggestions of how things should be done).

(a) *Ethos.* The focus should be on the patient, with his or her family included. This means placing the patients and their quality of life at the centre. Not only at the centre, but also in command; the patient is the authority responsible for his or her health. For the doctors, nurses and other employees, a culture of highest professional values, including unconditional commitment to patients, is the imperative to be adhered to. The principle of excellent care must pervade all activities. Goals have to be high, and leaders need to energize the organization to attain them.

(b) *Systemic approach.* Systemic thinking is a way of dealing competently with complex wholes. It enables systemic design for better organizations. The proposed thinking at the levels of both content and context forms a braid that enables a systemic process reaching from modelling to assessment to design to change. Unorthodox thinking, as our case showed, can produce management innovations that, although unwelcome at the beginning, go on to breed (unexpected) positive results and are adopted by the organization. To initiate and manage these innovations, powerful change agents are a *sine qua non.*[20]

(c) *Theory and practice.* Practitioners are often theory-averse. Therefore, we pursued another path, operating on a strong theoretical (and methodological) orientation, combining it with pragmatic, flexible implementation and – most importantly – with enduring dedication to the issues of health. This alternative path proved to be at the core of the success of Oncology Carinthia. The combination of theoretical, conceptually driven design and reflexive, committed practice is mandatory for dealing effectively with complex organizational issues in a social system.

(d) *Methodology.* We have tried to catalyse the dialectics of strategy and organization, proposing two methodologies (SD and VSM) that facilitate dealing with the complex issues under study. These are not the only methodologies available, but they appear to be mature devices on which one can rely in the face of complexity. And they are complementary. Much as with engineering, the methods chosen rely on the cyclic pair of reasoning and experimentation (Goldman, 2017).

(e) *Holistic organization design.* The systemic approach provides highly effective heuristic devices and frameworks (e.g., VSM) for enhancing the viability and adaptiveness of organizations. Holistic design combines decentralization and centralization, as well as information flows from inside-out and outside-in, which is a better model than top-down and bottom-up. Structures can be nimble and robust at the same time, and a complex structure can be even more economical than a simplistic one. The systemic approach based on VSM and other cybernetics-based structural models not discussed here[21] is very potent in putting networks and virtual organizations in place, which absorb complexity pervasively. This proves to be the case here in an organization that is above all humanistic.

---

[20] In the case under study both an internal and external change agent were active, and in harmony. This may have been an important prerequisite for the successful performance of Oncology Carinthia (Birkinshaw *et al.*, 2008).
[21] For example, Team Syntegrity, a systemic protocol for the interaction in large groups (Beer, 1994; Schwaninger, 2003).

These are the answers to our research question posed at the outset: How must health-care systems be designed to provide holistic, patient-centred care, of excellent quality and at bearable cost? The main implication of our study is that these insights and teachings can also deliver value to other protagonists and 'engineers' of health-care systems. Even more, we trust that the organizational concepts discussed herein can convey lessons to organizers in any industry, showing them new paths of coping with complexity. As far as management scientists and students are concerned, the systemic approach – framework and methodology – documented in this reflexive case might potentially enhance their professional knowledge and repertoire.

All of the aspects synthesized above are becoming ever more important, as the complexity of systems grows. The immediate results secured by organizations are less important than their viability in the long run. We have relied on the VSM to structure the OCS because that model defines not only the necessary but also the sufficient preconditions for such viability. Therefore, the design we have proposed and implemented is not only successful, but also has great potential for some time to come.

There are limitations to both the case and this chapter. As far as the case is concerned, even though the OCS has bred remarkable results as we have reported, none the less Oncology Carinthia is not perfect; there is room for further improvement. For example, to date our successes at the preventive end are below our initial expectations.

As far as the chapter we have written is concerned, its chief limitation lies in the difficulty of capturing the richness of 30 years of experience in a short document. For example, we cannot account for all the scenarios, the various 'what ifs' and 'so whats' that emerged in the process. Also, little has been said about the relationships between Oncology Carinthia and its external stakeholders, etc. Although we could write a book, still a concise piece of work like this is more digestible.

Acknowledgements

The authors wish to express their gratitude to all the people who have taken part in developing Oncology Carinthia for their enormous dedication and wonderful care: from the cleaning staff to doctors, nurses, administrators and state authorities. Our thanks also go to Professors Garcia-Díaz and Olaya for their outstanding editorial support. Their insightful comments have helped us to improve this chapter. Finally, we are grateful to Dr John Peck and Professor Stefan Ott for their precious support in matters of formulation and formalization.

Appendix: Mathematical Representations for Figures 12.5, 12.6 and 12.7

A1: VSM, for any System-in-Focus (one level of recursion; ref. Figure 12.5)

$$S = f\left(B_1,\ldots,B_n,C_1,\ldots,C_5\right) \tag{A.1}$$

The system as a whole (S) is made up of basic units B, the number of which can vary, and a set of five[22] management components C, which are connected by a coupling function f.

---

[22] Component 3* is subsumed under component 3.

A2: Recursive Structure of the VSM (ref. Figure 12.6)

$$R_z \subset R_y \subset R_x \subset R_w \qquad\qquad (A.2)$$

Viable systems are structured recursively. The sequence $R_{x,\ldots,z}$ denotes levels of recursion ranging from x to z. Units of lower levels are embedded in units of higher levels:

$$R_z \sim R_y \sim R_x \sim R_w \qquad\qquad (A.3)$$

Viable systems are fractals (i.e., they are self-similar in that their basic structure repeats itself at different levels of recursion).

A3: Virtual Teams (ref. Figure 12.7)

The virtual teams are entities that form themselves flexibly: according to the issue at hand, different combinations of resources will be pooled together ad hoc. In the case of a tumour board (*TB*), representatives R of the oncology team of a main hospital *OMH*, oncology stations *OS*, specialized departments *SD* and therapy teams *TT*. We describe this with a vector:

$$TB = \begin{pmatrix} R(OMH_x) \\ R(OS_y) \\ R(SD_y) \\ R(TT_z) \end{pmatrix} \qquad\qquad (A.4)$$

A second vector specifies, in more detail, that several units $i = 1,\ldots,n$ of each category (e.g., more than one oncology station) can be represented in one and the same tumour board:

$$TB = \begin{pmatrix} R_1(OMH_x) \\ \ldots \\ R_{n1}(OMH_x) \\ R_1(OS_y) \\ \ldots \\ R_{n2}(OS_y) \\ R_1(SD_y) \\ \ldots \\ R_{n3}(SD_y) \\ R_1(TT_z) \\ \ldots \\ R_{n4}(TT_z) \end{pmatrix} \qquad\qquad (A.5)$$

References

Allan, J.D. and Hall, B.A. (1988) Challenging the focus on technology: A critique of the medical model in a changing health care system. *Advances in Nursing Science*, **10**(3), 22–34.

Antonovsky, A. (1987) *Unraveling the Mystery of Health – How people manage stress and stay well*, Jossey-Bass, San Francisco, CA.

Ashby, W.R. (1956) *An Introduction to Cybernetics*, Chapman & Hall, London.

Beer, S. (1979) *The Heart of Enterprise*, John Wiley & Sons, Chichester.

Beer, S. (1981) *The Brain of the Firm*, John Wiley & Sons, Chichester.

Beer, S. (1984) The Viable System Model: Its provenance, development, methodology and pathology. *Journal of the Operational Research Society*, **35**(1), 7–25.

Beer, S. (1985) *Diagnosing the System for Organizations*, John Wiley & Sons, Chichester.

Beer, S. (1994) *Beyond Dispute. The invention of Team Syntegrity*, John Wiley & Sons, Chichester.

Birkinshaw, J., Hamel, G. and Mol, M.J. (2008) Management innovation. *Academy of Management Review*, **33**(4), 825–845.

Boos, F. and Mitterer, G. (2014) *Einführung in das systemische Management*, Carl-Auer-Systeme Verlag, Heidelberg.

Brownlee, S. (2007) *Overtreated. Why too much medicine is making us sicker and poorer*, Bloomsbury, New York, NY.

Crisan Tran, C.I. (2006) *Beers Viable System Model und die Lebensfähigkeit von Jungunternehmen: eine empirische Untersuchung*. PhD thesis, University of St. Gallen, Switzerland, no. 3201.

Espejo, R. (1993) Management of complexity in problem solving, in R. Espejo and M. Schwaninger (eds), *Organizational Fitness: Corporate fitness through management cybernetics*, Campus, Frankfurt, pp. 67–92.

Espejo, R. and Reyes, A. (2011) *Organizational Systems. Managing complexity with the* Viable System Model, Springer-Verlag, Berlin.

Espejo, R., Schuhmann, W., Schwaninger, M. and Bilello, U. (1996) *Organizational Transformation and Learning. A cybernetic approach to management*, John Wiley & Sons, Chichester.

Forrester, J.W. (1961) *Industrial Dynamics*, MIT Press, Cambridge, MA.

Goldman, S.L. (2017) Compromised exactness and the rationality of engineering (this volume).

Hontschik, B. (2010) Ich behandle ja keine Röntgenbilder. Interview, *Süddeutsche Zeitung*, **11** August 2010, p. 16.

Hoverstadt, P. (2008) *The Fractal Organization. Creating sustainable organizations with the Viable System Model*, John Wiley & Sons, Chichester.

Illich, I. (1976) *Medical Nemesis. The expropriation of health*, Pantheon Books, New York, NY.

Koen, B.V. (2003) *Discussion of the Method. Conducting the engineer's approach to problem solving*, Oxford University Press, Oxford.

Lipnack, J. and Stamps, J. (2000) *Virtual Teams* (2nd edn), John Wiley & Sons, New York, NY.

Loewit, G. (2010) *Der ohnmächtige Arzt. Hinter den Kulissen des Gesundheitssystems*, Haymon, Innsbruck.

Luhmann, N. (1995) *Social Systems*, Stanford University Press, Stanford, CA.

Malik, F. (2008) *Strategie des Managements komplexer Systeme. Ein Beitrag zur Management-Kybernetik evolutionärer Systeme* (10th edn), Haupt, Bern.

McCulloch, W. (1988) *Embodiments of Mind*, MIT Press, Cambridge, MA.

Nonaka, I. and Takeuchi, H. (1995) *The Knowledge-Creating Company*, Oxford University Press, New York, NY.

Pérez Ríos, J. (2012) *Design and Diagnosis for Sustainable Organizations: The viable system method*, Springer-Verlag, Berlin.

Pettigrew, A.M. (1985) *The Awakening Giant: Continuity and change in Imperial Chemical Industries*, Blackwell, Oxford.

Richardson, G. (1997) Problems in causal loop diagrams revisited, *System Dynamics Review*, **13**(3), 247–252.

Schwaninger, M. (1997) Integrative Systems Methodology: Heuristic for requisite variety, *International Transactions in Operational Research*, **4**, 109–123.

Schwaninger, M. (2003) A cybernetic model to enhance organizational intelligence, *System Analysis, Modelling and Simulation*, **43**(1), 53–65.

Schwaninger, M. (2004) Methodologies in conflict. Achieving synergies between System Dynamics and Organizational Cybernetics, *Systems Research and Behavioral Science*, **21**, 1–21.

Schwaninger, M. (2006) Design for viable organizations: The diagnostic power of the Viable System Model, *Kybernetes*, **35**(7/8), 955–966.

Schwaninger, M. (2009) *Intelligent Organizations: Powerful models for systemic management* (2nd edn), Springer-Verlag, Berlin.

Schwaninger, M. (2013) An Integrative Systems Methodology for dealing with complex issues, in J. Zelger, J. Müller and S. Plangger (eds), *GABEK VI – Sozial verantwortliche Entscheidungsprozesse*, StudienVerlag, Innsbruck, pp. 177–196.

Schwaninger, M. and Scheef, C. (2016) Testing the Viable System Model. Theoretical claim versus empirical evidence, *Cybernetics and Systems*, **47**(7), 544–569.

Senge, P.M. (1990) *The Fifth Discipline. The art and practice of the learning organization*, Century Business, London.

Sterman, J.D. (2000) *Business Dynamics. Systems thinking and modeling for a complex world*, Irwin/McGraw-Hill, Boston, MA.

Weber, M. and Schwaninger, M. (2002) Transforming an agricultural trade organization: A system-dynamics-based intervention. *System Dynamics Review*, **18**(3), 381–401.

Yin, R.K. (2014) *Case Study Research: Design and methods* (5th edn), Sage, Thousand Oaks, CA.

Ziegler, A., Lange, S. and Bender, R. (2007) Überlebenszeitanalyse: Der Log-Rang Test, *Deutsche Medizinische Wochenschrift*, **132**, e39–e41.

13

Reinforcing the Social in Social Systems Engineering – Lessons Learnt from Smart City Projects in the United Kingdom

Jenny O'Connor, Zeynep Gurguc and Koen H. van Dam

13.1 Introduction

Humankind and the natural world are increasingly governed by technology. In the era of big data, the digital economy, virtual reality and ubiquitous sensing, it is timely to explore whether these artificial systems can be better designed and managed to combat major global challenges and positively affect citizens' lives and livelihoods. Social systems engineering (SSE) speaks directly to this question, in that it proposes that a purposive, design-led approach has significant merit in comparison to uncoordinated, laissez-faire innovation. This chapter seeks to use real-world examples to identify the challenges and opportunities of taking a SSE approach, and to this end utilizes the 'smart cities' phenomenon as an example of 'SSE in action'.

'Smart cities' have become hugely popular in recent years [see, for example, the comparison by de Jong *et al.* (2015) of citations in academic literature], and there have been a number of useful explorations of the smart-city phenomenon from a theoretical perspective (e.g., Hollands, 2008; Townsend, 2013; Vanolo, 2013). However, in contrast, this chapter will offer a more functional review, in order to explore how smart-city aspirants interpret and actuate smart-city visions, and implement a SSE approach in this context. To this end, we evaluate how a number of smart-city projects in the United Kingdom are conceived, configured and deployed, and from this analysis assess what lessons are pertinent to the wider SSE enterprise.

Social Systems Engineering: The Design of Complexity, First Edition. Edited by César García-Díaz and Camilo Olaya.
© 2018 John Wiley & Sons Ltd. Published 2018 by John Wiley & Sons Ltd.

13.1.1   Cities as Testbeds

Cities have evolved over millennia in response to different stimuli, whether man-made or caused by nature; hence, they offer a useful example of an environment where both the natural and the artificial collide. Demographic changes brought about by factors such as increases in life expectancy and migration have been at the forefront of encouraging a more systematic approach to tackling some of these complex social issues – a 'new science of cities' (Batty, 2013) – and while rapid urbanization is generally regarded as a major global problem, cities are increasingly being perceived as having significant potential in facilitating citizens to live more successfully and sustainably together (Glaeser, 2011; Mitlin and Satterwaite, 1996). Thus, cities are reimagining themselves as territories with embedded logics of power and claim-making (Sassen, 2008); academics, industry and governments have become interested in how a systemic approach can be used to moderate and even capitalize upon the consequences of ever-greater urban growth. For example, Batty (2011) presents an overview of various city models that can be used to support decision-makers in achieving diverse outcomes, but stresses that a generic approach should be eschewed in favour of a locally tailored urban strategy. Hence, the main challenge facing cities is to develop a framework that will allow them to design interventions in a holistic way, and embrace both social and technical systems and their complex interactions. Social systems engineering potentially provides just such a perspective.

The term 'smart city' can be interpreted in diverse ways, ranging from the platitudinous – 'improving urban living' – to the highly technical. Attempts at definition have focused on a number of approaches. Smart cities are often defined by the possession of key attributes, constituents or initiatives in various forms of public/private partnership. For example, a triple-helix model of smart cities that hinges on local governance and policy reform, academic input and corporate initiatives (Leydesdorff and Deakin, 2011). Another approach sees smart cities described through a taxonomy of application domains, such as natural resources and energy, transport and mobility, government and economy, buildings, living and people (Neirotti et al., 2014). Improving the efficiency of these systems – often by increasing integration, interoperability and optimization – is then considered the 'smart' thing to do. A more holistic view focuses on a city being 'smart' when 'investments in human and social capital and traditional (transport) and modern (ICT) communication infrastructure fuel sustainable economic growth and a high quality of life, with a wise management of natural resources, through participatory governance' (Caragliu et al., 2011, p. 50). Alternatively, a more narrow interpretation suggests that the term is better understood as a purposeful and integrated approach to urban challenges, grounded in ubiquitous sensing, real-time information, (big) data analysis, centralized control through platforms and process optimization (Naphade et al., 2011). Whatever the case, the term 'smart city' is routinely acknowledged as an ambiguous and slippery concept (Tranos and Gertner, 2012), which does not easily lend itself to generic attempts at definition outside a highly specific context (Caragliu et al., 2011). Hence, evolution patterns remain highly dependent on local context and individual city composition. Moreover, the existence of many smart-city initiatives does not ensure that a city is better or more liveable in than others (Neirotti et al., 2014).

13.1.2   Smart Cities as Artificial Systems

While cities themselves are composed of both the natural and the artificial, *smart cities* – as a philosophical paradigm – are a purely artificial way of imagining urban futures. A systemic approach purports to tackle these complex, socially organized, large-scale, multi-organizational service systems, however they are riven with significant emergent challenges, and their very complexity and scale renders it impossible to make design or management decisions based on sufficient individual knowledge (Jones, 2014). Great interest exists in the potential of technology to address these 'wicked' problems (Rittel and Webber, 1973) within the urban domain; however, the world – or a city – is often a messier place than architects, engineers or designers would like it to be (Graham and Marvin, 2001; Jacobs, 1961).

Cities are socio-technical systems (Hughes, 1987) that evolve over time (Dennet, 1996), where multiple stakeholders work together, compete or otherwise influence each other's environment (Bruijn and Heuvelhof, 2008). Cities are also complex systems (Kominos *et al.* 2013), and in a complex system inferring the link between cause and effect is problematic. Assessing how interventions in one domain can lead to (intended or unintended) emergent behaviour at the systemic level is difficult, and in certain circumstances interventions can lead to the opposite outcome to what was initially intended.

Furthermore, it has been postulated that the self-congratulatory and purposive claim-making imbued within the term 'smart city' insidiously undermines and inhibits alternative imaginaries. The irrefutability latent within the term – for who cannot be in favour of cities becoming smarter? – neutralizes any alternatives as fundamentally unscientific and 'anti-progress'. This debate-shrivelling framing of smart cities inhibits alternative discourses on urban futures, many of which envision interventions that go beyond mere technological innovation (Hollands, 2008; Vanolo, 2013). In addition, while smart-city visions may offer powerful and attractive mechanisms for change, all technology poses both opportunity and risk (Beck, 1992; Giddens, 1991). The dualistic nature of technology has the potential to transform access to services and information, but is also inherently connected to various forms of privilege.[1] Simon (1969, p. 27) eloquently describes the difficulties in trying to develop technical solutions to real-world systems:

> To find optimal solutions with reasonable expenditures of effort when there are hundreds or thousands of variables, the powerful algorithms associated with OR (operations research) impose a strong mathematical structure on the decision problem. Their power is bought at the cost of shaping and squeezing the real-world problem to fit their computational requirements: for example, replacing the real-world criterion function and constraints with linear approximations […] of course the decision that is optimal for the simplified approximation will rarely be optimal in the real world.

13.1.3 Chapter Structure

The rest of this chapter is structured as follows. First we discuss the methodology used (Section 13.2) and provide the background context and approach taken by each case study city (Section 13.3). We subsequently analyse and discuss our case studies (Section 13.4), to determine the city's systemic approach from the practical initiatives undertaken, the extent to which the social components were actively represented, and whether the projected impact was sufficiently broad to include the wider social context and externalities. Finally, our conclusions

---

[1] The values embedded in the technology can often fail to represent diverse socio-economic groups and it is also unclear how conflicts are resolved in these systems.

(Section 13.5) reflect on the wider implications of the smart-city paradigm to social systems engineering. This analysis illustrates the complex factors that need to be explicitly acknowledged when taking a social systems engineering approach.

13.2   Methodology

To illustrate the role of social systems engineering in the smart-city context, we use a case-based approach in which we review and contrast a number of smart-city visions and outcomes. Based on these case studies, we address some of the opportunities and challenges inherent within the social systems engineering methodology. We selected four UK-based case studies: Glasgow, London, Bristol and Peterborough. The cities selected are respectively the winner and the three shortlisted cities of a major future cities competition conducted by the Technology Strategy Board[2] (TSB): the *Future Cities Demonstrator* (FCD). This nationwide competition was run in 2012 and promised £24m in funding to the winner and £3m to the runners-up (Technology Strategy Board, 2012). Applicants had to apply strategic thinking to integrating urban systems in order to 'improve the local economy, increase quality of life and reduce impact on the environment' (InnovateUK, 2012), and the bid documents describe their challenges, vision and approach. The FCD competition was used in this analysis for three key reasons. Firstly, these proposals were made publicly available, facilitating a review of a number of high-profile and well-documented smart-city projects. Secondly, by their participation in the competition, these cities can be considered as subscribers to a social systems engineering approach. Thirdly, all applicants described their developmental constraints, visions and future plans using a comparable format, at the same point in time (2012), with the same time span for implementation (2012–2014).

Our case studies explore the unique approach that each city has taken to implement its vision and address its specific challenges. We show that each of them is a different iteration of a SSE approach, and we critically review the proposals and evaluate how cities have translated their vision into impact. In addition, in order to extract lessons that are more widely applicable to the social systems engineering approach, we explore the opportunities and challenges inherent in top-down, design-led strategies that are common to the smart-city paradigm.

For each case study, we consider and describe the following in our review:

- A brief introduction to the city, including the challenges it is facing, as well as any unique local opportunities.
- The vision and objectives of the city, as described in their FCD application. These are the targets for the city derived from its challenges, leading to each city's future vision.
- The future city strategy, describing how the city will try to achieve the goals set out above and realize their vision and objectives.
- Implementation of the projects, summarizing what is currently happening and how the proposals from the FCD competition are maturing.
- The outcomes of local city initiatives, including but not limited to the FCD projects, and our interpretation of how they potentially impact the city.

[2] The TSB is an agency responsible for supporting innovation in the United Kingdom. The TSB has rebranded itself and is now known as InnovateUK.

13.3   Case Studies

In this section, we describe the urban initiatives for the four case study cities: Glasgow (the winner of a £24M TSB FCD grant), London, Bristol and Peterborough (the shortlisted candidates, which each received £3M for their scaled down initiatives).

13.3.1   Glasgow

Glasgow is one of the nine major cities in the United Kingdom, with a population of approximately 600,000 that increases to c. 1.2 million in the Greater Glasgow and Clyde area (Office for National Statistics, 2016). Glasgow benefits from a stable political leadership that is committed to a long-term vision for the city; this leadership has been supportive of the smart-city programme from its inception. The city is the commercial capital of Scotland and is well known for its heritage of trade and shipbuilding. This legacy of trade constitutes a key feature of the smart-city initiative that features business models designed to attract investment. In recent years, Glasgow has invested significantly in infrastructure projects, such as the New South Glasgow Hospitals Project,[3] and has also implemented substantive improvements to the city for the 2014 Commonwealth Games.[4] However, Glasgow faces some serious challenges, particularly in respect to inequality: in 2012 it had the lowest life expectancy of any city in the United Kingdom (Office for National Statistics, 2016); unemployment is a chronic problem, with 30.2% of households – the highest percentage in the United Kingdom in 2012 – described as 'workless'; furthermore, antisocial behaviour has been a long-term and intractable issue (ARUP, 2013).

(1)   Vision and Objectives

Glasgow details its future plans in the report *Future Glasgow 2011–2061: A Fifty Year Vision for the Future* (Glasgow City Council, 2012). It has ambitions to become the United Kingdom's first official smart city, and intends to achieve this by combining expertise from across the public, private and academic sectors. Glasgow's chief objective is to become a sustainable and efficient city, with 'citizens at the heart of every action taken'; hence, it is committed to developing and integrating technologies and applications in public safety, transport, health, technology and energy (Smart Cities Forum, 2015). Glasgow's ambition is to showcase how cities can grow their economies, improve their physical environment and enhance citizens' lives by implementing new technologies. Accordingly, the city's smart-city initiative focuses on:

- increasing employment prospects and attracting new businesses;
- providing quality local public services, especially in healthcare and education;
- decreasing the city's carbon footprint;
- reducing crime.

---

[3] £842 m to create advanced hospitals, taking into account energy consumption and carbon emissions.
[4] Including £500 m in district heating network, games venues, games transport network and waste management for the athlete's village.

(2) Smart-City Strategy

Future City Glasgow values collaboration and dialogue, hence it focuses on cultivating a strong relationship with diverse stakeholders such as Glasgow City Council, the Scottish Government and leading organizations within Glasgow across the public, private and academic sectors. To achieve smart-city outcomes and increase citizen engagement, the city wishes to improve the transparency and utility of open city data. To this end, their approach focuses on demonstrating the value of data within a specific context. For example, in the energy sector, comparative household usage data can be employed to promote citizen engagement and community energy projects can encourage the optimization of local energy resources. Future City Glasgow believes that the city can chiefly be enabled to become more efficient and sustainable via insight derived from the optimal use of increasingly available local data (Smart Cities Forum, 2015). Hence, it aspires to increase the richness of data available to users by supplementing existing data with the gathering of relevant new data. Moreover, the authorities support the view that the value of data does not necessarily depend on the volume of data, but rather its availability as a public good. In addition, they note that data protection and liability issues demand attention. Specific strategic initiatives include (Glasgow City Council, 2016):

- Glasgow Integrated Operations Centre (GIOC), which will monitor and control the city's new network of CCTV cameras and provide facial analysis and intends to improve decision-making, earlier reaction to events and real-time response to incidents around the city.
- The City Technology Platform, which will integrate the data streams and visualize data in a meaningful format. This will be publicly accessible via MyDashboard (e.g., energy use monitoring, real-time traffic information and transport information) and MyGlasgow (an incident-reporting service).
- Other FCD projects in areas including energy (demand-side management, city energy modelling, understanding the real-time impacts of insulation on homes, renewable energy opportunity mapping, behavioural change and fuel poverty reduction), health (active travel – encouraging cycling/walking through the use of open data and smartphone/tablet applications), transport (integrating mobile technologies to increase the efficiency of social transport in the city) and safety (application of intelligent street lighting in two demonstrator areas of the city).

(3) Implementation

As the winner of the FCD competition, Glasgow has started implementing key initiatives. Glasgow City Management System was established to manage the future city of Glasgow and the city launched the Future City Glasgow website to publish the achievement of key milestones (Glasgow City Council, 2016). GIOC acts as a state-of-the-art integrated traffic and public safety management system; it provides CCTV for public spaces, security for city council museums, police and traffic intelligence, as well as video analytics. The prototype MyGlasgow app – initially developed to report road issues – was expanded to include incident reporting in other domains such as parking, street lighting and waste collection.

Specific city demonstrator projects are also being implemented. For example, Glasgow's Active Travel Demonstrator offers the 'Glasgow Cycling' and 'Glasgow Walking' apps, which aim to transform the city into a more cycle- and pedestrian-friendly location with the goal of decreasing congestion and improving public health.[5] The city's energy demonstrator project promotes energy-efficient behaviour by encouraging residents[6] to use the 'Glasgow Energy App'; users can enter relevant data such as their property characteristics, home improvements and energy usage, and then receive information suggesting physical improvements and potential savings.[7] Glasgow is also pioneering the adoption of intelligent street lighting using real-time data on noise, movement and pollution in order to improve both lighting and safety throughout the city. Data from city projects is publicly available,[8] so citizens, stakeholders and other cities can benefit. A range of online interactive maps and correlations were also made publicly accessible to interested citizens.[9,10] Moreover, 'My Dashboard' allows residents and tourists to experience the power of open data first hand, and visualize urban data in a unique way by creating a personalized snapshot of Glasgow.

(4) Outcomes and Impact

Glasgow has made significant progress with its FCD initiatives; projects are active in many areas. GIOC is gathering data from across the city, the proposed apps are online and MyDashboard is fully operational. Given that many of the initiatives are still being rolled out, it is unclear what impact the smart strategies will have on some of the chronic and intractable issues currently facing the city. However, the research and learning programme 'GoWell', launched in 2005,[11] may offer some useful metrics of success over the coming years. This study gathers survey evidence on housing, health and neighbourhoods.

Glasgow took a very data-centric strategy to driving improvements within the city and tackling its key challenges. The city demonstrator projects were also chosen to create data and indicators that can facilitate impact evaluation. The city authorities took the view that data needed to be open, and its utility should be maximized via new tools. Glasgow authorities were concerned regarding the unavailability of some data, as well as data protection and liability issues.[12] However, despite the council's commitment to open data, they have done little to publish information on the progress of the various FCD schemes and initiatives that are currently being implemented. Moreover, the information that is publicly available – for example, the reviews of apps such as MyGlasgow in the iOS app store[13] – suggests low user

---

[5] 'Glasgow Cycling' offers a platform for cyclists to share their experiences of cycling within the city. Likewise, 'Glasgow Walking' demonstrates heritage walks and allows citizens and tourists to explore Glasgow on foot.

[6] It is available for every property type, including flats, houses, offices, schools and even leisure centres.

[7] Information can be viewed collectively across the city, using interactive 3D mapping to see how areas compare. It is also possible to measure how individual properties perform against each other.

[8] open.glasgow.gov.uk/.

[9] map.glasgow.gov.uk/.

[10] As said by Councillor Gordon Matheson, leader of Glasgow City Council and Chair of Future City Glasgow: 'This data hub is a fascinating piece of technology which empowers citizens by giving them easy access to a vast range of information about their city [...] whether you are an entrepreneur looking for new premises, residents hoping to set up a community energy project or an IT developer looking to create an innovative new pro' (InnovateUK, 2015).

[11] www.gowellonline.com/.

[12] For example, combining datasets may lead to unforeseen consequences and contribute to data misuse.

[13] The MyGlasgow app only receives 2/5 stars in the iOS AppStore.

utility. Indeed, given that there are some very similar, well-diffused apps that cover the same ground already supplied by market mechanisms, it seems that Glasgow may be somewhat guilty of reinventing the wheel.[14]

13.3.2 London

London is Europe's only mega city, with a population of approximately 8.6 million, which grows to approximately 14 million in the greater metropolitan area (Office for National Statistics, 2016). It is a centre for global finance and connectivity; it can compete on a scale that no other UK city and few global cities can rival according to the Greater London Authority (2012, p. 6):

> London offers unrivalled access to markets, talent and cultural diversity, and is a globally renowned destination for both big and small business. [...] It has the highest concentration of creative firms in any city or region in the world and is home to the largest concentration of IT software and services companies in Europe.

The city is the economic heart of the United Kingdom, however London enjoys only limited devolved power, and does not have full fiscal autonomy and control of local tax-raising and spending.[15] Although the London economy is growing significantly faster than that of the rest of the United Kingdom,[16] London faces significant pressing issues in respect to liveability, sustainability and managing a projected population growth of some 3 million people by 2050 (Greater London Authority Intelligence, 2013). These demographic changes will lead to challenges in a number of critical areas: they place additional strain on health and social care budgets; increase pressure on urban infrastructure, already struggling with demand;[17] increase air pollution, which is often linked to negative health outcomes (Walton et al., 2015);[18] and drive demand in an already heated property market, making access to housing difficult for many citizens. Furthermore, the impact of increased living costs, combined with stagnating earnings, affects the ability of both lower- and medium-income households to survive, and chronic inequality can lead to social unrest (Barro, 2000; Gupta, 1990).

(1) Vision and Objectives

London's objectives are focused on delivering tangible and integrated solutions for its citizens and businesses – to make London 'the best big city in the world to live and work' – while capitalizing on its connectivity, scale and complementarities. The mayor wishes to promote global business opportunities for UK-based firms and develop practical solutions for emerging economic, environmental and social challenges, particularly in East London. The Smart

---

[14] See, for example, the FixMyStreet project by mySociety at www.fixmystreet.com/.
[15] The devolved power of London is less than that of Wales (population 3 million) and Scotland (population 5 million), which have their own parliament or assembly. However, the Greater London Authority has greater powers than other local authorities, for example on transport.
[16] London's economy is growing at an annual rate of 3.3% compared with the national rate of 2.5% (RBS, 2015).
[17] Congestion in London is estimated to cost the economy £2 billion (London Assembly Transport Committee, 2011).
[18] Air pollution in London is currently above the European Union (EU) target.

London Plan adds: 'a smarter London is not a single, definitive solution, but a series of evolving interventions in response to our changing needs' (Smart London Board, 2013). Accordingly, the London FCD bid focused on a strategic area in the east of the city, which lies within four local authorities (Newham, Hackney, Tower Hamlets and Islington) and includes the enterprise zone near Old Street, sometimes referred to as 'Silicon Roundabout'.[19] Challenges faced by the area include population growth, large inequality, overburdened infrastructure, congestion, air pollution, resource inefficiency and unemployment. London wishes to develop initiatives for the local community, but also showcase how its smart-city solutions can benefit the United Kingdom more generally. Hence it is focusing on:

- attracting funding for key industry sectors;
- investing in infrastructure;
- creating new jobs and supporting skills training and development;
- improving energy efficiency of building stock;
- reducing carbon emissions and air pollutants to meet EU levels.

(2)   Smart-City Strategy

To this end, London's bid – *Linked London* – focuses on the use of testbeds to demonstrate the value of integrating city systems, showcases outcomes and impact to a global audience and proposes *data* as the new infrastructure. The feasibility study advises an integrated approach across four dimensions: the physical, the digital, the organizational and the social. A 'Digital Design Authority' will integrate data from across the workstreams and allow access by local people and businesses through open protocols. The strategy focuses on the following streams of work:

- urban transport logistic interventions such as route planning, dynamic pricing and use of electric vehicles to reduce congestion and pollution;
- real-time visualizations of infrastructure in East London to reduce disruption, pollution and economic inefficiencies;
- an intelligent power grid reutilizing rejected heat from London's underground system to improve air quality and dynamically balance real-time heat and power demands;
- sensors, smart communications infrastructure and a local services platform for households to facilitate assisted living, mobility services and fuel poverty reduction;
- a micro-work platform and new business models supported by local employment agencies, volunteer groups and business alliances to promote employment, volunteering and entrepreneurialism.

(3)   Implementation

London was active within the smart-city space for a number of years before the FCD bid, and hence has been working on various initiatives that precede and have a wider scope than is contained within the bid document. For example, Tech City, the technology start-up cluster in

---

[19] Silicon Roundabout is an area with a high concentration of technology companies that would like to be compared with, for example, Silicon Valley in the Bay Area of San Francisco.

East London, has been scaling naturally since 2010. CrossRail, a new line traversing central London, will start operating from 2018 with full services expected in 2019. Citymapper, a local start-up, provides highly accurate real-time transport information to commuters through a smartphone app using open data.[20] The London DataStore[21] has provided curated open data-sets since 2010. London is also the location for many meet-ups and hackathons, and provides a forum that allows innovators to network and collaborate on developing smart applications for the local market, usually with the intention to scale.

Initiatives such as the Smart London Board[22] have chosen to focus on citizens – 'Londoners at the core' – as the key metric in respect to smart development. Hence, digital inclusion, out-reach and engagement will seek to act as a rising tide where 'smart' citizens are facilitated through computer science education in schools, technology apprenticeships in businesses and online research communities, such as the mayoral initiative 'Talk London'. The city is also working on improving its network of public charging points for electric vehicles[23] and a number of electric vehicle car-sharing schemes are currently being rolled out. To reduce CO_2 emissions and further improve energy efficiency in buildings, new energy sensors and demand-side response was trialled in the 'Low Carbon London' research project. This involved a col-laboration with local energy companies and academic partners.

(4)   Outcomes and Impact

There has been little tangible development of many of the FCD plans, suggesting low com-mitment to the goals of the FCD document. To date, it appears that the London Digital Design Authority has not yet been created, nor other initiatives in the study instituted, and the Linked London programme has little visible presence since the proposal stage. To some extent, this is understandable given that the funds involved in the FCD competition repre-sent trifling sums in respect to the scale at which London infrastructural projects operate. However, many of the pressing issues outlined in the bid document continue to go unad-dressed, such as inequality and the consequences of a booming property market that chiefly attracts overseas investors. Current trends regarding the cost of living may radically remodel London and the aforementioned factors may significantly constrain London's achievement of smart-city goals.

However, unlike smaller conurbations in the United Kingdom, London is in the somewhat unique position that market mechanisms perhaps drive the most useful smart innovation. London has a scale and market size that is very attractive to both start-up innovators and large corporate actors, and hence some of the most successful 'pull-based'[24] smart innovation is either driven by the private sector or occurs through community-based social entrepreneur-ship. Furthermore, it is also important to note that while efficient services may ease, and even

---

[20] Using an application programming interface to access Transport for London's real-time data on the status of the transport network, including anything from the location of buses to planned engineering works.

[21] data.london.gov.uk/.

[22] The Smart London Board started work in 2013 and consists of a team of specialists and experts from different dis-ciplines, including from academia as well as the private sector, who advise the Mayor of London.

[23] www.sourcelondon.net/.

[24] In innovation literature, a 'push' occurs when a new invention is pushed by the supplier without proper consideration of whether or not it satisfies user need. A 'pull' occurs when a firm develops a product to fulfil a specific market need.

improve, the lives of Londoners, they are potentially of small consequence in terms of the myriad attractions that London offers. Despite the expense and frequent travails involved in living in London, it continues to act as a magnet for people from all over the world. Preserving these factors is key. As the Centre for London (2014, p. 17) notes:

> Some of the things that attract people to London are its tolerant and welcoming culture, rich public realm and cultural offer and buzzy social life [...] The city will need to find new ways of preserving the qualities that make it the alluring, successful place it is.

13.3.3 Bristol

Bristol is a city of some 437,500 inhabitants in the south west of the United Kingdom, with a further 200,000 residents within the greater metropolitan area (although these are not under the administration of Bristol City Council) (Office for National Statistics, 2016). It is one of the Core Cities[25] and has been enthusiastically engaged within the smart-cities space for quite some time. As the most prosperous city within the Core Cities grouping, it has many advantages: it is a regional powerhouse with diverse local industry; it is situated within a high-tech corridor with access to investment networks and creative hubs; it has three universities and a directly elected mayor, leading to a more visible local leadership and some devolved power in transport and city planning. It also possesses a relatively educated and liberal citizenry who are broadly engaged by sustainability goals and cultural development. However, like other cities, Bristol has a number of chronic problems. Although relatively prosperous, the city has a deeply problematic transport system[26] and has been significantly affected by austerity, inequality and the challenges of increasing urbanization. As a runner-up in the FCD competition, Bristol is well placed to deliver many of the targets within the feasibility study, despite receiving the smaller grant of £3 m.

(1) Vision and Objectives

The elected mayor – an architect and urbanist – emphasized the fact that cities should be perceived as solutions rather than problems; he is a proponent of technology as a powerful route to city prosperity, and wishes to promote Bristol as a 'testbed for urban innovation' and a 'laboratory for change', so the proposal for the feasibility study (Bristol City Council, 2012, p. 2) states:

> By the year 2020, Bristol will become one of Europe's top 20 cities; it will appear at, or close to, the very top of the league tables measuring sustainability, quality of life and achievement among European cities.

---

[25] The Core Cities group is a partnership between the ten largest cities in the United Kingdom, excluding London. The group advocates on behalf of its members, sets a research agenda and facilitates knowledge-sharing on best practice, efficiencies and solutions.

[26] Worker commutes in Bristol are longer than in any city outside London (estimated at an average of 38km).

This vision has the following components and direct objectives:

- Making our prosperity sustainable – economic growth (short-term 2020 target of 3500 new jobs, 40 new businesses).
- Reducing health and wealth inequalities.
- Building stronger and safer communities.
- Raising the aspirations and achievement of our people and families.

(2) Smart-City Strategy

In general, much of the local government discourse in respect to the future of Bristol is very focused on sustainable social outcomes, such as reducing inequality, increasing citizen engagement and cohesion, and fostering resilient communities. The FCD bid (Bristol City Council, 2012, p. 24) noted that solutions to the key challenges faced by Bristol would primarily be achieved through creating 'environmentally and socially sustainable jobs and growth':

> Connect Bristol will create the infrastructure and framework for the City to become agile-by-design. It will be poised and ready to move swiftly to capitalise on emerging trends, technology advances, modernised practices and new business models.

(3) Implementation

A cornerstone of Bristol's FCD proposal related to citizen engagement. Accordingly, the £3 m FCD funds were primarily dedicated to this. The Kwest living lab (an ENOLL[27] partner) will develop a city dashboard, among other projects. This involves working with local people and organizations to identify data relevant to their lives and the city, ranging from social media updates to readings taken from motion sensors within the home. Data streams will be used to develop online services and apps that are functional, engaging, innovative and relevant to the community and wider citizenry. Bristol has also been at the forefront of utilizing the city itself as a canvas or backdrop for artistic and cultural interventions, particularly those that fall under the rubric of the 'Playable City': this seeks to engage a wide audience through novel and creative digital approaches, such as the 'Hello Lampost' project.

Bristol is the European Green Capital City 2015, the first British city to be awarded this honour. Connect Bristol, the organization responsible for the FCD bid, has also developed a smart-city business model and created a dedicated innovation unit – Bristol Catapult – to actively promote city objectives. The 'Futures' department has specific remit to drive tech innovation, sustainability, EU integration and urban resilience. Invest Bristol and Bath, a new inward investment office that will serve all four local authorities jointly, was also created.

Bristol has pledged to reduce its CO_2 emissions by 40% by 2020[28] and produced a Bristol Low Carbon Smart City Roadmap. It has also taken a number of practical steps, such as

---

[27] European Network of Living Labs.
[28] From a 2012 baseline.

conducting smart-metering trials across 100 households and the creation of a local energy provider with a social mandate – Bristol Energy. The 'Bristol Is Open' project will monitor housing stock and promises to address smarter social care and housing through an Internet of Things (IoT) sensor network run through radio space. As previously noted, congestion is one of Bristol's most pressing problems, and there are a number of interventions aimed at tackling this issue – such as traffic restrictions, the creation of resident parking zones (RPZs) and a city-wide speed limit of 20 mph. There are also further plans to deploy a system of sensors to monitor the levels of pollution within the city.

Bristol is highly invested in reimagining itself as a pioneering tech innovation centre. In a £75m partnership with the University of Bristol, it is redeveloping the rediffusion network of underground cable television ducts to house a network of superfast, high-capacity fibre. It is expected to be the largest testbed of its kind in Europe, covering an area of 40 miles and potentially serving 1 million users (Brown, 2014). The city intends to charge telecommunications equipment firms to test new applications and services on the network at city scale under live conditions. In concert with this, a city network operating system is under development, which will be open, agnostic, programmable and future-proof. The council also intends to release a number of public datasets to promote service innovation by local small and medium enterprises.

(4) Outcomes and Impact

Bristol has made significant progress in delivering upon its future city ambitions. High-profile projects such as the software-defined network delivering ultrafast fibre and wireless canopy has certainly brought the city to a wider global audience, however this will have little direct impact for citizens, given that it is to be reserved for researchers and developers in order to avoid a conflict with local residential broadband suppliers. Although there has been a dominant technology-first approach, there is evidence of a commitment to incentivizing and inspiring citizens to participate through experimental and unorthodox methods such as play and grassroots initiatives. However, in common with many other cities, the constraints on a coherent and comprehensive urban strategy are often political, such as lack of authority over urban hinterlands or internal conflicts within authorities. Questions also remain regarding the downstream outcomes of transport and pollution interventions. In addition, Bristol's housing market is already highly buoyant, and urban improvements often have the potential to exacerbate inequality by attracting aspirational middle-income professionals to previously undesirable neighbourhoods. Increases in rents and property prices may drive pre-existing local communities from these areas.

13.3.4 Peterborough

Peterborough is a city in Cambridgeshire with a population of nearly 190,000 (Office for National Statistics, 2016). In the 1960s Peterborough was officially given the status of 'New Town', which boosted its growth.[29] In the process of growing, the city swallowed up several

---

[29] From 1968 to 1988 the city was designated for expansion to create space for the growing population of England.

smaller villages and an old airfield to enable the population to rapidly double from around 80,000 to 160,000, and today it is still the second fastest growing city in the United Kingdom. It was named one of four UK 'Environment Cities'[30] in 1994 and the Peterborough Environment City Trust was founded to continue this work, aiming to become the United Kingdom's 'Environment Capital'. Peterborough has the largest cluster of environmental businesses in the country, with 350 companies, but the economy also thrives in manufacturing, food and drink, digital media and financial services. Peterborough's economy grew faster than that of the rest of the United Kingdom in the late 1990s and early 2000s, with relatively low unemployment figures. Salaries, however, are on average lower than in the rest of the United Kingdom, which can partly be explained by the fact that the population is relatively young (Office for National Statistics, 2016). The city has several deprived areas, which rank among the lowest in the country on multiple indices of deprivation, and high levels of immigration – combined with fast population growth – leads to challenges in healthcare, quality of life and social cohesion. The health of citizens is general lower than average, with life expectancy of both men and women lower than the national average (Public Health England, 2013).

(1) Vision and Objectives

Given these challenges and opportunities for the city, in its FCD feasibility study Peterborough listed the following four priorities that have been adopted as the 'Sustainable Community Strategy':

- Creating opportunities and addressing inequalities, with a focus on improving health and education.
- Creating strong communities, with an emphasis on empowering local communities and improving community cohesion.
- Becoming the environmental capital of the United Kingdom, making the city greener and cleaner to create an example for other cities with its drive to reduce the dependence on natural resources and impact on the environment.
- Delivering growth that is sustainable, with economic prosperity and sustainable infrastructure.

Peterborough already adopted a single delivery plan and expects stakeholders to follow a shared agenda towards a sustainable city. The vision to become a smart city is thus part of an ongoing process in the city which started over 20 years ago and it is already working to address these priorities, including the building of nearly 300 low-carbon homes in what is the United Kingdom's largest zero-carbon housing development (Morris, 2016), developing the Peterborough Gateway industrial park and investing in the city's education system.

---

[30] This was a reward for the achievements of the city council in creating a green and environmentally friendly place to live and work.

For the FCD, the overall vision was summarized as follows by Peterborough City Council (2012, p. 3):

> The vision for Peterborough is for a bigger and better city that grows the right way through truly sustainable development: improving the quality of life for all its people and communities and creating a city which is a healthy, safe and exciting place to live, work and visit, and is famous as the environment capital of the UK.

(2) Smart-City Strategy

The result of a series of workshops and engagement events in which the city's citizens, local businesses, public-sector bodies and other stakeholders were invited to contribute to help shape the plans is the Peterborough DNA programme. This programme is composed of five key activities, which will stand on their own while also contributing to the overall objectives and vision by linking with the other activities:

- *Living Data* collects real-time urban data to enable user-friendly visualization and interaction to aid decision-making at city level.
- *Innovation Pool* builds on collaborative local networks, such as the EnviroCluster, to bring diverse stakeholders together to link challenges with ideas and solutions via open innovation.
- *City Metabolism* will deliver real-world testbeds to demonstrate the initiatives of other streams and improve resource efficiency.
- *Skills for our Future* emphasizes education and offers a platform to link employment demand and supply, as well as providing skills and careers advice to young people.
- *Transport Intelligence* addresses traffic congestion, parking and air-quality issues by installing a dense network of traffic sensors to monitor infrastructure and vehicles, and provide real-time data to optimize transport decisions.

(3) Implementation

Peterborough only received £3m as a FCD runner-up, however it was able to implement most of the proposed programme and launch the Peterborough DNA[31] portal. It launched the Living Data, Innovation Pool and Skills for the Future programmes, as well as the City Metabolism business area demonstrator. Both the Innovation Pool and Skills for the Future programmes focus on creating networks, knowledge-sharing and engaging different stakeholders in the city to induce creativity and improve collaboration. Additionally, the City Metabolism demonstrator has a sustainability focus, and aims to link challenges faced by local businesses with possible solutions coming from the Innovation Pool. By trying to improve resource efficiency at the macro level, this demonstrator highlights the social systems engineering approach and the city's high-level holistic perspective on urban systems. Among all the Peterborough DNA

---

[31] www.peterboroughdna.com/.

activities, the Transport Intelligence programme was the only intervention that was abandoned in the face of reduced FCD funding, although it appears that other ambitions were also scaled back to some extent.

In addition to delivering its FCD programme, Peterborough has instigated many other initiatives, policies and activities that coincide with a smart-cities approach, and a systems engineering outlook more generally. Under the rubric of the global 'One Planet Living' theme, collaboration with companies and organizations across Peterborough resulted in a city-wide proposal document that sets interim targets for the development of a sustainable city. Furthermore, the city's *Integrated Growth Study* (with ARUP) is rooted in consultation with the local community and offers a detailed vision and strategy for the city towards 2021. Innovative funding arrangements aiming to stimulate the local economy have also been employed in the city, with notable examples of public/private collaboration, such as the local mortgage and insurance scheme and the Peterborough energy service company. A 'bond-holder group'[32] of local businesses also offer local stakeholders the chance to meet at networking events.

The 'Peterborough Model'[33] is a 3D/Google Earth visualization for open data in the context of the city. This portal supports user engagement and data-sharing, while also providing insights for better decision-making. This now forms part of the Living Data stream. In addition to data visualization, the city also has its own Open Data platform.[34] Finally, Opportunity Peterborough,[35] the city's urban regeneration company, has been active since 2005. It runs a number of projects to support the local economy, primarily focusing on making the city more attractive to investment, facilitating engagement with local companies and coordinating skills development.

(4)   Outcomes and Impact

The impact of these smart-city initiatives was summarized by the Climate Change Manager of Peterborough City Council in an interview (Informa, 2015) as follows:

> There's no doubt that talking car parks or solar bins have merit in some areas, but in Peterborough we are focused on real challenges and real solutions that real people can and do use that make a difference.[36]

Peterborough's future city approach utilizes technical systems such as the Open Data store, data visualization and web platforms for collaboration and innovation. This highlights the city's approach, focusing on collaboration and skills rather than technology as an end in itself. For example, although ostensibly a technical solution, the chief objective of the Skills for our Future programme is to support local innovation and facilitate citizen access to in-demand skills and jobs. Furthermore, it is interesting to note that Transport Intelligence, the most

---

[32] www.opportunitypeterborough.co.uk/bondholder-network/about/.
[33] The Peterborough Model was implemented in collaboration with IBM, Royal Haskoning and Green Venture.
[34] data.peterborough.gov.uk/.
[35] www.opportunitypeterborough.co.uk/.
[36] www.brainwaveinnovations.co.uk/upload/cmspage/peterborough-dna/10-09-14_11-58-35_peterborough_dna_online.pdf.

technology focused project in the FCD bid, was the only intervention to be cut. The performance of Peterborough DNA is to be measured by impact on resident quality of life and improved operational conditions for local companies, however, as of yet, there is little data available on performance indicators to evaluate success.

13.4   Discussion

Smart-city projects are a useful example of social systems engineering in action. At a purely conceptual level, these initiatives and interventions seek to take a design-led and strategic approach to one of the most pressing global issues: rampant urbanization. Since 2000, over half the world's population lives in cities, and by 2050 this is expected to increase to 66% (United Nations, 2014). Cities are hubs of national production and allow citizens to co-exist with a lower environmental footprint; thus, city leaders are keen to capitalize upon key assets, achieve efficiencies and optimize the use of precious resources. However, as noted earlier, cities are extremely complex; designing interventions is far from straightforward, and due care for wider ecosystemic consequences is critical, as stated by UN-Habitat (2008, p. 182):

> Cities are not just brick and mortar: they symbolise the dreams, aspirations and hopes of societies. The management of a city's human, social, cultural and intellectual assets is, therefore, as important for harmonious urban development as is the management of a city's physical assets.

The smart-cities movement takes a SSE approach to determining how cities can best be designed and improved. The preceding section reviewed four UK case studies to understand how different cities seek to shape their urban future, and reconcile complex and occasionally conflicting values and challenges. The four exemplars underline the huge complexity and heterogeneity involved in attempting to innovate at the city scale; each city presents a unique set of challenges and there are few interventions free of political or social risk.

Below we discuss the key challenges that arise from taking a social systems engineering approach to future city design and operation. We illustrate the challenges with examples from our four case studies. The challenges are discussed under the following seven categories: push/pull adoption model; civic engagement; solutions and problems; metrics, quantification and optimization; project scope and lifecycle; collaboration and multidisciplinarity; knowledge-sharing.

13.4.1   Push/Pull Adoption Model

Despite the somewhat utopian character of smart-city discourse, little has changed in respect to how cities develop new urban systems. Procurement mostly occurs through traditional 'push' techniques, rather than through bottom-up scoping exercises of user need (i.e., 'pull'). Designing projects on a traditional technology-push model can result in focus being drawn away from moderate, demand-driven improvements in favour of grand, eye-catching schemes that may fail to have a core market or user base, and little benefit for the local community. For example, Bristol's high-profile redevelopment of its rediffusion network will not be accessible to users from the local community. Furthermore, a mismatch between output and demand can result in pilot schemes failing to translate into bona fide marketable services or products.

Cities may also fail to create the necessary incentives or environment for business models to translate initiatives from testbed to market.[37] However, in exceptional cases (such as London) – where a city has sufficient scale, resources and an entrepreneurial culture – the best innovation often occurs through laissez faire market mechanisms, and city leaders are perhaps often best advised to simply 'get out of the way' (Von Hippel *et al.*, 1999) and just create the right environment to let it happen. While London has several high-profile partnerships with industry addressing, for example, smart innovation in infrastructure, its short-term strategy concerns digital inclusion and fostering a local culture of innovation. This primarily focuses on developing knowledge-sharing networks between city boroughs, enhancing technology skills and coding education at local schools, and facilitating internships or apprenticeships for local young people within the digital sector. It is hoped that such initiatives will provide a solid foundation for engaging the local community and local authorities in the possibilities and limitations of technology, and thus generate a supportive environment for pull-driven adoption.

13.4.2   Civic Engagement

Cities are ostensibly interested in delivering new forms of participatory democracy and encouraging citizens to reduce their impact on city services, particularly at critical peak times. All four case studies highlight the importance of civic engagement and user behaviour change in achieving successful outcomes. However, as has been widely recognized in the academic literature, civic engagement and behaviour change are both extremely difficult to engineer and sustain (Norris, 2002; Thaler and Sunstein, 2008), and it is challenging to transfer solutions directly from one context to another. Moreover, the premise that desirable urban imaginaries are directly predicated upon a strong element of civic input and participation can easily be obscured by the perception that future direction can simply be 'revealed by large data' (Penn *et al.*, 2013). Hence, while there is a great deal of rhetoric in respect to engaging users, in practice it is unclear to what extent cities deliver upon these commitments. Indeed, while Bristol has had significant success through associated but independent initiatives based around cultural events and digital outreach such as the Playable City,[38] the success of city council efforts to engage users through its dedicated smart-city strategy is unclear. If cities commission software to be developed as part of their smart-city activities, it is worth considering doing this more closely with citizens so that the software product better reflects the needs of the potential users (Hovmand, 2014), but also offers the opportunity to communicate ideas and increase participants' knowledge (Vennix, 1996; Voinov and Bousquet, 2010). Failing to substantively address citizen engagement potentially results in a number of perverse outcomes, such as: (a) there is a dataset available, so developers produce a service for which there is no real market; (b) there is an unmet need, but the data is presented badly so users don't use the interface – and hence the service; (c) cities 'reinvent the wheel' and reproduce what has already been done well in the private sector;[39] or (d) there is a fundamental misapprehension of the degree of citizen interest in greater access to curated urban data.

---

[37] Notably, the InnovateUK approach tries to improve by funding projects which bring innovation from the academic to the private sector or from ideas and concepts to prototypes and services.
[38] http://www.watershed.co.uk/about/about-us
[39] For example, MyGlasgow app seems similar to the popular FixMyStreet app.

13.4.3   Solutions and Problems

Many cities aspire to utilize a holistic approach to achieve their future vision. For example, Peterborough's City Metabolism programme is based on the concept of 'urban metabolism'. This approach highlights a SSE outlook on city systems, and is described as 'the sum total of the technical and socio-economic processes that occur in cities, resulting in growth, production of energy, and elimination of waste' (Kennedy *et al.*, 2007, p. 1965). However, despite good intentions, it is often difficult to translate these visions into practice, and can result in overly technical solutions with narrow applications. Moreover, even supposedly 'holistic' approaches often shy away from intractable or diffuse macro-social problems, due to the intense uncertainty that goes hand-in-hand with any interventions in complex, living environments (cf. Pitt, Chapter 5 in this volume, for a useful discussion on the challenges of technological interventions in techno-social domains). For example, all four cities reviewed in this chapter noted inequality as one of the chief constraints facing sustainable urban development. Indeed, it is commonly noted as one of the chief problems facing society more generally (Lansley, 2012; Stiglitz, 2012; Wilkinson and Pickett, 2009). However, smart-city imaginaries – where they do directly acknowledge inequality and other associated urban ills – generally focus on 'economic development' as the panacea, which may be an overly simplistic solution. This rationale reflects a rather naive logic where economic development is presented as a 'black box', with few details on how it will counter deep-rooted and complex ills such as social exclusion and chronic poverty. Hence, for many cities, it seems easier to focus on projects with more tangible outcomes – such as reducing congestion or air pollution – than seeking to enumerate strategies to combat the macro 'wicked problems' or investigate the externalities involved in these complex projects.

13.4.4   Metrics, Quantification and Optimization

It has become common for metrics assessing cost of living, pollution and congestion to be interpreted as signifiers for city 'liveability', and hence to dominate the discourse. However, choosing what is important over what is not, and its quantification and measurement within social systems engineering, immediately involves questions of ethics and values. Metrics devolving from these choices are frequently culturally or socially ascribed (Friedman *et al.*, 2006), and a hierarchy built on metrics may focus attention away from other more diffuse, but important, outcomes. Even in the case where cities do expressly evaluate fuzzier goals such as citizen engagement, there is scant cause for optimism; for example, Glasgow's 'GoWell' survey project assessed community engagement among other metrics, and research conclusions questioned 'whether sufficient attention was being given to ensuring the ability of community organisations to follow democratic procedures and to represent the diversity of views within communities'. It also found that processes of community engagement and consultation were open to manipulation by both sides (GoWell, 2012). Hence, it is important to recognize that the goals of the city government may not be aligned with the goals of citizens. Outcomes such as efficiency, optimization, predictability, convenience and security – while potentially useful goals in themselves – may fail to capture what makes a city a truly great place to live and work.

13.4.5   Project Scope and Lifecycles

Given that cities face such complex challenges, it is common for strong leadership to be regarded as essential for driving change. However, despite the attractions of a centralized decision-making authority such as a city mayoralty, there are a number of associated concerns: the risk of totalitarianism; the lack of requisite political or technical expertise; excessive influence on the part of corporates; and the privileging of expensive technical solutions over more low-key strategies such as policy changes[40] or changing behaviour through raising awareness. Furthermore, given the brevity of local electoral cycles in the United Kingdom (every 4 years), committed decision-makers may not be in a position to oversee a 5- or 10-year future city roadmap. When using a purposive and design-led SSE approach, it is important to continuously address the reality of limits to decision-making power and the need for measurable impact in shorter time scales. Evidence from testbeds and other smart-city initiatives should help local actors make a case for long-term investment and commitment to projects beyond the short-term election cycles. Furthermore, myopic thinking around election cycles can impact the scope and adventure of urban projects with little accountability in respect of long-term consequences. This situation is frequently compounded by decision-makers lacking jurisdiction over a city's hinterland, such as is the case with Bristol. Operating within an artificial city boundary can mean that problems are simply moved from the centre to the periphery, and this compromises the implementation of initiatives at sufficient scale.

13.4.6   Collaboration and Multidisciplinarity

Smart-city frameworks commonly propose a collaborative and multidisciplinary approach to social systems engineering. While the participatory aspect of such projects is laudable, in practice it is unclear how well teams function together, and how values are determined or conflicts reconciled. Project goals may be imbued with problematic preferences, which are frequently inconsistent or ill-defined. Moreover, the focus on a certain agenda as 'optimal' or 'evidence-based' allows city managers to say: 'It wasn't me who made the decision, it was the data' (Haque, as cited in Poole, 2014). Moreover, data is collected directly from many technical systems (e.g., transport nodes, electricity meters, planning systems) with relative ease and accuracy, however social data often relies on self-reporting (e.g., of incidents to local authorities or to travel app providers, or through interviews and surveys deployed in research studies). Hence, there is a risk that recalled data may be biased, inadequate or simply fallacious. Participation by collaborators in projects can also be fluid; the amount of time and effort they can devote often varies, so involvement may fluctuate. Moreover, participants' attention is often highly bounded, and they can struggle to remain objective in respect to solutions outside their own specific professional interests, and hence may fail to duly recognize a 'greater good' (cf. Goldman, Chapter 1 in this volume, for a more substantive discussion of the challenges faced by engineers attempting to implement idealized, universal theories within real-world contexts). While this may be of little individual significance in terms of team relations, the

---

[40] Since the creation of a directly elected mayoralty in Bristol, only 19% of councillors now feel they can get involved in important decision-making vs. 61% before the mayor came to power (Niranjan, 2015).

aggregate impact can be profound, particularly in a sensitive context such as urban development. Peterborough attempts to address this issue in an organic way through its Innovation Pool model, which seeks to capitalize upon diverse, local expertise to address common 'push/pull' innovation issues and the 'solutions in search of problems' phenomenon. It is crucial that cities foster collaboration in a substantive way – one that goes beyond mere rhetoric – by bringing interested parties together in meaningful ways.

13.4.7   Knowledge-Sharing

One of the most trumpeted attractions of smart-city projects relates to knowledge-sharing among various cohorts and networks. Although cities do seek to compete with each other in attracting investment and highly skilled workers, there has been a widespread recognition of the need to share learning and experiences. Partnerships with academia facilitate a philosophical reinterpretation of cities – where they can reimagine themselves as solutions rather than problems – and partnerships with the private sector can allow cities to gain expertise on urban technologies and the state of scientific development more generally. However, a lack of rigorous focus on all outcomes – be they positive, of little merit, or malign – can result in little learning across the sector. Indeed, in common with testbeds in other sociotechnical domains, the smart-city space is characterized by little robust investigation of interventions that went awry or failed to achieve intended goals. Furthermore, there is little scoping of the impact of negative externalities outside a clearly delineated project scope. For example, improving services in a low-income urban area often drives gentrification, which in turn may result in low-income workers having to live further from their places of work, creating additional strain on the already stressed transport system and potentially interfering with other urban strategies such as anti-obesity initiatives based around physical activity. Hence, in order to promote inclusion while optimizing outcomes, SSE approaches should strive to truly leverage the outcomes of knowledge-sharing and benefit from documented best practice, robust risk assessments and honest reports about unintended or negative consequences.

13.5   Conclusion

Smart-city visions are often presented as the future of urban living, where optimized use of resources and increased automation go hand-in-hand with inclusive societies in which citizens have a greater role in decision-making through direct democracy. However, any intervention in the urban environment – be it new technology or campaigns for behaviour change – imposes varying degrees of transformation on the social system and hence is imbued with risk. Medium- to long-term objectives are all extremely important in social systems engineering, however these need to be complemented with bottom-up, incremental and achievable short-term outcomes. 'A paradoxical, but perhaps realistic, view of design goals is that their function is to motivate activity which will in turn generate new goals' (Simon, 1969, p. 162). This research suggests that many contemporary smart-city projects tend to focus on high-level outcomes such as open data, smart-city standards and privacy governance for the promotion of sustainable city goals, however these need to be complemented with a good mix of short-term and community-based objectives.

Future city projects must also be extremely cautious in respect of uncertain or unintended outcomes. Many critics are now coming to believe that informal and serendipitous social processes are the 'genius' of the city (Sennett, 2011), and a smart city should therefore emulate 'a shifting flock of birds or school of fish' (Ratti and Townsend, 2011, p. 42), where individuals respond to subtle cues from their fellow citizens rather than a regimented, top-down, deterministic socio-technical structure. Like individuals and organizations – and complex systems more generally – cities are subject to powerful, intangible phenomena and harmless-seeming interventions can have unpredictable or malign consequences for the delicate mechanisms that govern social life (Merton, 1936). Given that many smart innovations are designed with a view to being deployed across mega agglomerations that contain millions of citizens, the consequences of error or unintended conflicts are hugely magnified in such sensitive and potentially volatile ecosystems. The smart-city movement – and social systems engineering – needs to tackle this issue more thoughtfully; we believe that the social systems engineering approach needs to be grounded within a sensitive and wide-ranging analysis of the many factors, some possibly unquantifiable, that are key to making cities great places to live and work.

Technology has enormous potential to address issues of sustainability and liveability, however the values and privileges it inherently enshrines need to be actively interrogated and managed. There is little 'low-hanging fruit' and, in an era of shrinking budgets and aging populations, cities feel that they must aim for maximum efficiency in use of resources. Hence, in the search for solutions – or even improved problem definition – there has been a turn to technology and in particular, new insight has been sought through data analytics. However, it is crucial to understand that while data can provide evidence in respect of specific questions, it should not overtake the importance of the 'social' component in SSE.

References

ARUP (2013) Solutions for cities: An analysis of the feasibility studies from the Future Cities Demonstrator programme. Available at: publications.arup.com/Publications/S/Solutions_for_Cities.aspx (retrieved 27 May 2016).

Barro, R.J. (2000) Inequality and growth in a panel of countries. *Journal of Economic Growth*, **5**(1), 5–32.

Batty, M. (2011) A generic framework for computational spatial modelling, in A.J. Heppenstall, A.T. Crooks, L.M. See and M. Batty (eds), *Agent-Based Models of Geographical Systems*, Springer-Verlag, London.

Batty, M. (2013) *The New Science of Cities*, MIT Press, Boston, MA.

Beck, U. (1992) *Risk Society: Towards a new modernity*, Sage, New Delhi.

Bristol City Council (2012) Connect Bristol feasibility study, November 2012.

Brown, J.M. (2014) Bristol to become a smart city laboratory. *Financial Times*, 30 October 2014.

Bruijn, J.D. and Heuvelhof, E.T. (2008) *Management in Networks. On multi-actor decision making*, Routledge, London.

Caragliu, A., Del Bo, C. and Nijkamp, P. (2011) Smart cities in Europe. *Journal of Urban Technology*, **18**(2), 65–82.

Centre for London (2014) The brightest star: A manifesto for London. Available at: centreforlondon.org/publication/brightest-star-manifesto-london/ (retrieved 26 May 2016).

de Jong, M., Joss, S., Schraven, D., Zhan, C. and Weijnen, M. (2015) Sustainable–smart–resilient–low carbon–eco–knowledge cities; making sense of a multitude of concepts promoting sustainable urbanization. *Journal of Cleaner Production*, **109**, 25–38.

Dennett, D.C. (1996) *Darwin's Dangerous Idea: Evolution and the meanings of life*, Simon & Schuster, New York, NY.

Friedman, B., Kahn, P. and Borning, A. (2006) Value sensitive design and information systems, in P. Zhang and D. Galletta (eds), *Human–Computer Interaction and Management Information Systems: Foundations*, M.E. Sharpe, New York, NY.

Giddens, A. (1991) *The Consequences of Modernity*, Polity Press, Cambridge.

Glaeser, E. (2011) *Triumph of the City: How our greatest invention makes us richer, smarter, greener, healthier and happier*, Penguin, New York, NY.

Glasgow City Council (2012) Glasgow city management system: final report, November 2012.

Glasgow City Council (2016) Future City Glasgow. Available at: futurecity.glasgow.gov.uk/ (retrieved 27 May 2016).

GoWell (2012) Empowerment. Available at: www.gowellonline.com/research_and_findings/key_findings/empowerment (retrieved 26 May 2016).

Graham, S. and Marvin, S. (2001) *Splintering Urbanism: Networked infrastructures, technological mobilities and the urban condition*, Routledge, Oxford.

Greater London Authority (2012) Linked London – TSB future cities demonstrator feasibility report. Available at: connect.innovateuk.org/documents/3130726/3794125/Feasibility+Study+-+Greater+London+Authority.pdf/4a645968-6f89-46bb-aafb-7286bbe77d2d (retrieved 23 October 2015).

Greater London Authority Intelligence (2013) Population and employment projections to support the London infrastructure plan 2050. Available at: www.london.gov.uk/sites/default/files/gla_migrate_files_destination/Population%20and%20employment%20projections%20to%20support%20the%20London%20Infrastructure%20Plan%202050.pdf (retrieved 23 October 2015).

Gupta, D. (1990) *The Economics of Political Violence*, Praeger, New York, NY.

Hollands, R. (2008) Will the real smart city please stand up? Intelligent, progressive or entrepreneurial? *City*, **12**(3), 303–320.

Hovmand, P. (2014) *Community Based System Dynamics*, Springer-Verlag, New York, NY.

Hughes, T.P. (1987) The evolution of large technological systems, in W. Bijker, T. Hughes and T. Pinch (eds), *The Social Construction of Technological Systems*, MIT Press, Cambridge, MA.

Informa (2015) Q&A with Charlotte Palmer, Climate Change Manager, Peterborough City Council. Available at: smarttofuture.com/qa-with-charlotte-palmer-climate-change-manager-peterborough-city-council/ (retrieved 26 May 2016).

InnovateUK (2012) City council's feasibility studies. Available at: connect.innovateuk.org/web/future-cities-special-interest-group/feasibility-studies (retrieved 27 May 2016).

InnovateUK (2015) Glasgow launches world leading city data hub. Available at: connect.innovateuk.org/web/future-cities-special-interest-group/article-view/-/blogs/glasgow-launches-world-leading-city-data-hub (retrieved 27 May 2016).

Jacobs, J. (1961) *The Death and Life of Great American Cities*, Random House, New York, NY.

Jones, P. (2014) Systemic design principles for complex social systems, in G. Metcalf (ed.), *Social Systems and Design*, Springer-Verlag, Tokyo.

Kennedy, C.A., Cuddihy, J. and Engel Yan, J. (2007) The changing metabolism of cities. *Journal of Industrial Ecology*, **11**, 43–59.

Kominos, N., Pallot, M. and Schaffer, H. (2013) Special issue on smart cities and the future internet in Europe. *Journal of the Knowledge Economy*, **40**(2), 119–134.

Lansley, S. (2012) *The Cost of Inequality: Why economic equality is essential for recovery*, Gibson Square, London.

Leydesdorff, L. and Deakin, M. (2011) The triple-helix model of smart cities: A neo-evolutionary perspective. *Journal of Urban Technology*, **18**(2), 53–63.

London Assembly Transport Committee (2011) The future of road congestion in London, June 2011. Available at: www.london.gov.uk/sites/default/files/The%20future%20of%20road%20congestion%20in%20London%20June%202011.pdf (retrieved 26 May 2016).

Merton, R.K. (1936) The unanticipated consequences of purposive social action. *American Sociological Review*, **1**(6), 894–904.

Mitlin, D. and Satterthwaite, D. (1996) Sustainable development and cities, in C. Pugh (ed.), *Sustainability, the Environment and Urbanization*, Earthscan Publications, London.

Morris (2016) Vista – Peterborough. Available at: morrishomes.co.uk/find-your-perfect-home/cambridgeshire/peterborough/vista/ (retrieved 26 May 2016).

Naphade, M., Banavar, G., Harrison, C., Paraszczak, J. and Morris, R. (2011) Smarter cities and their innovation challenges. *Computer*, **44**(6), 32–39.

Neirotti, P., De Marco, A., Cagliano, A., Mangano, G. and Scorrano, F. (2014) Current trends in smart cities initiatives: Some stylized facts. *Cities*, **38**, 25–36.

Niranjan, A. (2015) Did having an elected mayor benefit Bristol? *CityMetric*, 26 March 2015. Available at: www.citymetric.com/politics/did-having-elected-mayor-benefit-bristol-882 (retrieved 27 May 2016).

Norris, P. (2002) *Digital Divide: Civic engagement, information poverty and the Internet worldwide*, Cambridge University Press, Cambridge.

Office for National Statistics (2016) Office for national statistics website. Available at: www.ons.gov.uk/ (retrieved 26 May 2016).

Penn, A.S., Knight, C.J., Lloyd, D.J., Avitabile, D., Kok, K., Schiller, F. *et al.* (2013) Participatory development and analysis of a fuzzy cognitive map of the establishment of a bio-based economy in the Humber region. *PLoS One*, **8**(11), e78319.

Peterborough City Council (2012) Future Cities Demonstrator Feasibility Study – Peterborough DNA. Final Report, FCD/DNA/Final, TSB Reference No. 23443-162343.

Poole, S. (2014) The truth about smart cities: 'In the end, they will destroy democracy'. *The Guardian*, 17 December 2014.

Public Health England (2013) Public health observatories – Area: Peterborough UA. Available at: www.apho.org.uk/resource/item.aspx?RID=50363 (retrieved 26 May 2016).

Ratti, C. and Townsend, A. (2011) The social nexus. *Scientific American*, **305**, 42–48.

RBS (2015) Regional growth figures released for Q4 2014. Available at: www.rbs.com/news/2015/february/regional-growth-tracker-q4-2014.html (retrieved 26 May 2016).

Rittel, H.W.J. and Webber, M.M. (1973) Dilemmas in a general theory of planning. *Policy Sciences*, **4**, 155–169.

Sassen, S. (2008) *Territory, Authority, Rights: From medieval to global assemblages*, Princeton University Press, Princeton, NJ.

Sennett, R. (2011) The stupefying smart city. Available at: lsecities.net/media/objects/articles/the-stupefying-smart-city/en-gb/ (retrieved 23 September 2016).

Simon, H.A. (1969) *The Sciences of the Artificial* (1st edn), MIT Press, Cambridge, MA.

Smart Cities Forum (2015) Smart cities forum energy and sustainability task group report – recommendations to the government. Available at: www.glasgow.gov.uk/councillorsandcommittees/viewSelectedDocument.asp?c=P62AFQUTNTT10G2U (retrieved 23 October 2015).

Smart London Board (2013) Smart London plan, 19 September 2013. Available at: www.london.gov.uk/sites/default/files/smart_london_plan.pdf (retrieved 26 May 2016).

Stiglitz, J. (2012) *The Price of Inequality: How today's divided society endangers our future*, W.W. Norton & Co., New York, NY.

Technology Strategy Board (2012) Future Cities Demonstrator – competition for large-scale demonstrator project funding, February 2012, T12/068.

Thaler, R.H. and Sunstein, C.R. (2008) *Nudge: Improving decisions about health, wealth and happiness*, Yale University Press, New Haven, CT.

Townsend, A.M. (2013) *Smart Cities: Big data, civic hackers, and the quest for a new utopia*. W.W. Norton & Co., London.

Tranos, E. and Gertner, D. (2012) Smart networked cities? *Innovation: The European Journal of Social Science Research*, **25**(2), 175–190.

UN-Habitat (2008) *State of the World's Cities 2008/2009*, UN-Habitat, Nairobi.

United Nations (2014) World's population increasingly urban with more than half living in urban areas, 10 July 2014. Available at: www.un.org/development/desa/en/news/population/world-urbanization-prospects.html (retrieved 26 May 2016).

Vanolo, A. (2013) Smartmentality: The smart city as disciplinary strategy. *Urban Studies*, **51**(5), 883–898.

Vennix, J.A.M. (1996) *Group Model Building*, John Wiley & Sons, Chichester.

Voinov, A. and Bousquet, F. (2010) Modelling with stakeholders. *Environmental Modelling & Software*, **25**, 1268–1281.

Von Hippel, E., Thomke, S. and Sonnack, M. (1999) Creating breakthroughs at 3M. *Harvard Business Review*, **77**, 47–57.

Walton, H., Dajnak, D., Beevers, S., Williams, M., Watkiss, P. and Hunt, A. (2015) Understanding the health impacts of air pollution in London, 14 July 2015 for Transport for London and the Greater London Authority. Available at: www.london.gov.uk/sites/default/files/HIAinLondon_KingsReport_14072015_final_0.pdf (retrieved 26 September 2016).

Wilkinson, R.G. and Pickett, K. (2009) *The Spirit Level: Why more equal societies almost always do better*, Allen Lane, London.

Index

abstraction, 47, 60, 118–120, 122, 125–127, 134, 140, 147, 184–186, 190–192
abstraction-aggregation hierarchy, 118–120, 127
action, 1–4, 6, 14, 21, 23, 27–28, 35, 40, 48–49, 52, 59, 66–68, 70–77, 79, 83, 86, 134, 138, 140–141, 143–145, 153–154, 158–163, 165–166, 170–172, 180–181, 218, 228, 238, 267, 271, 283
  research, 163, 173, 216–217, 220, 222, 225, 227, 229–230
actor-network theory (ANT), 7, 157–158, 160–163, 165–166, 168, 172–175
actor, 1, 7, 29, 46, 49, 52, 66–67, 69–71, 73–75, 79–81, 83, 85–86, 108, 137–141, 143, 146–147, 149, 157–166, 168–172, 174–175, 180, 184, 212, 218–219, 222, 228–230, 237, 256, 276, 286
adaptation, 5, 37, 41–42, 52, 61, 77, 159, 216
adaptive, 5, 22, 31–33, 37–40, 42, 53, 57–58, 61, 78, 81, 161, 182–183, 216, 259
adopters, 200–207, 210–212
adoption
  dynamics, 202–203, 205–211
  probability, 205–207
agent-based modelling, 81, 108, 133, 139, 157–159, 161, 163, 165, 167, 169, 171, 173–175, 187, 200
agents, 31–32, 34, 40–41, 49, 51–52, 55–57, 78, 81, 110, 134–135, 137, 139–142, 145, 151, 153, 157–65, 172–173, 180, 187–188, 190, 200, 203–207, 210–212, 229, 258–260, 262
agent template, 138, 141–142, 147
air pollution, 274–275, 285
app, 272–273, 276, 284, 286
application design, 137, 142, 149

applied science, 3, 13–14, 16, 24–25
Aristotle, 23, 50, 94–95, 97
artefact, 2–5, 7, 56, 66, 70–80, 160, 162, 166, 174, 180–181, 184–186, 190–192, 194, 229
automobile industry, 104, 119

Bacon, Francis, 15, 20
behaviour, 4–5, 7–8, 16, 29, 31–42, 46–49, 51–52, 55, 58, 60, 66–68, 72, 74, 78, 82, 85–86, 103–104, 119, 127, 134–142, 144–145, 147–148, 151, 154, 157–159, 161–163, 172, 175, 187, 199–203, 207, 218, 222, 237, 243, 248–250, 258, 260, 269, 271, 273, 284, 286–287
black swan, 32, 36–37, 40
boundary objects, 229
Bristol, 270–271, 277–279, 283–284, 286

citizens, 14, 16, 71–72, 74, 78, 81–82, 92, 140–141, 143–149, 151–154, 267–268, 271, 273–274, 276, 279–281, 283–285, 287–288
climate change, 40, 82–83, 282
closeness, 201, 203
codification, 184–186, 190–192
cognitive scaffolding, 185
Colombia, 165–171
community, 4, 7–8, 36, 45, 55, 60–61, 66, 92, 134–135, 139, 169, 179, 183, 194, 201, 203, 210–212, 217, 272–273, 275–276, 278–285, 287
complex
  adaptive system, 31–33, 35–40, 69, 81, 85
  socio-technical systems (CSS), 45–52, 54–55, 58–61, 103, 128, 180
  system, 31–33, 35–39, 46–47, 55, 57–58, 199, 219, 269, 288